Studies in Computational Intelligence 407

Editor-in-Chief

Prof. Janusz Kacprzyk
Systems Research Institute
Polish Academy of Sciences
ul. Newelska 6
01-447 Warsaw
Poland
E-mail: kacprzyk@ibspan.waw.pl

For further volumes:
http://www.springer.com/series/7092

Germano Resconi

Geometry of Knowledge for Intelligent Systems

Springer

Author
Prof. Germano Resconi
Catholic University
Dept. Mathematics and Physics
Brescia
Italy

ISSN 1860-949X e-ISSN 1860-9503
ISBN 978-3-642-27971-3 e-ISBN 978-3-642-27972-0
DOI 10.1007/978-3-642-27972-0
Springer Heidelberg New York Dordrecht London

Library of Congress Control Number: 2011945322

Printed on acid-free paper

Springer is part of Springer Science+Business Media (www.springer.com)

Preface

There is a tremendous interest among researchers and application engineers in agent-based systems. These systems are increasing used virtually in every area right from healthcare to the computer games. The book is on the geometry of agent knowledge. Chapter 1 presents an introduction as well as the significance of the area of research presented in the book.

Chapter 2 presents the connection between tensor calculus and formal concepts definition. The definition of the tensor calculus is evolved to model geometry transformations to simplify formal description of a given problem. Tensor calculus gives us invariant forms for any possible change of the reference in the geometric description of problems. Conversely, formal concept analysis presents the fundamental elements such as object and attribute to describe the concepts in the mind of the agent. Thus, the main idea behind the book is to show that is possible to present a geometric image of the concept and introduce invariants in the agent mind.

Chapter 3 describes the geometric coherence of the agent expressed by the minimum path in a given manifold. Agents describe concepts in a geometric way and use the geometry to navigate in the space. In chapter 4, we introduce symmetry and rules to express the meaning of the agent in a geometric way. In chapter 5 we describe the logic as a part of a geometric space. Logic is not only true and false but new topics in logic as fuzzy set and many value logic extend the simple idea of true and false to values or coordinates in geometric space where the true and false are positions in this space. Thus, the holistic approach to Fuzzy and many value logic of the agent can be well represented by geometry of agent knowledge. In chapters 7 and 8 we introduce uncertainty of agent and geometry representation.

In conclusion the geometry gives us the main reference for agent concepts and associate agent knowledge at any level. For example, at the level of neural network, at the level of logic, at the level of computation and at the level of uncertainty.

We believe that this book will prove useful to the researchers, professors and students of various disciplines including, physics, computer science, engineering and philosophy. We are grateful to the Springer-Verlag for their editorial assistance.

Germano Resconi
Italy

Contents

Chapter 1
An Introduction to the Geometry of Agent Knowledge

Abstract. This chapter provides an introduction to the geometry of knowledge. It briefly introduces important concepts and presents a summary of the contents of the book.

1.1 Introduction

The essential nature of things which was called *physics* by the Greeks, is a unifying concept. These mystical philosophers viewed the world as a *kind of living organism which was subject to change*. At that time there was considered to be no difference between concepts and physical objects. As commonly understood now, the Greeks had unified the concepts in nature by the use of geometry.

We will show in this book that we can restore this previous unity by the use of a suitable geometric representation of the knowledge from a mathematical and physical point of view. An opposite that of the Greek theory appeared soon and dominated western thought. The philosophers who created this argued that the universe is composed by a set of passive dead particles moving in a void. The movement of these particles is regulating by a number of "god-given" physical laws which were discovered by Newton and can explain the entire physical universe, including human beings. Descartes philosophy contends that nature is formed by two separate parts. The mind and matter. Using this concept Newton and Descartes built a structure of the universe without inclusion of human concepts or knowledge.

In the Newton's mechanics nature is independent of human concepts or perception so it can be predicted and totally controlled. The knowledge of the observer or Agent is not included in natural phenomena.

A series of scientific revolutions occurred in the twentieth century. One of the first began with the discovery of the electrical and magnetic Fields by Michael Faraday and Clark Maxwell. An electrical or magnetic field consists of an infinite number of forces located at every point of space. The forces are not the arbitrary physical entities of Newton, but are contained in fields under laws that are true for all of the infinite forces of the field. The field covers all the space which contains the forces. Every particle in the universe is not solitary but is connected to the

G. Resconi: Geometry of Knowledge for Intelligent Systems, SCI 407, pp. 1–3.
springerlink.com

others in a holistic way by the physical field. For the infinite number of the forces in the field we are unable to devise an experiment to test the properties of the field which encompasses the entire Universe. The field is basically a mental concept which forms a physical entity. The mental concept forms part of the physical description of the Universe.

Maxwell realized the profound importance of the field and declared that the field could not be explained mechanically as local effect of one force on a single particle in an empty space. The sources of the field are not physically in contact with the particles. They generated fields which at the velocity of the light cover all empty space and generate forces that move the particles. Every elementary source generates a basic field of forces. The field in the universe is the superposition of all elementary fields for any one particle that cover the entire Universe. The forces that move the particles are generated by the superposition of all the forces generated by all the sources, particles, in the universe. All the sources (particles) unify to result in the movement of a particle. The universe is a body that acts as an indivisible unity. The entire universe is conceptually a unity that guides the movement of the particle. The separation of the universe into parts is an artificial concept since every part of the universe is involved in the construction of the field. Feedback exists between the particles and the field. We cannot separate one from the other. In the field image of the universe the concept of space and time is independent of the dynamics of the particles. In Einstein's image the geometry of the space time concept changes with the observer's velocity and gravity. Time is no longer absolute but is transformed in accord with the observer's physical state. Einstein discovered that the gravitational field can be represented as a deformation of the space-time reference geometry. He substituted the forces with the deformation of space-time where the particle moves in a complex way under the inertial law. The physical model is connected in a strong way with the perceived conceptual space – time structure. A much higher degree of conceptualisation is obtained by the use of quantum mechanics. Here the states of the particles (velocity, position, energy and so on) are no longer numbers but are operators that change the probability of having a particle in a particular state (position, velocity, energy and so on) in the universe. Particles in different positions of the space are not independent from one to another but correlate to each other in a geometric way as indicated by the Hilbert geometric space.

A matter field that controls the probability without the use of any energy guides the movement of the particles. There are two types of fields in the universe. One is the classical field of forces which transmit energy at the velocity of light. The other is the non-energetic field of the probability that guides, at a microscopic level the movement of the particles. The non-locality of the particles is a new type of uncertainty introduced by quantum mechanics. The perception by one observer (Adaptive Agent) in quantum mechanics is part of the definition of the probability that guides the movement of the particles.

The only variable that the observer can change is the probability of finding the particle in a particular state. Future experiments depend on the past experiments. The most conceptual physical model of the Universe is given by quantum field theory using the geometric representation given by Hilbert space. In a quantum computer we have computation using fields. In the classical computation the unit of information is the bit which can assume the value of one or zero. In quantum mechanics we may substitute at any bit, which is a number, a Qubit or Quantum bit. The qubit is the equivalent of the bit in computer science. The qubit was introduced in the quantum computer. The difference between the bit and the quantum bit is that the bit is able to assume only the value of 1 or 0. The qubit is a probability field and may have either the bit 1 or 0.

In classical computation by numbers only two states are possible for the bit. That is state one or state zero. In quantum mechanics a superposition with different weights of states one and zero are possible. In this way from states One and Zero may have infinite numbers of intermediate states for different pairs of values of weights. Using normal intuition it is difficult to understand that one particle can rotate simultaneously the right (bit one) and to the left direction (bit zero) with different degrees or probabilities. The new frontier of the quantum mechanics forces the logic to differ from the true dichotomy in that it generates new concepts to describe the quantum phenomena. This new logic is termed quantum logic. This logic attempts to describe, in a conceptual way, the Superposition Phenomena.

The important concept which we study in this book is the Field and its Geometric Representation. To develop a geometric image of the gravity, Einstein used Tensor Calculus but this is very different from the knowledge instruments that used now. For instance techniques of data mining, neural networks, formal concept analysis, quantum computer and other topics are quite different from tensor calculus. Our aim in this book is to rebuild the tensor calculus and develop a new representation. By using new geometric image we can unify all the topics that have been studied in recent years to create a bridge between the traditional physical phenomena based on an individual and subjective approach to the knowledge. In conclusion as the universe is a unity we postulate that we must include in this unity the conceptual level of individual. The conceptual level should not be confused with particles in physics. It must be considered as an integral part of the universe. The introduction of the observer to quantum mechanics and relativity indicates that the universe must be integrated with the conceptual part of the individual. The conceptual part lies beyond the traditional physical domain and extends the human knowledge of the universe.

[illegible]

Chapter 2
Tensor Calculus and Formal Concepts

Abstract. This chapter introduces tensor calculus using formal concepts which aid the better understanding of intelligence.

2.1 Vectors and Superposition of Attributes

In a given space a using Cartesian orthogonal reference system a vector can be represented by its n Cartesian components. When we define the axis as:

$$H_1, H_2, \ldots, H_p \tag{2.1}$$

The Cartesian components are

$$w_1, w_2, \ldots, w_p \tag{2.2}$$

Using the context in Table 2.1

Table 2.1 Context Representation

Attributes

	Context	H_1	H_2	...	H_p
	A_1	h_{11}	h_{12}	...	h_{1p}
Objects	A_2	h_{21}	h_{22}	....	h_{2p}
	...	...	...	...	...
	A_q	h_{n1}	h_{n2}	....	h_{np}

When $h_{h,k} \in \{0, 1\}$ H_k are attributes and A_h are objects. Table 2.1 represents a context in a formal concept analysis. When $h_{h,k}$ are Boolean numbers we have an ordinary context table in the formal concept analysis.

The values in Table 2.1 can be represented by the matrix H as follows:

$$H = \begin{bmatrix} h_{11} & h_{12} & \cdots & h_{1p} \\ h_{21} & h_{22} & \cdots & h_{2p} \\ \cdots & \cdots & \cdots & \cdots \\ h_{q1} & h_{q2} & \cdots & h_{qp} \end{bmatrix} \tag{2.3}$$

G. Resconi: Geometry of Knowledge for Intelligent Systems, SCI 407, pp. 5–29.
springerlink.com

The transformation of the components **W** by using H gives a new set of components **X**:

$$\mathbf{Y} = \mathrm{H}\,\mathbf{W} \tag{2.4}$$

or

$$Y = HW = \begin{bmatrix} h_{11} & h_{12} & \dots & h_{1p} \\ h_{21} & h_{22} & \dots & h_{2p} \\ \dots & \dots & \dots & \dots \\ h_{q1} & h_{q2} & \dots & h_{qp} \end{bmatrix} \begin{bmatrix} w_1 \\ w_2 \\ \dots \\ w_p \end{bmatrix} = \begin{bmatrix} y_1 \\ y_2 \\ \dots \\ y_q \end{bmatrix} \tag{2.5}$$

and

$$Y = w_1 \begin{bmatrix} h_{11} \\ h_{21} \\ \dots \\ h_{q1} \end{bmatrix} + w_2 \begin{bmatrix} h_{12} \\ h_{22} \\ \dots \\ h_{q2} \end{bmatrix} + \dots.. + w_p \begin{bmatrix} h_{1p} \\ h_{2p} \\ \dots \\ h_{qp} \end{bmatrix} = w_1 H_{\alpha,1} + w_2 H_{\alpha,2} + \dots. + w_p H_{\alpha,p}$$

We now consider the geometric image of the matrix H. For simplicity we use a three dimension space. We have

$$H = \begin{bmatrix} h_{11} & h_{12} & h_{13} \\ h_{21} & h_{22} & h_{23} \\ h_{31} & h_{32} & h_{33} \end{bmatrix}$$

This represents the context in the formal concept analysis Table 2.2.

Table 2.2 Context Representation for Three Objects A and Three Attributes H

Context	H_1	H_2	H_3
A_1	h_{11}	h_{12}	h_{13}
A_2	h_{21}	h_{22}	h_{23}
A_3	h_{31}	h_{32}	h_{33}

The column vectors

$$H_1 = \begin{bmatrix} h_{11} \\ h_{21} \\ h_{31} \end{bmatrix}, \ \mathrm{H}_2 = \begin{bmatrix} h_{12} \\ h_{22} \\ h_{32} \end{bmatrix}, \mathrm{H}_3 = \begin{bmatrix} h_{13} \\ h_{23} \\ h_{33} \end{bmatrix}$$

In H are a set of three vectors in a three dimensional space of the objects given in Table 2.2. In Figure 2.1 we show the image of the vectors in H for the one attribute H_i (i = 1,2,3) in Table 2.2. It also represents the space of the objects A_i (i = 1,2,3).

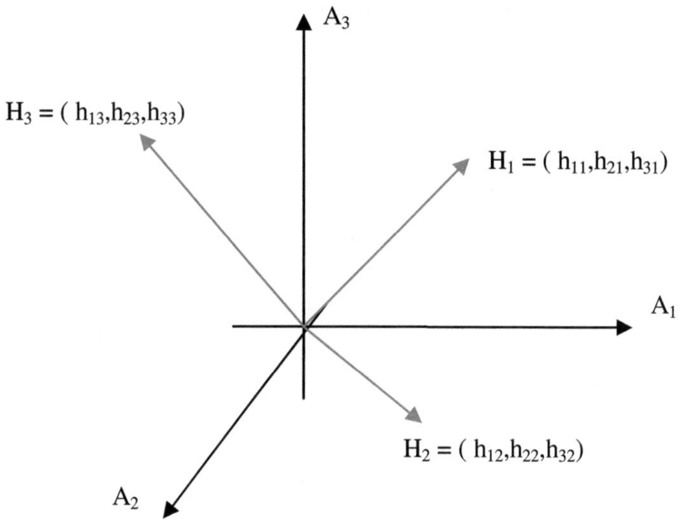

Fig. 2.1 ObjectsA$_k$ and Attributes H_j representing the column vectors in the matrix H

The vector X has three components w_1 , w_2 , w_3 on the reference given by the three vectors H_1 , H_2 , H_3 as shown in figure 2.2.

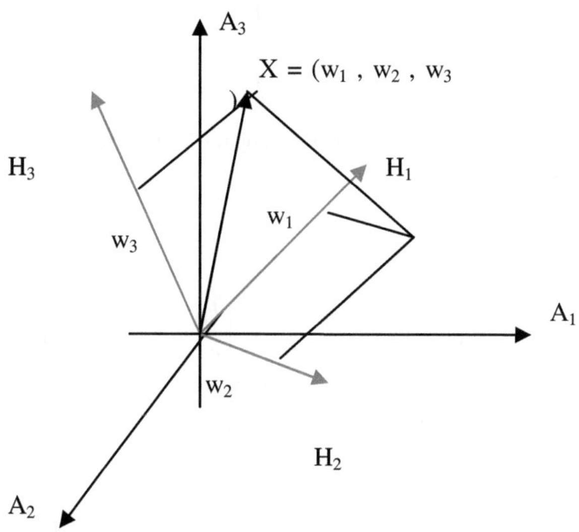

Fig. 2.2 Objects A_k and Attributes H_j represented by the column vectors in the matrix H and the coordinates **W** for the vector **X**.

In the classical tensor calculus equation (1.5) becomes

$$y_k = \sum_{j=1}^{p} h_k^j w_j = h_k^j w_j \tag{2.6}$$

The superposition given by the (1.5) or (1.6) can be represented by the neural scheme shown in figure 2.3.

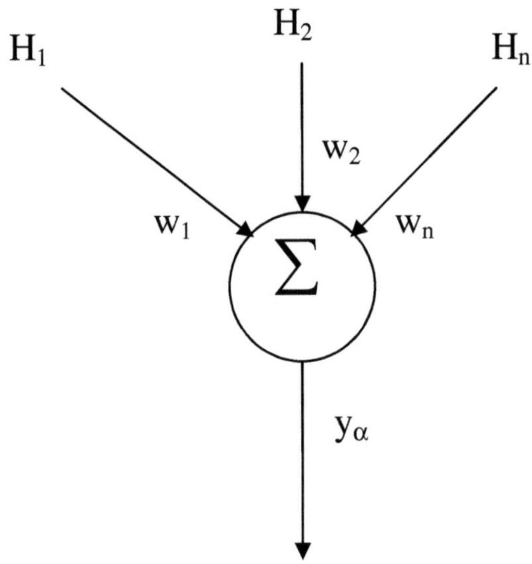

Fig. 2.3 The Neural Image of the Superposition by the Transformation of the Vector **W** into the Vector **Y**

When

$$H^T H = \begin{bmatrix} h_{11} & h_{12} & \cdots & h_{1p} \\ h_{21} & h_{22} & \cdots & h_{2p} \\ \cdots & \cdots & \cdots & \cdots \\ h_{q1} & h_{q2} & \cdots & h_{qp} \end{bmatrix}^T \begin{bmatrix} h_{11} & h_{12} & \cdots & h_{1p} \\ h_{21} & h_{22} & \cdots & h_{2p} \\ \cdots & \cdots & \cdots & \cdots \\ h_{q1} & h_{q2} & \cdots & h_{qp} \end{bmatrix} = \underset{\delta_{h,k}}{\begin{bmatrix} 1 & 0 & \cdots & 0 \\ 0 & 1 & \cdots & 0 \\ \cdots & \cdots & \cdots & \cdots \\ 0 & 0 & \cdots & 1 \end{bmatrix}} = \tag{2.7}$$

The set of vectors H_k are orthogonal to the others. The attribute reference B is a Cartesian orthogonal reference. We note that every vector relation is

$$\mathbf{V}_1 = \mathbf{V}_2 \tag{2.8}$$

When $\mathbf{V}_1$ becomes H $\mathbf{V}_1$ and B becomes H $\mathbf{V}_2$. H is a quadratic matrix and H^{-1} exists. We obtain

$$H\,\mathbf{V}_1 = H\,\mathbf{V}_2 \tag{2.9}$$

When we multiply equation (2.9) for H^{-1} we return to (2.8). It is possible to prove that any vector relation is invariant using the transformation H.

It can be proved that the value of the scalar product of two vectors $\mathbf{Z}_1$, $\mathbf{Z}_2$ for equation (2.7) is the same after the transformation H. In fact we have

$$(H\,\mathbf{Z}_1)^T (H\,\mathbf{Z}_2) = \mathbf{Z}_1^T H^T H\,\mathbf{Z}_2 = \mathbf{Z}_1^T \mathbf{Z}_2 \tag{2.10}$$

Where $\mathbf{Z}_1^T \mathbf{Z}_2$ is the scalar product of the vector $\mathbf{Z}_1$ and the vector $\mathbf{Z}_2$.

Computation of the Weights in the Superposition by Covariant and Controvariant Components of Vectors in Object Space

Now because

$$Y = H\,W, \quad \text{we have} \quad W = H^{-1} Y \tag{2.11}$$

When exist the inverse of H or H^{-1}.

Because the equation (2.11) can be written in this way

$$W = H^{-1} X = (H^T H)^{-1} H^T X \tag{2.12}$$

The identity in (2.12) is true only when H^{-1} exists.

When H^{-1} does not exist, but $(H^T H)^{-1}$ exists we can compute the weight W by the expression

$$W = (H^T H)^{-1} H^T X \tag{2.13}$$

Given the vector X in the space of the objects, equation (2.13) using the tensor calculus is

$$x^h = g^{-1} x_k \tag{2.14}$$

In tensor calculus g is the metric tensor and

$$g = H^T H = \sum_k h_{kj} h_{ki} = h^k_j h_{ki} \tag{2.15}$$

The covariant components of X are

$$x_k = H^T X = h_k^j A_j \tag{2.16}$$

The controvariant components of X are

$$x^i = g^{-1} x_j = g^{i,j} x_j = W \tag{2.17}$$

We note that *when H is a square matrix* we obtain

$$HW = h_{j,i} x^i \qquad h_{j,i} x^i = A_j \tag{2.18}$$

When H is a rectangular matrix with q > p, we have

$$QX = HW \qquad HW = H(H^T H)^{-1} H^T X = h_{j,i} x^i = y_j \tag{2.19}$$

With the property

$$Q^2 X = H(H^T H)^{-1} H^T H(H^T H)^{-1} H^T X = H(H^T H)^{-1} H^T X = QX \tag{2.20}$$

Where

$$Q x_k = h_{j,i} x^i = (h_{j,i} g^{iq} h_q^k) x_k = h_{j,i} (h_{pi} h_{pq})^{-1} h_q^k x_k \tag{2.21}$$

and

$$\begin{aligned} Q^2 x_k &= (h_{j,i} g^{ik}) h_q^k (h_{j,i} g^{ik}) h_q^k x_k = \\ h_{j,i} (h_{pi} h_k^p)^{-1} h_q^k \, h_{j,i} (h_{pi} h_k^p)^{-1} h_q^k x_k &= \\ h_{j,i} (h_{pi} h_k^p)^{-1} h_q^k x_k &= Q x_k \end{aligned} \tag{2.22}$$

Where Q is a projection operator. In figure 2.4 we show the geometric image of the projection operator for three dimensional object space and the two dimensional attribute space.

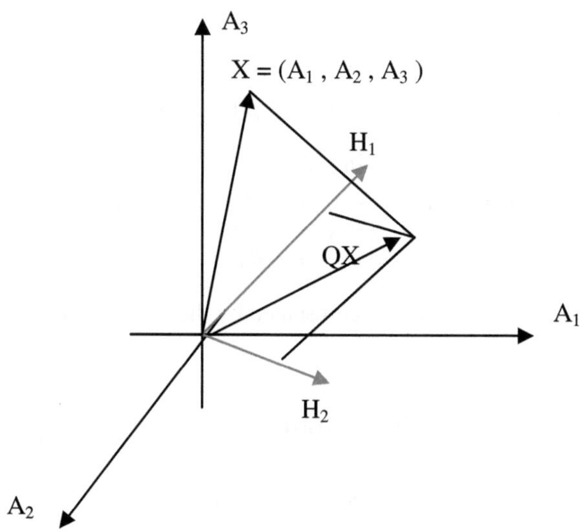

Fig. 2.4 The projection operator Q in the tensor calculus and in the formal concept analysis

Example

$$\begin{aligned} H &= \begin{bmatrix} \cos(\alpha) & \cos(\beta) \\ \sin(\alpha) & \sin(\beta) \end{bmatrix} \\ g &= H^T H = \\ &\begin{bmatrix} 1 & \cos(\alpha-\beta) \\ \cos(\alpha-\beta) & 1 \end{bmatrix} \end{aligned} \tag{2.23}$$

Given the vectors $X = \begin{bmatrix} A_1 \\ A_2 \end{bmatrix}$ in the space of the objects of table 2.1, using equation (2.16) we have

$$\begin{aligned}
&x_k = \begin{bmatrix} x_1 \\ x_2 \end{bmatrix} = H^T X = \\
&\begin{bmatrix} \cos(\alpha) & \sin(\alpha) \\ \cos(\beta) & \sin(\beta) \end{bmatrix} \begin{bmatrix} A_1 \\ A_2 \end{bmatrix} = \\
&\begin{bmatrix} A_1 \cos(\alpha) + A_2 \sin(\alpha) \\ A_1 \cos(\beta) + A_2 \sin(\beta) \end{bmatrix} \\
&x^h = g^{-1} x_k = \\
&\begin{bmatrix} \dfrac{(A_1 \cos(\alpha) + A_2 \sin(\alpha)) - \cos(\alpha - \beta)(A_1 \cos(\beta) + A_2 \sin(\beta))}{\sin^2(\alpha - \beta)} \\ \dfrac{(A_1 \cos(\beta) + A_2 \sin(\beta)) - \cos(\alpha - \beta)(A_1 \cos(\alpha) + A_2 \sin(\alpha))}{\sin^2(\alpha - \beta)} \end{bmatrix}
\end{aligned} \tag{2.24}$$

The two components of X can be represented by the graph in figure 2.5.

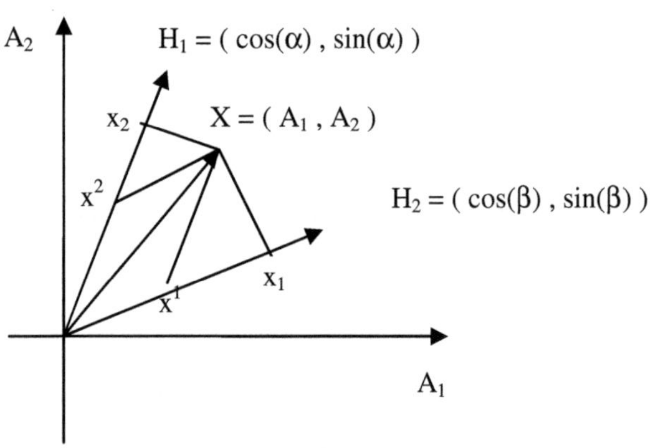

Fig. 2.5 General coordinates H_1, H_2, Object coordinates A_1, A_2. The components of X, the Covariant components x_k and controvariant components x^k.

Another example is the general transformation of the coordinates.

$$\begin{cases} \bar{z}_1 = \bar{z}_1(z_1, z_2, \ldots\ldots, z_n) \\ \bar{z}2 = \bar{z}2(z_1, z_2, \ldots\ldots, z_n) \\ \ldots \\ \bar{z}n = \bar{z}n(z_1, z_2, \ldots\ldots, z_n) \end{cases} \tag{2.25}$$

Where the matrix H is

$$H = \begin{bmatrix} \frac{\partial z_1}{\partial \bar{z}1} & \frac{\partial z_2}{\partial \bar{z}1} & \ldots & \frac{\partial z_n}{\partial \bar{z}1} \\ \frac{\partial z_1}{\partial \bar{z}2} & \frac{\partial z_2}{\partial \bar{z}2} & \ldots & \frac{\partial z_n}{\partial \bar{z}2} \\ \ldots & \ldots & \ldots & \ldots \\ \frac{\partial z_1}{\partial \bar{z}n} & \frac{\partial z_2}{\partial \bar{z}n} & \ldots & \frac{\partial z_n}{\partial \bar{z}n} \end{bmatrix} \tag{2.26}$$

A simple system $X = x_k$ is covariant when the transformation (2.25) changes as shown

$$\bar{x}k = H^T X = \sum_j \frac{\partial z^k}{\partial \bar{z}j} x_j \tag{2.27}$$

Where x_j is in the Euclidean geometry and in this theory is located in the object space. A system is controvariant when the transformation (2.25) changes as shown

$$\begin{gathered} \bar{x}^k = (H^T H)^{-1} H^T X = \\ H^{-1} X = \sum_j \frac{\partial \bar{z}^k}{\partial z_j} x^j = \\ g^{-1} x_j = g^{k,j} \bar{x} j \end{gathered} \tag{2.28}$$

Note that for the Jacobian property we have

$$H^{-1} = \begin{bmatrix} \frac{\partial \bar{z}1}{\partial z_1} & \frac{\partial \bar{z}2}{\partial z_1} & \ldots & \frac{\partial \bar{z}n}{\partial z_1} \\ \frac{\partial \bar{z}1}{\partial z_2} & \frac{\partial \bar{z}2}{\partial z_2} & \ldots & \frac{\partial \bar{z}n}{\partial z_2} \\ \ldots & \ldots & \ldots & \ldots \\ \frac{\partial \bar{z}1}{\partial z_n} & \frac{\partial \bar{z}2}{\partial z_n} & \ldots & \frac{\partial \bar{z}n}{\partial z_n} \end{bmatrix} \tag{2.29}$$

Here the projection operator Q X is the identity QX = X . The transformation (2.25) is an internal isomorphism the dimensions of the space are unchanged after the transformation.

The interpretation of the tensor calculus by formal concept analysis is an extension of classical tensor calculus of differential geometry.

2.2 Invariants

The space of the objects includes the vectors of the attributes. Here, any vector X is projected in the object subspace defined by the vectors of the attributes. Given QX = y_k, we compute the scalar

$$s^2 = (QX)^T (QX) = \sum_k y_k y_k = y^k y_k \tag{2.30}$$

Note that the space of the objects is a Cartesian Space with the Euclidean geometry, and

$$y^k = y_k$$

In general coordinates the transformation of the components QX into the space of the objects the components of the attributes are

$$\bar{y}_k = H^T QX = h_k^j y_j \tag{2.31}$$

and

$$\bar{y}^k = (H^T H)^{-1} H^T QX = g^{-1} h_k^j y_j \tag{2.32}$$

The definition of the projection operator, equations (2.31) and (2.32) can be written as

$$\bar{y}_k = H^T H(H^T H)^{-1} H^T X = H^T X = h_k^j x_j = \bar{x}_k \tag{2.33}$$

and

$$\bar{y}^k = (H^T H)^{-1} H^T H (H^T H)^{-1} H^T X = (H^T H)^{-1} H^T X = g^{-1} h_k^j x_j = \bar{x}^k \tag{2.34}$$

The scalar can be computed using the new components as

$$
\begin{aligned}
& s^2 = \bar{y}^k \bar{y}_k = \\
& ((H^T H)^{-1} H^T X)^T H^T X = \\
& X^T H (H^T H)^{-1} (H^T H)(H^T H)^{-1} H^T X = \\
& X^T (H (H^T H)^{-1} H^T) H (H^T H)^{-1} H^T X = \\
& X^T Q^T Q X = \\
& (QX)^T (QX) = y^k y_k
\end{aligned}
\tag{2.35}
$$

Now solve the equation

$$QX = X \tag{2.36}$$

When H is a quadratic matrix we have

$$
\begin{aligned}
& QX = H(H^T H)^{-1} H^T X = \\
& H(H)^{-1} (H^T)^{-1} H^T X = X
\end{aligned}
\tag{2.37}
$$

When $H = \begin{bmatrix} A \\ B \end{bmatrix}$ and A is a quadratic matrix with the p dimension, we can write $B = C\,A$, where $C = B\,A^{-1}$. When $X = \begin{bmatrix} X_A \\ X_B \end{bmatrix}$ and X_A is a vector with p dimensions , and $X_B = C\,X_A$. We can say that $Q\,X = X$.

Now we have

$$HS = X$$

That can write in this way

$$H^T H S = H^T X$$

Or

$$S = (H^T H)^{-1} H^T X$$

So we have

$$\begin{bmatrix} A \\ B \end{bmatrix} S = \begin{bmatrix} A \\ CA \end{bmatrix} S = \begin{bmatrix} AS \\ CAS \end{bmatrix} = \begin{bmatrix} X_A \\ CX_A \end{bmatrix} = \begin{bmatrix} X_A \\ X_B \end{bmatrix}$$

And we have the INVARIANT form

$$CX_A = X_B \tag{2.38}$$

All the vector X that cannot satisfy at the previous invariant has the projection operator Q X different from X.

We obtain

$$
\begin{aligned}
QX &= \begin{bmatrix} A \\ CA \end{bmatrix} \left(\begin{bmatrix} A \\ CA \end{bmatrix}^T \begin{bmatrix} A \\ CA \end{bmatrix} \right)^{-1} \begin{bmatrix} A \\ CA \end{bmatrix}^T \begin{bmatrix} X_A \\ CX_A \end{bmatrix} = \\
&\begin{bmatrix} A \\ CA \end{bmatrix} \left(\begin{bmatrix} A & CA \end{bmatrix} \begin{bmatrix} A \\ CA \end{bmatrix} \right)^{-1} \begin{bmatrix} A \\ CA \end{bmatrix}^T \begin{bmatrix} X_A \\ CX_A \end{bmatrix} \\
&= \begin{bmatrix} A \\ CA \end{bmatrix} (A^T A + (CA)^T (CA))^{-1} \begin{bmatrix} A \\ CA \end{bmatrix}^T \begin{bmatrix} X_A \\ CX_A \end{bmatrix} \\
&= \begin{bmatrix} A \\ CA \end{bmatrix} (A^T A + A^T C^T CA))^{-1} \begin{bmatrix} A \\ CA \end{bmatrix}^T \begin{bmatrix} X_A \\ CX_A \end{bmatrix} \\
&= \begin{bmatrix} A \\ CA \end{bmatrix} (A^T (A + C^T CA))^{-1} \begin{bmatrix} A \\ CA \end{bmatrix}^T \begin{bmatrix} X_A \\ CX_A \end{bmatrix} \\
&= \begin{bmatrix} A \\ CA \end{bmatrix} (A^T (I + C^T C)A)^{-1} \begin{bmatrix} A \\ CA \end{bmatrix}^T \begin{bmatrix} X_A \\ CX_A \end{bmatrix} \\
&= \begin{bmatrix} A \\ CA \end{bmatrix} A^{-1} (I + C^T C)^{-1} (A^T)^{-1} \begin{bmatrix} A \\ CA \end{bmatrix}^T \begin{bmatrix} X_A \\ CX_A \end{bmatrix} \\
&= \begin{bmatrix} I \\ C \end{bmatrix} (I + C^T C)^{-1} \begin{bmatrix} I & C^T \end{bmatrix} \begin{bmatrix} X_A \\ CX_A \end{bmatrix} \\
&= \begin{bmatrix} I \\ C \end{bmatrix} (I + C^T C)^{-1} (I + C^T C) X_A = \begin{bmatrix} I \\ C \end{bmatrix} X_A = \begin{bmatrix} X_A \\ CX_A \end{bmatrix}
\end{aligned}
\tag{2.39}
$$

When QX = X, we have

$$s^2 = \bar{y}^k \bar{y}_k = x^k x_k$$

Because x is a vector in the Euclidean Space of the objects, $x^k = x_k$ and

$$s^2 = \bar{y}^k \bar{y}_k = x^k x_k = \sum_k x_k^2$$

In this case s^2 is invariant for any context H where Q X = X.

Example

For $H = \begin{bmatrix} 1 & 0 \\ 1 & -1 \\ 0 & 1 \end{bmatrix}$ we have $A = \begin{bmatrix} 1 & 0 \\ 1 & -1 \end{bmatrix}$, $B = \begin{bmatrix} 0 & 1 \end{bmatrix}$

$$C = BA^{-1} = \begin{bmatrix} 1 & -1 \end{bmatrix}$$

Now for

$$X = \begin{bmatrix} x_1 \\ x_2 \\ C\begin{bmatrix} x_1 \\ x_2 \end{bmatrix} \end{bmatrix} = \begin{bmatrix} x_1 \\ x_2 \\ x_1 - x_2 \end{bmatrix}$$

We have

$$Q\begin{bmatrix} x_1 \\ x_2 \\ x_1 - x_2 \end{bmatrix} = \begin{bmatrix} x_1 \\ x_2 \\ x_1 - x_2 \end{bmatrix}$$

Now for $X = \begin{bmatrix} X_A \\ X_B \end{bmatrix}$

$$\begin{aligned} QX &= \begin{bmatrix} I \\ C \end{bmatrix} (I + C^T C)^{-1} \begin{bmatrix} I & C^T \end{bmatrix} \begin{bmatrix} X_A \\ X_B \end{bmatrix} = \\ &= \begin{bmatrix} I \\ C \end{bmatrix} (I + C^T C)^{-1} (X_A + C^T X_B) \\ &= \begin{bmatrix} I \\ C \end{bmatrix} (I + C^T C)^{-1} (X_A + C^T (CX_A + D)) \qquad (2.40) \\ &= \begin{bmatrix} I \\ C \end{bmatrix} X_A + \begin{bmatrix} I \\ C \end{bmatrix} (I + C^T C)^{-1} C^T D \\ &= \begin{bmatrix} X_A + (I + C^T C)^{-1} C^T D \\ C(X_A + (I + C^T C)^{-1} C^T D \end{bmatrix} = \begin{bmatrix} X_A + E \\ C(X_A + E) \end{bmatrix} \end{aligned}$$

Example

When

$$X = \begin{bmatrix} x_1 \\ x_2 \\ x_3 \end{bmatrix}, X_A = \begin{bmatrix} x_1 \\ x_2 \end{bmatrix}, X_B = \begin{bmatrix} x_3 \end{bmatrix}$$

$$H = \begin{bmatrix} 1 & 0 \\ 1 & -1 \\ 0 & 1 \end{bmatrix}$$

We have

$$C = \begin{bmatrix} 1 & -1 \end{bmatrix}$$

$$X_B = CX_A + D$$

$$D = \begin{bmatrix} x_3 \end{bmatrix} - \begin{bmatrix} x_1 - x_2 \end{bmatrix}$$

and

$$QX = \begin{bmatrix} X_A \\ CX_A \end{bmatrix} + \begin{bmatrix} (I+C^T C)^{-1} C^T D \\ C(I+C^T C)^{-1} C^T D \end{bmatrix} =$$

$$\begin{bmatrix} x_1 \\ x_2 \\ x_1 - x_2 \end{bmatrix} + \begin{bmatrix} \frac{1}{3}(x_3 - x_1 + x_2) \\ \frac{1}{3}(-x_3 + x_1 - x_2) \\ \frac{2}{3}(x_3 - x_1 + x_2) \end{bmatrix}$$

Where

$$\frac{1}{3}(x_3 - x_1 + x_2) - \frac{1}{3}(-x_3 + x_1 - x_2) = \frac{2}{3}(-x_3 + x_1 - x_2)$$

or

$$\begin{bmatrix} \frac{1}{3}(x_3 - x_1 + x_2) \\ \frac{1}{3}(-x_3 + x_1 - x_2) \\ \frac{2}{3}(x_3 - x_1 + x_2) \end{bmatrix} = \begin{bmatrix} E \\ CE \end{bmatrix}$$

Where

$$E = \begin{bmatrix} \frac{1}{3}(x_3 - x_1 + x_2) \\ \frac{1}{3}(-x_3 + x_1 - x_2) \end{bmatrix}$$

In (2.35) we use the relation

$$Qx_k = H(H^T H)^{-1} H^T X =$$

$$Q^T x_k = (H(H^T H)^{-1} H^T)^T x_k$$

for

$$(H^T H)^T = H^T H \tag{2.41}$$

we have

$$(H(H^T H)^{-1} H^T)^T x_k = H(H^T H)^{-1} H^T X$$

$$Q\, x_k = Q^T x_k$$

In conclusion

$$\begin{aligned} s^2 &= \bar{y}^k \, \bar{y}_k = \\ \bar{x}^k \, \bar{x}_k &= y^k \, y_k \end{aligned} \tag{2.42}$$

The scalar s^2 is invariant when we move from the space of the objects to the space of the attributes. Equation (2.37) can be written as

$$\begin{aligned} s^2 &= \bar{y}^k \, \bar{y}_k = \\ \bar{y}^k \, g \, \bar{y}^k &= \bar{x}^k \, g \, \bar{x}^k \\ \bar{y}_k g^{-1} \bar{y}_k &= \bar{x}_k \, g^{-1} \bar{x}_k \end{aligned} \tag{2.43}$$

The equation (2.37) with the weights can be written as

$$\begin{aligned} s^2 &= W^T g \, W \\ W &= (H^T H)^{-1} H^T X = \bar{x}^k \end{aligned} \tag{2.44}$$

For the projection operator Q we have the property

$$(X - Q\,X)^T Q\,X = X^T Q\,X - (Q\,X)^T Q\,X = X^T Q\,X - X^T Q^T Q\,X$$

but as $Q^T Q = Q$ so we have

$$(X - QX)^T QX = X^T Q\,X - X^T Q\,X = 0 \tag{2.45}$$

The vector $(X - Q\,X)$ is orthogonal to the vector $Q\,X$

Projection Space

When we have n attribute spaces

$$H^1 = \begin{bmatrix} h_{11}^{\;\;1} & h_{12}^{\;\;1} & \dots & h_{1p}^{\;\;1} \\ h_{21}^{\;\;1} & h_{22}^{\;\;1} & \dots & h_{2p}^{\;\;1} \\ \dots & \dots & \dots & \dots \\ h_{q1}^{\;\;1} & h_{q2}^{\;\;1} & \dots & h_{qp}^{\;\;1} \end{bmatrix}$$

$$H^2 = \begin{bmatrix} h_{11}^{\;\;2} & h_{12}^{\;\;2} & \dots & h_{1p}^{\;\;2} \\ h_{21}^{\;\;2} & h_{22}^{\;\;2} & \dots & h_{2p}^{\;\;2} \\ \dots & \dots & \dots & \dots \\ h_{q1}^{\;\;2} & h_{q2}^{\;\;2} & \dots & h_{qp}^{\;\;2} \end{bmatrix}$$

....

$$H^n = \begin{bmatrix} h_{11}^{\;\;n} & h_{12}^{\;\;n} & \dots & h_{1p}^{\;\;n} \\ h_{21}^{\;\;n} & h_{22}^{\;\;n} & \dots & h_{2p}^{\;\;n} \\ \dots & \dots & \dots & \dots \\ h_{q1}^{\;\;n} & h_{q2}^{\;\;n} & \dots & h_{qp}^{\;\;n} \end{bmatrix} \tag{2.46}$$

The projection space is given by the vectors

$$P = [Q_1 X, Q_2 X, \dots\dots, Q_n X] \tag{2.47}$$

Where Q_k are the k projection operators into the different attribute spaces in (2.40).

We now compute the scalar s_k^2 for the different projections. We obtain

$$\begin{aligned} & X^T [Q_1 X, Q_2 X, \dots\dots, Q_n X] = [X^T Q_1 X, X^T Q_2 X, \dots\dots, X^T Q_n X] = \\ & [X^T Q_1^{\;T} Q_1 X, X^T Q_2^{\;T} Q_2 X, \dots\dots, X^T Q_n^{\;T} Q_n X] = \\ & [(Q_1 X)^T Q_1 X, (Q_2 X)^T Q_2 X, \dots\dots, (Q_n X)^T Q_n X] = \\ & [s_1^2, s_2^2, \dots\dots, s_n^2] = S \end{aligned} \tag{2.48}$$

Then

$$X^T P = S \tag{2.49}$$

and

$$\begin{aligned} P^T X &= S^T \\ X &= (P^T)^{-1} S^T \end{aligned} \tag{2.50}$$

When the inverse $(P^T)^{-1}$ exists we can compute X from its projection P.

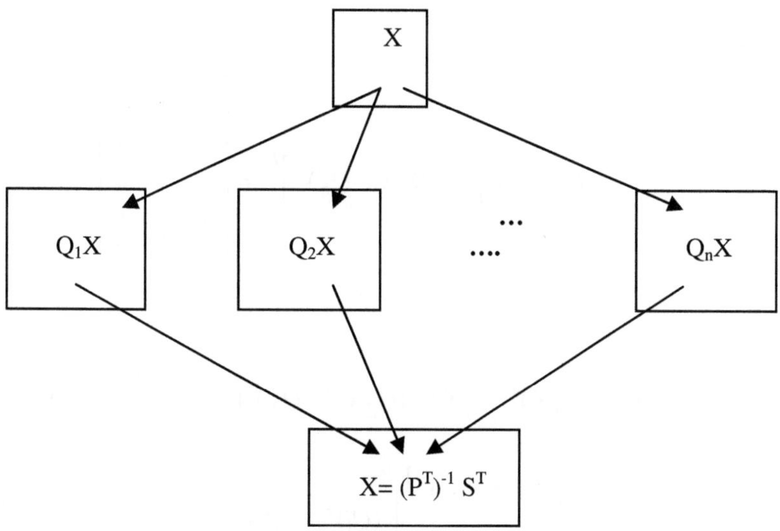

Fig. 2.6 The projection of V on the different genome spaces is used to obtain the projections Q_1X ,Q_2X,...., Q_n. X. Using the projections we can rebuilt the original vector X

The scheme of a multimedia connection between contexts and super context is shown in figure 2.6.

For example given three abstract spaces

$$H^1 = \begin{bmatrix} 1 & 1 \\ 1 & 0 \\ 0 & 1 \end{bmatrix}, \; H^2 = \begin{bmatrix} 1 & 0 \\ 1 & 1 \\ 0 & 0 \end{bmatrix}, \; H^3 = \begin{bmatrix} 1 & 0 \\ 0 & 1 \\ 1 & 0 \end{bmatrix}$$

and using the vector $X = \begin{bmatrix} 0 \\ 1 \\ 1 \end{bmatrix}$, we obtain the three projections

$$Q_1 X = \begin{bmatrix} \frac{2}{3} \\ \frac{1}{3} \\ \frac{1}{3} \end{bmatrix}, Q_2 X = \begin{bmatrix} 0 \\ 1 \\ 0 \end{bmatrix}, Q_3 X = \begin{bmatrix} \frac{1}{2} \\ 1 \\ \frac{1}{2} \end{bmatrix}$$

and

$$P^T = \begin{bmatrix} Q_1 X & Q_2 X & Q_3 X \end{bmatrix}^T = \begin{bmatrix} \frac{2}{3} & \frac{1}{3} & \frac{1}{3} \\ 0 & 1 & 0 \\ \frac{1}{2} & 1 & \frac{1}{2} \end{bmatrix}$$

for

$$S = \begin{bmatrix} (Q_1 X)^T (Q_1 X) & (Q_2 X)^T (Q_2 X) & (Q_3 X)^T (Q_3 X) \end{bmatrix}$$

$$X = (P^T)^{-1} S^T = \begin{bmatrix} \frac{2}{3} & \frac{1}{3} & \frac{1}{3} \\ 0 & 1 & 0 \\ \frac{1}{2} & 1 & \frac{1}{2} \end{bmatrix}^{-1} \begin{bmatrix} (Q_1 V)^T (Q_1 V) \\ (Q_2 V)^T (Q_2 V) \\ (Q_3 V)^T (Q_3 V) \end{bmatrix} = \begin{bmatrix} 3 & 1 & -2 \\ 0 & 1 & 0 \\ -3 & -3 & 4 \end{bmatrix} \begin{bmatrix} \frac{2}{3} \\ 1 \\ \frac{3}{2} \end{bmatrix} = \begin{bmatrix} 0 \\ 1 \\ 1 \end{bmatrix}$$

or

$$\begin{bmatrix} 0 \\ 1 \\ 1 \end{bmatrix} = \frac{2}{3} \begin{bmatrix} 3 \\ 0 \\ -3 \end{bmatrix} + 1 \begin{bmatrix} 1 \\ 1 \\ -3 \end{bmatrix} + \frac{3}{2} \begin{bmatrix} -2 \\ 0 \\ 4 \end{bmatrix}$$

Extension of the superposition of the attributes

Previously, we have defined a superposition process with the weights W. Now we can extend the superposition process in this way.

$$\begin{bmatrix} y_1 \\ y_2 \\ \dots \\ y_q \end{bmatrix} = \begin{bmatrix} h_{11} & h_{12} & \dots & h_{1p} \\ h_{21} & h_{22} & \dots & h_{2p} \\ \dots & \dots & \dots & \dots \\ h_{q1} & h_{q2} & \dots & h_{qp} \end{bmatrix} \begin{bmatrix} w_1 \\ w_2 \\ \dots \\ w_p \end{bmatrix}$$

or (2.51)

$$\begin{bmatrix} y_1 \\ y_2 \\ \dots \\ y_q \end{bmatrix} = w_1 \begin{bmatrix} h_{11} \\ h_{21} \\ \dots \\ h_{q1} \end{bmatrix} + w_2 \begin{bmatrix} h_{12} \\ h_{22} \\ \dots \\ h_{q2} \end{bmatrix} + \dots\dots + w_p \begin{bmatrix} h_{1p} \\ h_{2p} \\ \dots \\ h_{qp} \end{bmatrix}$$

The weights $\mathbf{w_j}$ here are vectors. The expression (2.45) has a meaning only when the components of the vectors are defined.

$$w_1, w_2, \dots\dots, w_p \tag{2.52}$$

For the neural image of the extended superposition we have

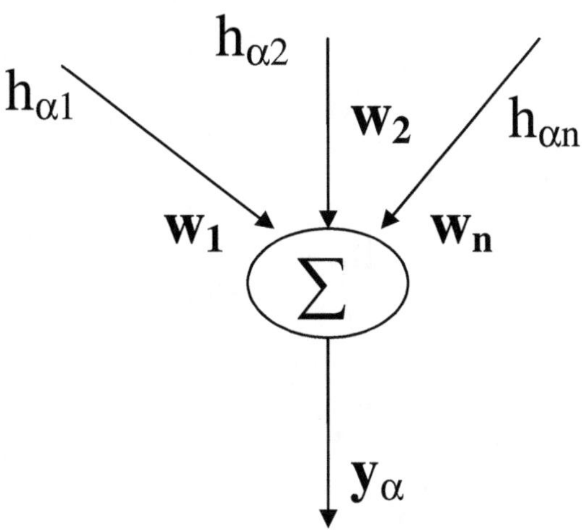

Fig. 2.7 Any weight in the new type of neuron is a vector

For equation (2.5) the components $Z_{i,j}$ of the vectors w_i on the axis

$$\begin{bmatrix} h_{11} \\ h_{21} \\ \dots \\ h_{q1} \end{bmatrix}, \begin{bmatrix} h_{12} \\ h_{22} \\ \dots \\ h_{q2} \end{bmatrix}, \dots\dots\dots, \begin{bmatrix} h_{1p} \\ h_{2p} \\ \dots \\ h_{qp} \end{bmatrix}$$

satisfy relation (2.13).

$$w_j = Hz_j = \begin{bmatrix} h_{11} & h_{12} & \dots & h_{1p} \\ h_{21} & h_{22} & \dots & h_{2p} \\ \dots & \dots & \dots & \dots \\ h_{q1} & h_{q2} & \dots & h_{qp} \end{bmatrix} \begin{bmatrix} z_{j,1} \\ z_{j,2} \\ \dots \\ z_{j,p} \end{bmatrix} \tag{2.53}$$

or

$$w_j = z_{j,1} \begin{bmatrix} h_{11} \\ h_{21} \\ \dots \\ h_{q1} \end{bmatrix} + z_{j,2} \begin{bmatrix} h_{12} \\ h_{22} \\ \dots \\ h_{q2} \end{bmatrix} + \dots\dots + z_{j,p} \begin{bmatrix} h_{1p} \\ h_{2p} \\ \dots \\ h_{qp} \end{bmatrix} =$$

$$z_{j,1}H_{\alpha,1} + z_{j,2}H_{\alpha,2} + \dots + z_{j,p}H_{\alpha,p}$$

When we substitute the (2.47) into (2.45), we have

$$y_{h,k} = \begin{bmatrix} y_{11} & y_{12} & \dots & y_{1p} \\ y_{21} & y_{22} & \dots & y_{2p} \\ \dots & \dots & \dots & \dots \\ y_{q1} & y_{q2} & \dots & y_{qp} \end{bmatrix} = \begin{bmatrix} h_{11} & h_{12} & \dots & h_{1p} \\ h_{21} & h_{22} & \dots & h_{2p} \\ \dots & \dots & \dots & \dots \\ h_{q1} & h_{q2} & \dots & h_{qp} \end{bmatrix} \begin{bmatrix} \sum_{j=1}^{p} z_{1.j}h_{k.j} \\ \sum_{j=1}^{p} z_{2.j}H_{k.j} \\ \dots \\ \sum_{j=1}^{p} z_{p.j}H_{k.j} \end{bmatrix} = \sum_i \sum_j h_{j,h}h_{i,k}z_{j,i} \tag{1.48}$$

In the neuron image we have

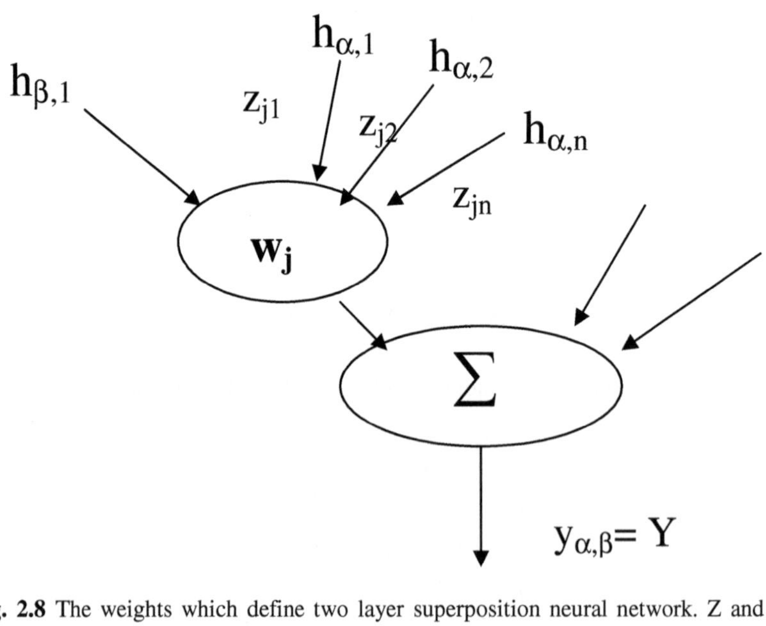

Fig. 2.8 The weights which define two layer superposition neural network. Z and Y are tensors.

In classical tensor calculus y is a tensor of the second order. Equation (2.48) can be written in this way

$$y = (H \oplus H)\, Z \tag{2.54}$$

where V and Z are matrices and $H \oplus H$ is the tensor product of X with itself.

For example given

$$H = \begin{bmatrix} h_{1,1} & h_{1,2} & h_{1,3} \\ h_{2,1} & h_{2,2} & h_{2,3} \\ h_{3,1} & h_{3,2} & h_{3,3} \end{bmatrix} \tag{2.55}$$

with the tensor product we have

$$H \oplus H = \begin{bmatrix} h_{1,1}h_{1,1} & h_{1,1}h_{1,2} & h_{1,1}h_{1,3} & h_{1,2}h_{1,1} & h_{1,2}h_{1,2} & h_{1,2}h_{1,3} & h_{1,3}h_{1,1} & h_{1,3}h_{1,2} & h_{1,3}h_{1,3} \\ h_{1,1}h_{2,1} & h_{1,1}h_{2,2} & h_{1,1}h_{2,3} & h_{1,2}h_{2,1} & h_{1,2}h_{2,2} & h_{1,2}h_{2,3} & h_{1,3}h_{2,1} & h_{1,3}h_{2,2} & h_{1,3}h_{2,3} \\ h_{1,1}h_{3,1} & h_{1,1}h_{3,2} & h_{1,1}h_{3,3} & h_{1,2}h_{3,1} & h_{1,2}h_{3,2} & h_{1,2}h_{3,3} & h_{1,3}h_{3,1} & h_{1,3}h_{3,3} & h_{1,3}h_{3,3} \\ h_{2,1}h_{1,1} & h_{2,1}h_{1,1} & h_{2,1}h_{1,1} & h_{2,2}h_{1,1} & h_{2,2}h_{1,2} & h_{2,2}h_{1,3} & h_{2,3}h_{1,1} & h_{2,3}h_{1,2} & h_{2,3}h_{1,3} \\ h_{2,1}h_{1,1} & h_{2,1}h_{1,1} & h_{2,1}h_{1,1} & h_{2,2}h_{2,1} & h_{2,2}h_{2,2} & h_{2,2}h_{2,3} & h_{2,3}h_{2,1} & h_{2,3}h_{2,2} & h_{2,3}h_{2,3} \\ h_{2,1}h_{1,1} & h_{2,1}h_{1,1} & h_{2,1}h_{1,1} & h_{2,2}h_{3,1} & h_{2,2}h_{3,2} & h_{2,2}h_{3,3} & h_{2,3}h_{3,1} & h_{2,3}h_{3,2} & h_{2,3}h_{3,3} \\ h_{3,1}h_{1,1} & h_{3,1}h_{1,1} & h_{3,1}h_{1,1} & h_{3,2}h_{1,1} & h_{3,2}h_{1,2} & h_{3,2}h_{1,3} & h_{3,3}h_{1,1} & h_{3,3}h_{1,2} & h_{3,3}h_{1,3} \\ h_{3,1}h_{1,1} & h_{3,1}h_{1,1} & h_{3,1}h_{1,1} & h_{3,2}h_{2,1} & h_{3,2}h_{2,2} & h_{3,2}h_{2,3} & h_{3,3}h_{2,1} & h_{3,3}h_{2,2} & h_{3,3}h_{2,3} \\ h_{3,1}h_{1,1} & h_{31}h_{1,1} & h_{3,1}h_{1,1} & h_{3,2}h_{3,1} & h_{3,2}h_{3,2} & h_{3,2}h_{3,3} & h_{3,3}h_{3,1} & h_{3,3}h_{3,2} & h_{3,3}h_{3,3} \end{bmatrix} \tag{1.51}$$

we have

$$Y = \begin{bmatrix} y_{1,1} & y_{1,2} & y_{1,3} \\ y_{2,1} & y_{2,2} & y_{2,3} \\ y_{3,1} & y_{3,2} & y_{3,3} \end{bmatrix} = z_{1,1} \begin{bmatrix} h_{1,1}h_{1,1} & h_{1,1}h_{2,1} & h_{1,1}h_{3,1} \\ h_{2,1}h_{1,1} & h_{2,1}h_{2,1} & h_{2,1}h_{3,1} \\ h_{3,1}h_{1,1} & h_{3,1}h_{2,1} & h_{3,1}h_{3,1} \end{bmatrix} + z_{1,2} \begin{bmatrix} h_{1,1}h_{1,2} & h_{1,1}h_{2,2} & h_{1,1}h_{3,2} \\ h_{2,1}h_{1,2} & h_{2,1}h_{2,2} & h_{2,1}h_{3,2} \\ h_{3,1}h_{1,2} & h_{3,1}h_{2,2} & h_{3,1}h_{3,2} \end{bmatrix} + \dots$$

$$+ z_{3,3} \begin{bmatrix} h_{1,3}h_{1,3} & h_{1,3}h_{2,3} & h_{1,3}h_{3,3} \\ h_{2,3}h_{1,3} & h_{2,3}h_{2,3} & h_{2,3}h_{3,3} \\ h_{3,3}h_{1,3} & h_{3,3}h_{2,3} & h_{3,3}h_{3,3} \end{bmatrix} \tag{1.52}$$

where

$$R_{j,i} = \begin{bmatrix} h_{1,j}h_{1,i} & h_{1,j}h_{2,i} & h_{1,j}h_{3,i} \\ h_{2,j}h_{1,i} & h_{2,j}h_{2,i} & h_{2,j}h_{3,i} \\ h_{3,j}h_{1,i} & h_{3,j}h_{2,i} & h_{3,j}h_{3,i} \end{bmatrix} \tag{2.56}$$

are the nine components of a vector in a space of nine dimensions. The values of the tensor Z are the components of the tensor Y on the reference whose axes are

$$H_1 = \begin{bmatrix} h_{11} \\ h_{21} \\ h_{31} \end{bmatrix}, \ H_2 = \begin{bmatrix} h_{12} \\ h_{22} \\ h_{32} \end{bmatrix}, \ H_3 = \begin{bmatrix} h_{13} \\ h_{23} \\ h_{33} \end{bmatrix} \tag{2.57}$$

Second Order Tensor and Relations

Given the set of relations shown in figure (2.9)

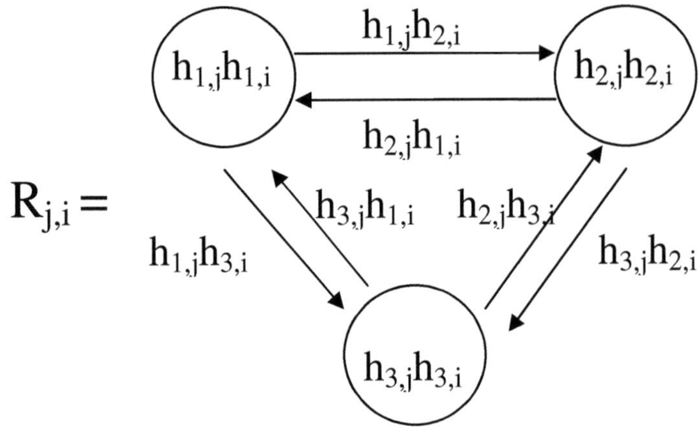

Fig. 2.9 The basic set of 9 relations between three elements. We move from one relation to another by the variation of the index j and the index i.

In the extension of the superposition for the (2.48) the components $z_{1,j}$ of the tensor y can be represented by Figure 2.10.

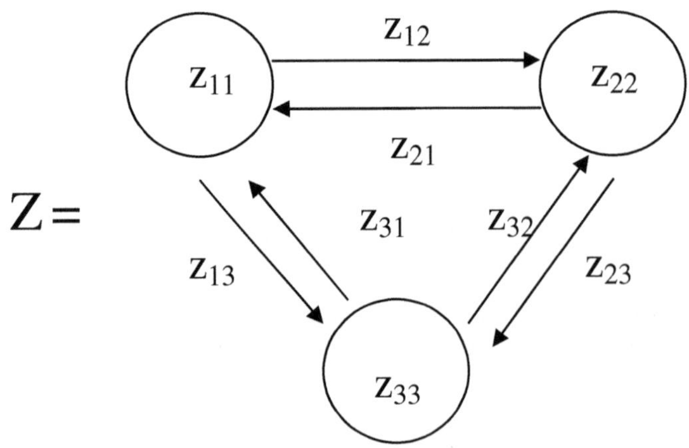

Fig. 2.10 The graph image of the components $z_{i,j}$ of the tensor y

Figure 2.11 shows the tensor y and the components in the objects space.

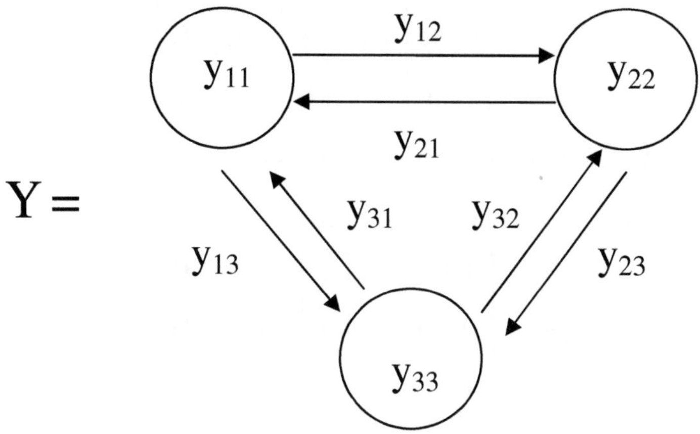

Fig. 2.11 Image of the tensor $y_{j,I}$ as a relationship

In conclusion using the three vectors H_1 , H_2 and h_3, it is possible to determine the nine components Z that shown in the graph in Figure 2.10 of the tensor Y which is given in Figure 2.11. This is done by using the components in the space of the objects in H.

Example

In table 2.3 there are two colours "Red" and "Yellow", The book has only one colour "Yellow" and the Pencil has two colours "Red" and "Green".

Table 2.3 Context representation of the three objects A and the three attributes B

Context	Red	Yellow	*Green*
Table	1	1	*0*
Book	0	1	*0*
Pencil	*1*	*0*	*1*

Here we have the reference

$$H = \begin{bmatrix} 1 & 1 & 0 \\ 0 & 1 & 0 \\ 1 & 0 & 1 \end{bmatrix} \tag{2.58}$$

For equation (2.5) we then have

$$Y = w_1 \begin{bmatrix} 1 \\ 0 \\ 1 \end{bmatrix} + w_2 \begin{bmatrix} 1 \\ 1 \\ 0 \end{bmatrix} + w_3 \begin{bmatrix} 0 \\ 0 \\ 1 \end{bmatrix} \tag{2.59}$$

The vector Y is formed by the superposition of the three vectors for any of the colours.

The components Z of the graph on the axis

$$H_1 = \begin{bmatrix} 1 \\ 0 \\ 1 \end{bmatrix}, H_2 = \begin{bmatrix} 1 \\ 1 \\ 0 \end{bmatrix}, H_3 = \begin{bmatrix} 0 \\ 0 \\ 1 \end{bmatrix} \tag{2.60}$$

These are the weights of the superposition of the 9 basic relations. We show one of these relations in 1.5.

$$R_{1,1} = \begin{bmatrix} h_{1,1}h_{1,1} & h_{1,1}h_{2,1} & h_{1,1}h_{3,1} \\ h_{2,1}h_{1,1} & h_{2,1}h_{2,1} & h_{2,1}h_{3,1} \\ h_{3,1}h_{1,1} & h_{3,1}h_{2,1} & h_{3,1}h_{3,1} \end{bmatrix} = \begin{bmatrix} 1 & 0 & 1 \\ 0 & 0 & 0 \\ 1 & 0 & 1 \end{bmatrix} \tag{2.61}$$

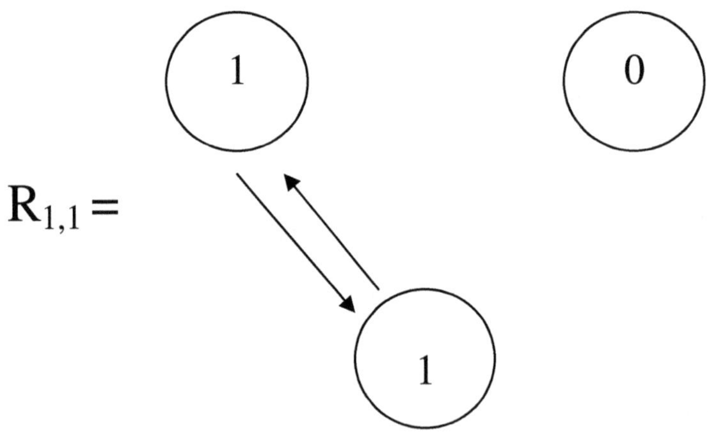

Fig. 2.12 The image of the basic relation R_{11} obtained by using of the data given in table 2.3

Using the second order tensor which is represented by its context in formal concept analysis, it is possible to generate a model of a relation as a weighted superposition of basic relations. This extends the vector decomposition for a basic set of vectors.

It is possible to extend beyond the second tensor and generate superposition of basic tensors with the weights as tensors. Here the components $Z_{h,k,n}$ of the tensor of the third order is more complex that a simple relationship between two points. The components are 3x3x3 = 27. Any basic tensor *is a relation on 9 objects* where H is 3 x 3 matrix.

In general, we have the basic relations.

$$R_{j,i,k} = h_{\alpha,j} h_{\beta,i} h_{\gamma,k} \tag{2.62}$$

Chapter 3
Geometry and Agent Coherence

This chapter presents the geometric coherence of the agent expressed by a graph of actions (flow) and nodes that can be sources or tasks (effort). Any graph source, action, tasks. Any agent of the first order can be considered to be a system, which by the use of resources can activate an action, which enables the task to be done. Resources can be physical resources, functions, tables of data or any type of information necessary to do the action. Action is a method or a set of methods by which we solve our problem. Initially, we can represent any simple agent using figure 3.1 and a network of simple agents.

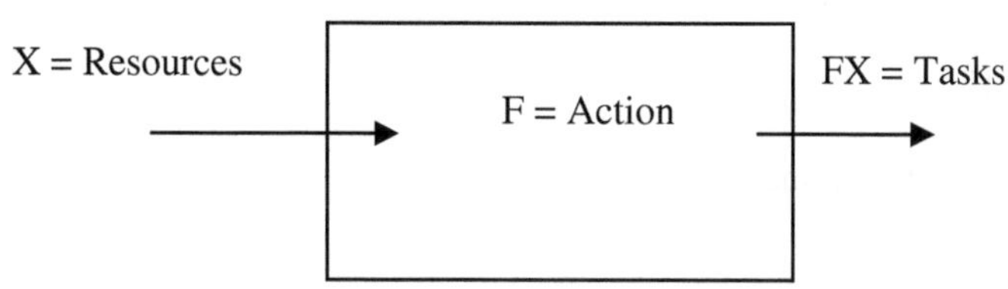

Fig. 3.1 Representing Resource, Action and Task of the Agents.

$$H = \begin{bmatrix} & V_1 & \dots & V_n \\ action_1 & v_{1,1} & \dots & v_{1,n} \\ action_2 & v_{2,1} & \dots & v_{2,n} \\ \dots & \dots & \dots & \dots \\ action_m & v_{m,1} & \dots & v_{m,n} \end{bmatrix}$$

where the values $v_{h,k} \in \{0,1\}$, The V_k are mashes of the graph. For example in the graph

G. Resconi: Geometry of Knowledge for Intelligent Systems, SCI 407, pp. 31–60.
springerlink.com

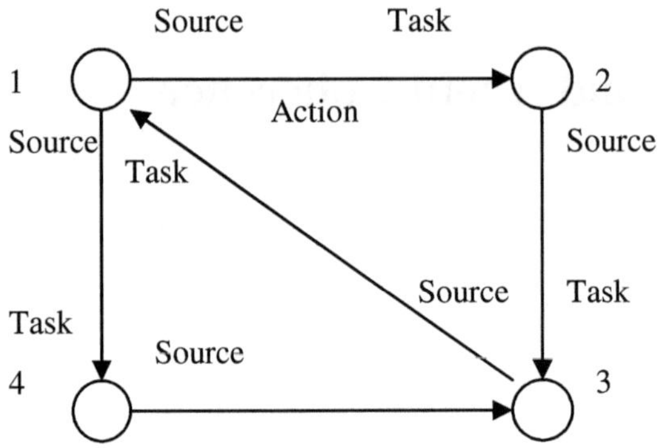

We have the matrix

$$H = \begin{bmatrix} & (1,2,3) & (1,3,4) \\ action_1 & 1 & 0 \\ action_2 & 1 & 0 \\ action_3 & 1 & 1 \\ action_4 & 0 & 1 \\ action_5 & 0 & 1 \end{bmatrix}$$

Where (1,2,3) is the first mash or V_1 and (1,3,4) is the second mash or V_2. Now we can compute the geometric elements in the chapter 2. So we have

$$G = H^T H$$

Where G is the tensor metric and

$$S = (H^T H)^{-1} H^T X$$

For which

$$S^T G S = D$$

Where D is the metric of the graph space given by sources, action and tasks. At the end we have the projection operator

$$Y = H S.$$

That give the coherent value of X respect to the graph H. Now we give more information on the meaning of the graph sources, action and tasks. X are external values of the sources and tasks that we want to introduce inside the network.

3.1 Agents and Coherence

An Agent of the First Order is a Resource. Tasks and Actions are defined by the use of other agents to include this particular type of agent. We term them Agents of the Second Order. We must then extend the traditional paradigm of input output structure shown in figure 3.3. To build the Second Order Agent we change the traditional graph image of actions to that shown in figure 3.4. In a new image we show both the task and the sources by the symbols Sk and Tk. We next obtain figure 3.3 where at any node of the graph the tasks and the sources with the same number are the same entity.

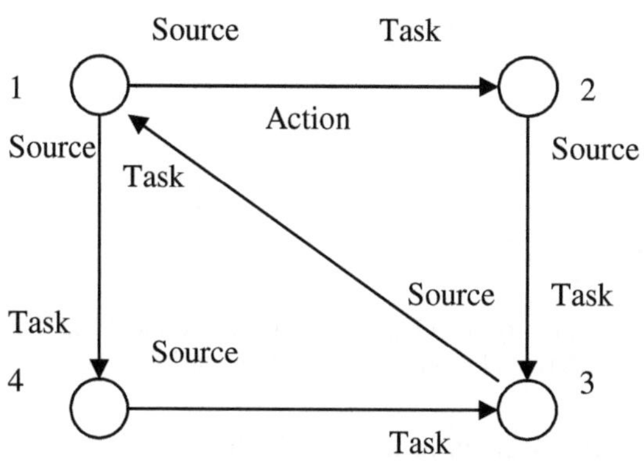

Fig. 3.2 Network showing the Action between Different sources and the task. For example from the source 1, we can obtain two different actions tasks 2 and the task 4

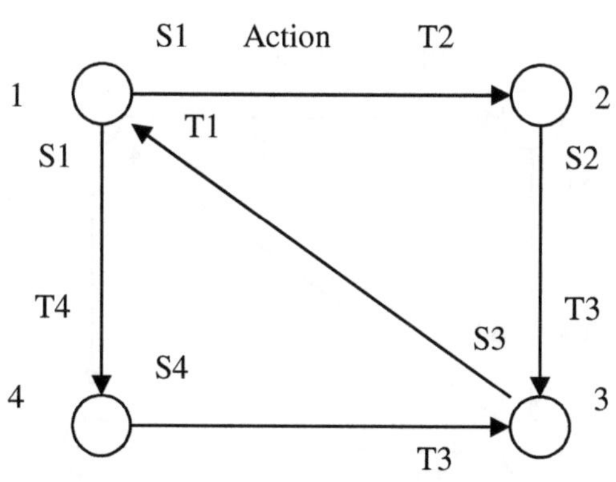

Fig. 3.3 At any node of the graph the tasks and the sources with the same number are the same entity. For example we have that $S_1 = T_1$.

The graph shown in figure 3.3 is the same entity in one of the nodes. It can be a source and may also be a task. This means that we have a fundamental symmetry.

That is "The task T in one node is the same as the source S at the same node". This is an example of the symmetry principle. The agent of the second order is responsible for the allocation of the source, the Task and the Actions. When we have this type of agent we can change the graph in figure 3.3.

The change in figure 3.4 is an example. We can change the task and the source using the same actions without changing the Symmetry Principle. The transformation is then as follows

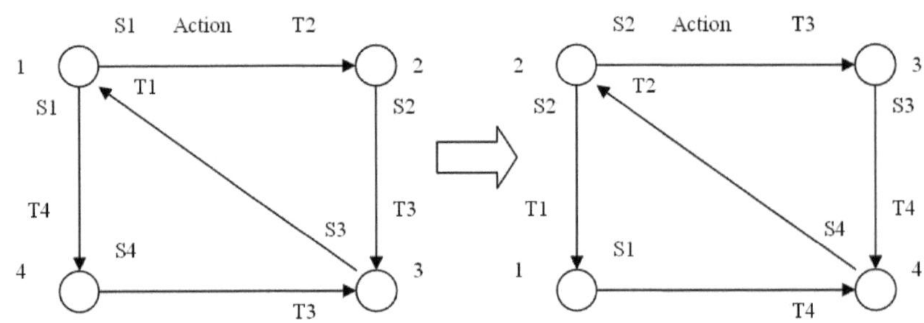

Fig. 3.4 The transformation of the Graph, the Source and the Task. The structure remains unchanged but has a different allocation of sources and tasks.

An example of this is given in figure 3.4. Here we show a possible change of tasks and sources using the same structure. The new graph is obtained by the use of the second order agents and is similar but is not equal to that given previously. Similarly it means that the form has the same properties but is not equal .The new graph given in figure 3.4 can also be shown in this way.

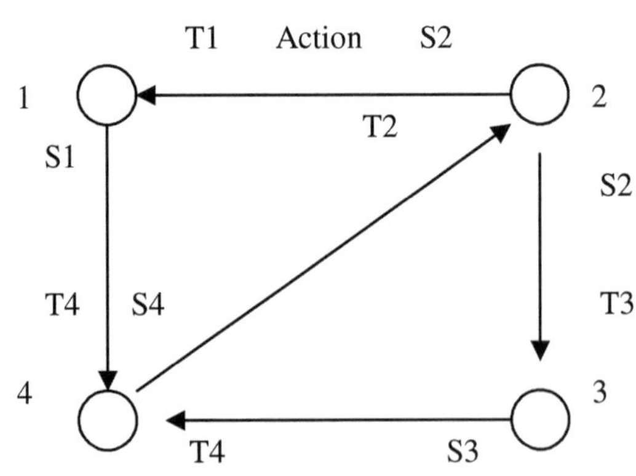

Fig. 3.5 After transformation the Graph changes its Image but the properties are unchanged. It is a Similar Graph.

An operator P which changes the sources and the tasks is

$$P[Sources] = P[Tasks] = \begin{pmatrix} 1 & 2 & 3 & 4 \\ 2 & 3 & 4 & 1 \end{pmatrix}$$

The transformation by the agent of the second order in figure 3.4 may be written as,

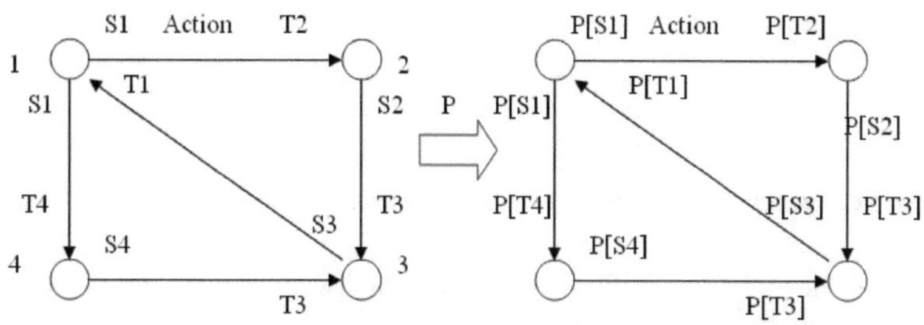

Fig. 3.6 Representation of the transformation of the Tasks and Sources by use of the Operator P

An agent of the Second Order can generate more complex transformations of the sources and the tasks. An agent of the Second Order can assemble tasks in clusters and then it is possible to consider all of the tasks in the same cluster as being equal. Here we have the transformation

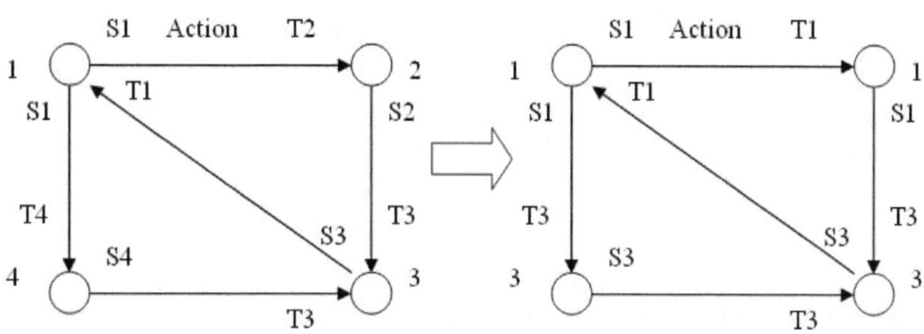

Fig. 3.7 Agents of the second order form clusters of tasks and sources in such a way that they have the same structure and repetition.

The cluster approach shown in figure 3.7 generates a new graph that is homomorphous to the original graph. Homomorphous means that the resulting new graph is equal to the previous but with cluster of nodes marked in the same way so that we are unable to separate them.

The part of the graph inside the clusters cannot be separated again. This is known as Granulation. The resulting graph in figure 3.7 can also be written in a similar way.

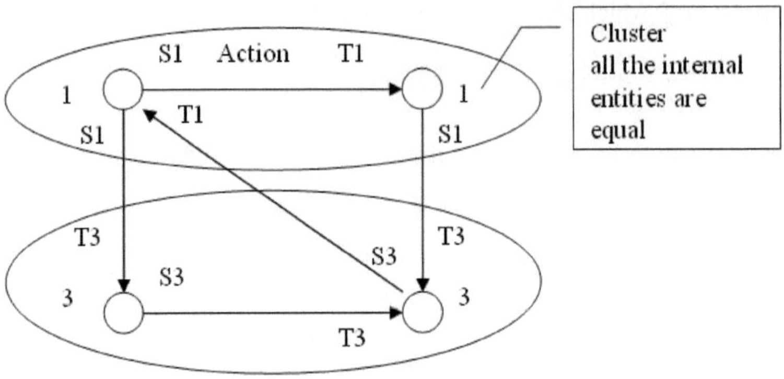

Fig. 3.8 The agent having the same structure creates similar clusters of tasks and sources

We also have the possibility of an agent of the second order. This possibility is the opportunity to avoid the symmetry principle between the sources and the tasks. Here are the two different operators

$$P[Sources] = \begin{pmatrix} 1 & 2 & 3 & 4 \\ 2 & 3 & 4 & 1 \end{pmatrix} \text{ and } Q[Tasks] = \begin{pmatrix} 1 & 2 & 3 & 4 \\ 2 & 1 & 4 & 3 \end{pmatrix}$$

we obtain this transformation.

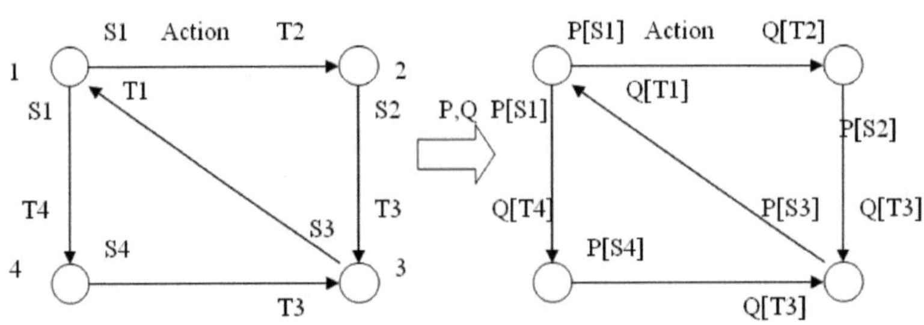

Fig. 3.9 An Agent of the second order can allocate the sources and the tasks in different ways using the same action graph

The new graph shown in figure 3.9 was obtained by using agents of the second order as the sources and the tasks allocation.

In the ordinary Action Sources Task Graph, it is assumed that at any point of the graph the task converges on the point of the previous actions. The intent is to show that the result is the same although the points are changed by the new actions used. In many cases this may not be true. We can consider that any node of the action graph is not a passive node but is an active node having its own internal actions. *For an internal actions the entity* which type of "task" is transformed into another entity which depends on the type of source.

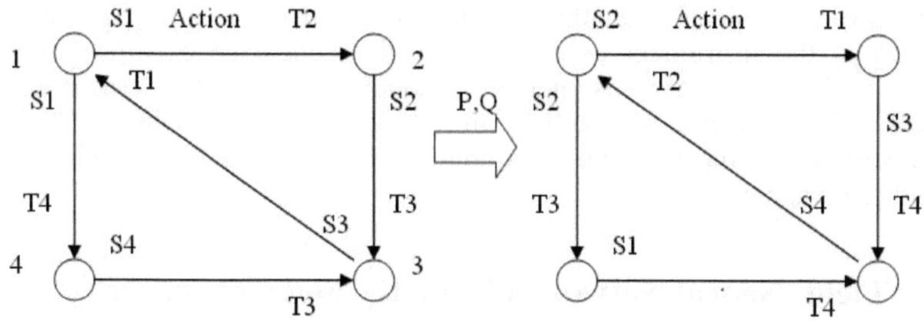

Fig. 3.10 The operators P and Q are used to allocate the tasks and the sources

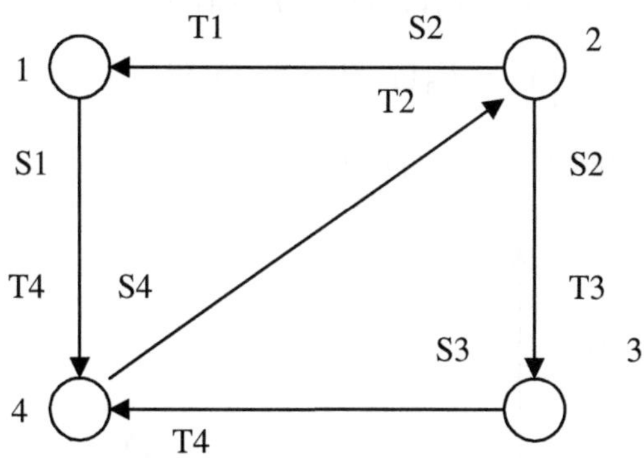

Fig. 3.11 Ordering the sources and tasks by the transformation of the second order agents obtained by using the operators P and Q. The actions are the same as those actions indicated in figure 3.5.

In figure 3.11 we always have an important symmetry for the same node. There is only one source which can be used in the different actions. For example in the node 2 the source S2 is repeated twice, One task is to obtain the task T1 and the other task is to obtain task T3. We can say that the transformation of the second order agent is a different allocation of tasks and sources using the same graph of actions. We illustrate this on figure 3.12.

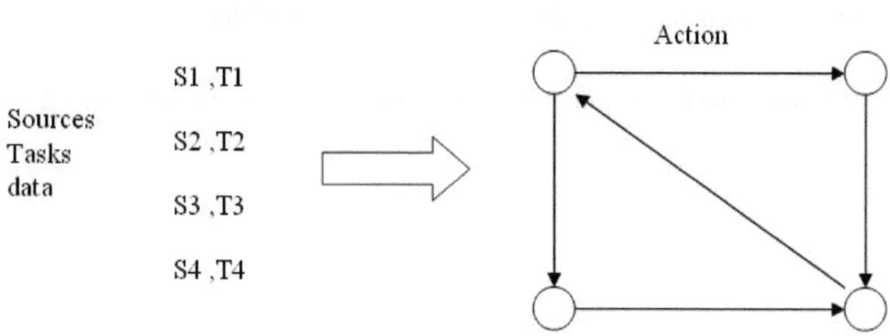

Fig. 3.12 The Ordering of the sources and tasks using the transformation of the Second Order Agents obtained by use of the operators P and Q. The graph is invariant for Second Order Agent transformations.

The invariant rule (for Graphs) for all possible changes of the Source – Task Network is the instrument used to define the coherence existing within the Second Order Agent. Agents that are incoherent change the structure of the sources and represent a task without an invariant rule. In conclusion it is noted that invariance is the key entity required to obtain coherent behaviour in the second order agent actions.

3.2 Field, Neural Network Geometry and Coherence

Given the transformation

$$\begin{cases} x_1 = F_1(\rho,\theta) = \rho\cos(\theta) \\ x_2 = F_2(\rho,\theta) = \rho\sin(\theta) \end{cases} \tag{3.1}$$

A geometric illustration for the equation is given in Figure 3.13.

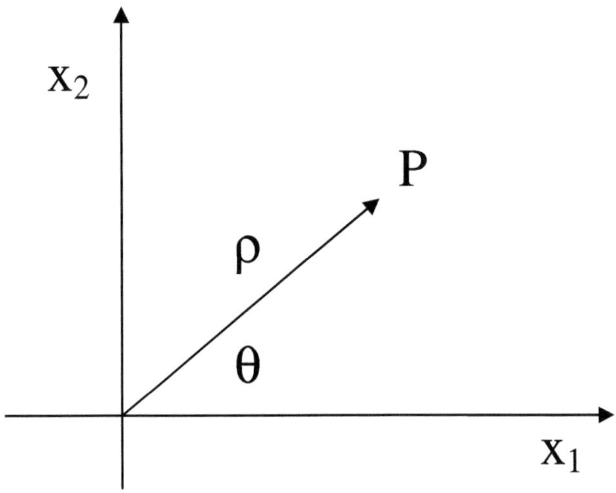

Fig. 3.13 Cartesian Coordinates and Polar coordinates for the Point P

The fields or transformations for in (3.1) are represented in this manner.

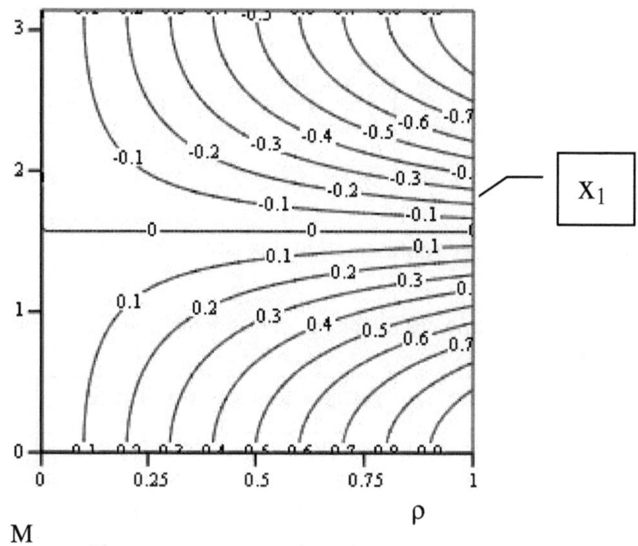

Fig. 3.14 Field . $x_1 = F_1(\rho,\theta) = \rho\cos(\theta)$ The field is the value of the coordinate y_1 when change the polar coordinates θ , ρ

For the second field we have

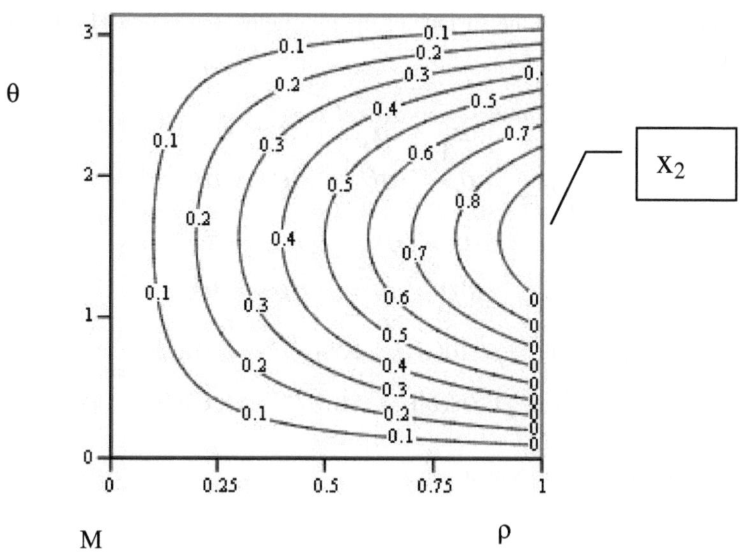

Fig. 3.15 Field $x_2 = F_2(\rho,\theta) = \rho\sin(\theta)$ the field is the value of the coordinate y_2 when changed the polar coordinates.

For the system (3.1) we have

$$H = \begin{bmatrix} \frac{\partial x_1}{\partial \rho} & \frac{\partial x_1}{\partial \theta} \\ \frac{\partial x_2}{\partial \rho} & \frac{\partial x_2}{\partial \theta} \end{bmatrix} = \begin{bmatrix} \cos(\theta) & -\rho\sin(\theta) \\ \sin(\theta) & \rho\cos(\theta) \end{bmatrix} \tag{3.2}$$

The matrix H can also be represented in this table.

Table 3.1 Objects and attributes for the derivatives of the fields

	Attribute 1	Attribute 2	X=dx
Object 1	$\frac{\partial x_1}{\partial \rho}$	$\frac{\partial x_1}{\partial \theta}$	dx_1
Object 2	$\frac{\partial x_2}{\partial \rho}$	$\frac{\partial x_2}{\partial \theta}$	dx_2

Where X is an infinitesimal vector in the space of the objects (x_1, x_2).

When we move on the surface of the two fields, the matrix H gives us the local metric in this way

$$G = H^T H = \begin{bmatrix} \frac{\partial x_1}{\partial \rho} & \frac{\partial x_1}{\partial \theta} \\ \frac{\partial x_2}{\partial \rho} & \frac{\partial x_2}{\partial \theta} \end{bmatrix}^T \begin{bmatrix} \frac{\partial x_1}{\partial \rho} & \frac{\partial x_1}{\partial \theta} \\ \frac{\partial x_2}{\partial \rho} & \frac{\partial x_2}{\partial \theta} \end{bmatrix} = \begin{bmatrix} \cos(\theta) & -\rho\sin(\theta) \\ \sin(\theta) & \rho\cos(\theta) \end{bmatrix}^T \begin{bmatrix} \cos(\theta) & -\rho\sin(\theta) \\ \sin(\theta) & \rho\cos(\theta) \end{bmatrix} = \begin{bmatrix} 1 & 0 \\ 0 & \rho^2 \end{bmatrix} \tag{3.3}$$

Where G is the metric tensor.

For (3.1) we have

$$dx=\begin{bmatrix} dx_1 \\ dx_2 \end{bmatrix}=\begin{bmatrix} \frac{\partial x_1}{\partial \rho} & \frac{\partial x_1}{\partial \theta} \\ \frac{\partial x_2}{\partial \rho} & \frac{\partial x_2}{\partial \theta} \end{bmatrix}\begin{bmatrix} d\rho \\ d\theta \end{bmatrix}=$$

$$\begin{bmatrix} \frac{\partial y_1}{\partial \rho}d\rho+\frac{\partial y_1}{\partial \theta}d\theta \\ \frac{\partial y_2}{\partial \rho}d\rho+\frac{\partial y_2}{\partial \theta}d\theta \end{bmatrix}=d\rho\begin{bmatrix} \frac{\partial x_1}{\partial \rho} \\ \frac{\partial x_2}{\partial \rho} \end{bmatrix}+d\theta\begin{bmatrix} \frac{\partial x_1}{\partial \theta} \\ \frac{\partial x_2}{\partial \theta} \end{bmatrix}= \tag{3.4}$$

$$HdS=H(H^T H)^{-1}H^T dx$$

Where

$$dS=\begin{bmatrix} dS_1 \\ dS_2 \end{bmatrix}=\begin{bmatrix} d\rho \\ d\theta \end{bmatrix}=\begin{bmatrix} \frac{\partial x_1}{\partial \rho} & \frac{\partial x_1}{\partial \theta} \\ \frac{\partial x_2}{\partial \rho} & \frac{\partial x_2}{\partial \theta} \end{bmatrix}^{-1}\begin{bmatrix} dx_1 \\ dx_2 \end{bmatrix}=$$

$$(\begin{bmatrix} \frac{\partial x_1}{\partial \rho} & \frac{\partial x_1}{\partial \theta} \\ \frac{\partial x_2}{\partial \rho} & \frac{\partial x_2}{\partial \theta} \end{bmatrix}^T\begin{bmatrix} \frac{\partial x_1}{\partial \rho} & \frac{\partial x_1}{\partial \theta} \\ \frac{\partial x_2}{\partial \rho} & \frac{\partial x_2}{\partial \theta} \end{bmatrix})^{-1}\begin{bmatrix} \frac{\partial x_1}{\partial \rho} & \frac{\partial x_1}{\partial \theta} \\ \frac{\partial x_2}{\partial \rho} & \frac{\partial x_2}{\partial \theta} \end{bmatrix}^T\begin{bmatrix} dx_1 \\ dx_2 \end{bmatrix}=$$

$$(H^T H)^{-1}H^T dx=G^{-1}H^T dx$$

Using the concept of we obtain for the expression (3.4), the following neural image,

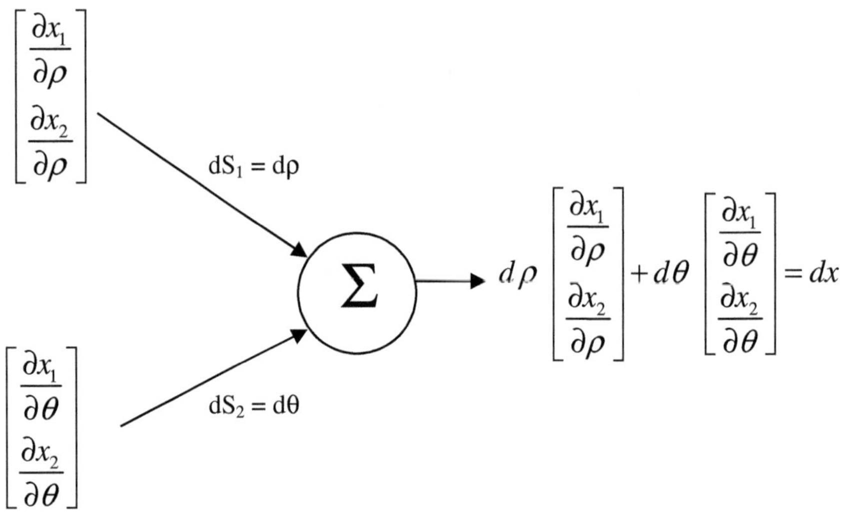

Fig. 3.16 The neural image of the superposition of the derivatives of the fields x with respect to the polar coordinates

Now we can show that

$$ds^2 = (dS)^T G(dS) = \begin{bmatrix} d\rho \\ d\theta \end{bmatrix}^T \begin{bmatrix} 1 & 0 \\ 0 & \rho^2 \end{bmatrix} \begin{bmatrix} d\rho \\ d\theta \end{bmatrix} = \\ d\rho^2 + \rho^2 d\theta^2 = dx_1^2 + dx_2^2 \tag{3.5}$$

When we take use conjugate sources or coordinates, we obtain

$$dS^C = \begin{bmatrix} dS^C{}_1 \\ dS^C{}_2 \end{bmatrix} = \begin{bmatrix} \frac{\partial x_1}{\partial \rho} & \frac{\partial x_1}{\partial \theta} \\ \frac{\partial x_2}{\partial \rho} & \frac{\partial x_2}{\partial \theta} \end{bmatrix}^T \begin{bmatrix} dx_1 \\ dx_2 \end{bmatrix} = H^T dx$$

Then

$$ds^2 = (dS^C)^T G^{-1}(dS^C) = (H^T \begin{bmatrix} dx_1 \\ dx_2 \end{bmatrix})^T (H^T H)^{-1} (H^T \begin{bmatrix} dx_1 \\ dx_2 \end{bmatrix}) =$$

$$\begin{bmatrix} dx_1 \\ dx_2 \end{bmatrix}^T H(H^T H)^{-1} H^T \begin{bmatrix} dx_1 \\ dx_2 \end{bmatrix} = \begin{bmatrix} dx_1 \\ dx_2 \end{bmatrix}^T HH^{-1}(H^T)^{-1} H^T \begin{bmatrix} dx_1 \\ dx_2 \end{bmatrix} = dx_1^2 + dx_2^2$$

For $dS = G^{-1}dS^C$, using the language of the tensor calculus we have

$$dS = G^{-1}dS^C$$

In the Einstein notation we have

$$dS^i = G^{-1}dS_j = G^{i,j}dS_j$$

Where

and $dS^i = dS$ *are the controvariant basis* (ρ,θ)

$dS_j = dS^C$ *are the covariant basis*

We will show geometric examples of the controvariant and covariant basis. Given the transformation

$$x_1 = x'_1 + x'_2 cos(\alpha)$$
$$x_2 = x'_2 sin(\alpha)$$

Geometrically, it can be shown as follows

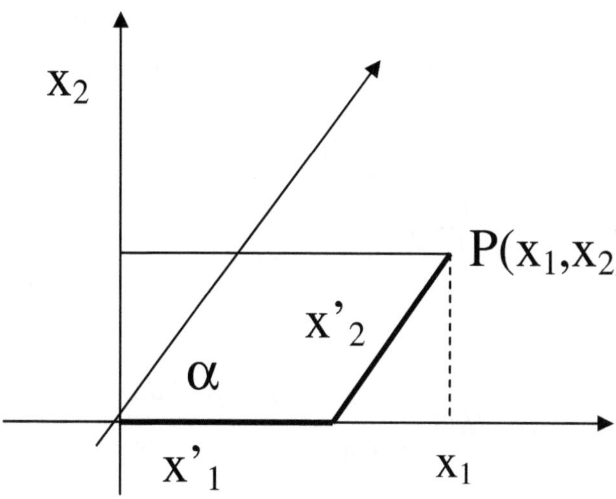

Fig. 3.17 Represents an example for general coordinates

Using the transformation (3.5) we obtain

$$H = \begin{bmatrix} \dfrac{\partial x_1}{\partial x'_1} & \dfrac{\partial x_1}{\partial x'_2} \\ \dfrac{\partial x_2}{\partial x'_1} & \dfrac{\partial x_2}{\partial x'_2} \end{bmatrix} = \begin{bmatrix} 1 & \cos(\alpha) \\ 0 & \sin(\alpha) \end{bmatrix}$$

The controvariant components are

$$dS = \begin{bmatrix} dS_1 \\ dS_2 \end{bmatrix} = \begin{bmatrix} dx'_1 \\ dx'_2 \end{bmatrix} = \begin{bmatrix} \dfrac{\partial x_1}{\partial x'_1} & \dfrac{\partial x_1}{\partial x'_2} \\ \dfrac{\partial x_2}{\partial x'_1} & \dfrac{\partial x_2}{\partial x'_2} \end{bmatrix}^{-1} \begin{bmatrix} dx_1 \\ dx_2 \end{bmatrix} = H^{-1} \begin{bmatrix} dx_1 \\ dx_2 \end{bmatrix}$$

$$\left(\begin{bmatrix} \dfrac{\partial x_1}{\partial x'_1} & \dfrac{\partial x_1}{\partial x'_2} \\ \dfrac{\partial x_2}{\partial x'_1} & \dfrac{\partial x_2}{\partial x'_2} \end{bmatrix}^T \begin{bmatrix} \dfrac{\partial x_1}{\partial x'_1} & \dfrac{\partial x_1}{\partial x'_2} \\ \dfrac{\partial x_2}{\partial x'_1} & \dfrac{\partial x_2}{\partial x'_2} \end{bmatrix} \right)^{-1} \begin{bmatrix} \dfrac{\partial x_1}{\partial x'_1} & \dfrac{\partial x_1}{\partial x'_2} \\ \dfrac{\partial x_2}{\partial x'_1} & \dfrac{\partial x_2}{\partial x'_2} \end{bmatrix}^T \begin{bmatrix} dx_1 \\ dx_2 \end{bmatrix} =$$

$$\begin{bmatrix} 1 & -\dfrac{\cos(\alpha)}{\sin(\alpha)} \\ 0 & \dfrac{1}{\sin(\alpha)} \end{bmatrix} \begin{bmatrix} dx_1 \\ dx_2 \end{bmatrix} = \begin{bmatrix} dx_1 - \dfrac{\cos(\alpha)}{\sin(\alpha)} dx_2 \\ \dfrac{dx_2}{\sin(\alpha)} \end{bmatrix}$$

We note that

$$dx = HdS = \begin{bmatrix} 1 & \cos(\alpha) \\ \cos(\alpha) & 1 \end{bmatrix} \begin{bmatrix} dS_1 \\ dS_2 \end{bmatrix}$$

The conjugate basis or covariant basis is given by

$$dS^C = \begin{bmatrix} dS^C{}_1 \\ dS^C{}_2 \end{bmatrix} = H^T \begin{bmatrix} dx_1 \\ dx_2 \end{bmatrix} = \begin{bmatrix} 1 & 0 \\ \cos(\alpha) & \sin(\alpha) \end{bmatrix} \begin{bmatrix} dx_1 \\ dx_2 \end{bmatrix} = \begin{bmatrix} dx_1 \\ dx_1 \cos(\alpha) + dx_2 \sin(\alpha) \end{bmatrix}$$

Geometrically, the basis and the conjugate basis are represented as follows:

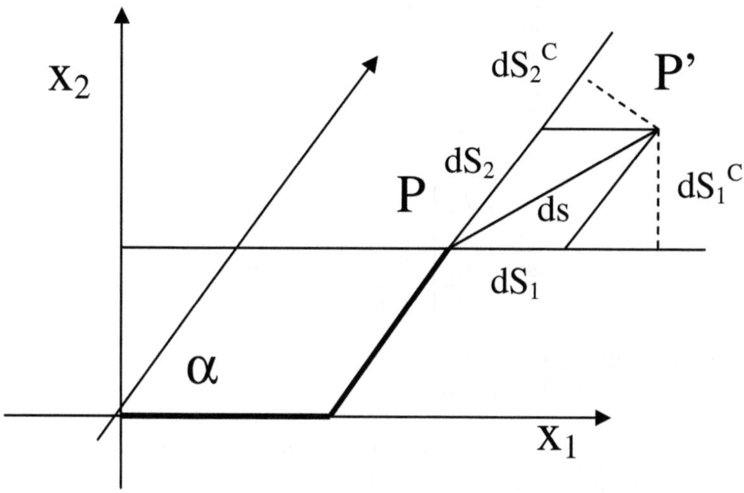

Fig. 3.18 The geometric representation of the controvariant components (sources dS_1 and dS_2) and the conjugate sources (covariant basis dS_1^C and dS_2^C)

The previous expressions results in the metric

$$ds^2 = (dS)^T G(dS) = \begin{bmatrix} dS_1 \\ dS_2 \end{bmatrix}^T \begin{bmatrix} 1 & \cos(\alpha) \\ \cos(\alpha) & 1 \end{bmatrix} \begin{bmatrix} dS_1 \\ dS_2 \end{bmatrix} =$$
$$dS_1^{\,2} + dS_2^{\,2} + 2\cos(\alpha) dS_1 dS_2 = dx_1^{\,2} + dx_2^{\,2}$$

From the previous considerations we observe

$$dx = HdS = \begin{bmatrix} \frac{\partial x_1}{\partial \rho} & \frac{\partial x_1}{\partial \theta} \\ \frac{\partial x_2}{\partial \rho} & \frac{\partial x_2}{\partial \theta} \end{bmatrix} \begin{bmatrix} d\rho \\ d\theta \end{bmatrix} = \begin{bmatrix} (d\rho \frac{\partial}{\partial \rho} + d\theta \frac{\partial}{\partial \theta}) x_1 \\ (d\rho \frac{\partial}{\partial \rho} + d\theta \frac{\partial}{\partial \theta}) x_2 \end{bmatrix} \tag{3.6}$$

When we have x = x(t) ; S = S(t) and t is time we obtain

$$dx = \frac{dx}{dt} dt \ , \ dS = \frac{dS}{dt} dt$$

We can write (3.6) as follows;

$$\frac{dx}{dt}dt = H\frac{dS}{dt}dt$$
$$and$$
$$\frac{dx}{dt} = H\frac{dS}{dt}$$
$$or \tag{3.7}$$
$$\frac{dx}{dt} = \begin{bmatrix} \frac{dx_1}{dt} \\ \frac{dx_2}{dt} \end{bmatrix} = \begin{bmatrix} \frac{\partial x_1}{\partial \rho} & \frac{\partial x_1}{\partial \theta} \\ \frac{\partial x_2}{\partial \rho} & \frac{\partial x_2}{\partial \theta} \end{bmatrix} \begin{bmatrix} \frac{d\rho}{dt} \\ \frac{d\theta}{dt} \end{bmatrix}$$

The derivatives of the basis coordinates.

We can reverse the (3.7) as follows,

$$\begin{bmatrix} \frac{dS_1}{dt} \\ \frac{dS_2}{dt} \end{bmatrix} = \begin{bmatrix} \frac{d\rho}{dt} \\ \frac{d\theta}{dt} \end{bmatrix} = \begin{bmatrix} \frac{\partial x_1}{\partial \rho} & \frac{\partial x_1}{\partial \theta} \\ \frac{\partial x_2}{\partial \rho} & \frac{\partial x_2}{\partial \theta} \end{bmatrix}^{-1} \begin{bmatrix} \frac{dx_1}{dt} \\ \frac{dx_2}{dt} \end{bmatrix} = H^{-1} \begin{bmatrix} \frac{dx_1}{dt} \\ \frac{dx_2}{dt} \end{bmatrix} =$$
$$(H^T H)^{-1} H^T \begin{bmatrix} \frac{dx_1}{dt} \\ \frac{dx_2}{dt} \end{bmatrix} = \begin{bmatrix} \frac{\partial \rho}{\partial x_1} & \frac{\partial \rho}{\partial x_2} \\ \frac{\partial \theta}{\partial x_1} & \frac{\partial \theta}{\partial x_2} \end{bmatrix} \begin{bmatrix} \frac{dx_1}{dt} \\ \frac{dx_2}{dt} \end{bmatrix} \tag{3.8}$$

The two relations establish a connection between the derivatives of the two basis. In (3.8) the input vector

$$X = \begin{bmatrix} \frac{dx_1}{dt} \\ \frac{dx_2}{dt} \end{bmatrix}$$

has derivatives which are free. We can impose a constraint to the derivative as follows;

$$\begin{cases} \frac{dx_1}{dt} = f_1(x_1, x_2) \\ \frac{dx_2}{dt} = f_2(x_1, x_2) \end{cases} \tag{3.9}$$

The sources are also limited by the relations

$$\begin{bmatrix} \frac{dS_1}{dt} \\ \frac{dS_2}{dt} \end{bmatrix} = \begin{bmatrix} \frac{d\rho}{dt} \\ \frac{d\theta}{dt} \end{bmatrix} = \begin{bmatrix} \frac{\partial \rho}{\partial x_1} & \frac{\partial \rho}{\partial x_2} \\ \frac{\partial \theta}{\partial x_1} & \frac{\partial \theta}{\partial x_2} \end{bmatrix} \begin{bmatrix} f_1(x_1, x_2) \\ f_2(x_1, x_2) \end{bmatrix} =$$

$$\begin{bmatrix} f_1(x_1, x_2)\frac{\partial \rho}{\partial x_1} + f_2(x_1, x_2)\frac{\partial \rho}{\partial x_2} \\ f_1(x_1, x_2)\frac{\partial \theta}{\partial x_1} + f_2(x_1, x_2)\frac{\partial \theta}{\partial x_2} \end{bmatrix} = \begin{bmatrix} D\rho \\ D\theta \end{bmatrix}$$

Where

$$D = f_1(x_1, x_2)\frac{\partial}{\partial x_1} + f_2(x_1, x_2)\frac{\partial}{\partial x_2}$$

is the Lie derivative. We also know that the solution of the equation (3.8) may be written as follows

$$\begin{bmatrix} x_1(t) \\ x_2(t) \end{bmatrix} = \exp(tD) \begin{bmatrix} x_1(0) \\ x_2(0) \end{bmatrix}$$

or

$$x(t) = e^{tD} x(0) = \sum_{\upsilon=0}^{\infty} \frac{t^\upsilon}{\upsilon!} D^\upsilon x(0) \tag{3.10}$$

Equation (3.10) is the Lie series.

3.3 Transformations in Which the General Quadratic Form Is Invariant

We know that

$$\left(\begin{bmatrix} \cos(\alpha) & \sin(\alpha) \\ -\sin(\alpha) & \cos(\alpha) \end{bmatrix} \begin{bmatrix} A \\ B \end{bmatrix}\right)^T \begin{bmatrix} \cos(\alpha) & \sin(\alpha) \\ -\sin(\alpha) & \cos(\alpha) \end{bmatrix} \begin{bmatrix} A \\ B \end{bmatrix} = A^2 + B^2$$

For the share transformation

$$\begin{bmatrix} x_1 \\ x_2 \end{bmatrix} = \begin{bmatrix} 1 & 0 \\ m & 1 \end{bmatrix} \begin{bmatrix} x_1' \\ x_2' \end{bmatrix}$$

we have the quadratic form

$$\rho^2 = (1+m^2)S_1^2 + S_2^2 + 2mS_1S_2$$

We can now compute the sources

$$\begin{bmatrix} S_1 \\ S_2 \end{bmatrix} = \begin{bmatrix} x_1' \\ x_2' \end{bmatrix} = \begin{bmatrix} 1 & 0 \\ m & 1 \end{bmatrix}^{-1} \begin{bmatrix} A \\ B \end{bmatrix} = \begin{bmatrix} 1 & 0 \\ -m & 1 \end{bmatrix} \begin{bmatrix} A \\ B \end{bmatrix}$$

We also want compute the transformation U of the sources as follows.

We have

$$U\begin{bmatrix} S_1 \\ S_2 \end{bmatrix} = U\begin{bmatrix} 1 & 0 \\ -m & 1 \end{bmatrix} \begin{bmatrix} A \\ B \end{bmatrix} = \begin{bmatrix} 1 & 0 \\ -m & 1 \end{bmatrix} \begin{bmatrix} \cos(\alpha) & \sin(\alpha) \\ -\sin(\alpha) & \cos(\alpha) \end{bmatrix} \begin{bmatrix} A \\ B \end{bmatrix}$$

for

$$U\begin{bmatrix} 1 & 0 \\ m & 1 \end{bmatrix}^{-1} = \begin{bmatrix} 1 & 0 \\ m & 1 \end{bmatrix}^{-1} \begin{bmatrix} \cos(\alpha) & \sin(\alpha) \\ -\sin(\alpha) & \cos(\alpha) \end{bmatrix}$$

we have

$$U = \begin{bmatrix} 1 & 0 \\ m & 1 \end{bmatrix}^{-1} \begin{bmatrix} \cos(\alpha) & \sin(\alpha) \\ -\sin(\alpha) & \cos(\alpha) \end{bmatrix} \begin{bmatrix} 1 & 0 \\ m & 1 \end{bmatrix} =$$

$$\begin{bmatrix} \cos(\alpha) - m\sin(\alpha) & -\sin(\alpha) \\ \sin(\alpha) + m^2\sin(\alpha) & m\sin(\alpha) + \cos(\alpha) \end{bmatrix} =$$

$$T^{-1}RT$$

We have the same invariance form

$$\rho^2 = (1+m^2)S_1^2 + S_2^2 + 2mS_1S_2$$

The change of the transformation $\begin{bmatrix} \cos(\alpha) & \sin(\alpha) \\ -\sin(\alpha) & \cos(\alpha) \end{bmatrix}$ into U is due to a second order agent. The first order agent changes the basis. The change of the transformation is a more complex operation and is given by agents of the second order.

We now show that for

$$G = \begin{bmatrix} 1+m^2 & m \\ m & 1 \end{bmatrix} = \begin{bmatrix} 1 & 0 \\ m & 1 \end{bmatrix}^T \begin{bmatrix} 1 & 0 \\ m & 1 \end{bmatrix} = T^T T$$

We have

$$\rho^2 = (U\begin{bmatrix} S_1 \\ S_2 \end{bmatrix})^T \begin{bmatrix} 1+m^2 & m \\ m & 1 \end{bmatrix} (U\begin{bmatrix} S_1 \\ S_2 \end{bmatrix}) = \begin{bmatrix} S_1 \\ S_2 \end{bmatrix}^T (U^T G U) \begin{bmatrix} S_1 \\ S_2 \end{bmatrix}$$

$$U^T G U = (T^{-1}RT)^T (T^T T)(T^{-1}RT) = T^T R^T (T^{-1})^T (T^T T)(T^{-1}RT) =$$
$$T^T R^T R T = T^T T = G$$

We can find other transformations U using the unitary matrix B where $B^T B = I$. In general we have

$$U = T^{-1} B\, T$$

With the transformation

$$\rho^2 = (U\begin{bmatrix} S_1 \\ S_2 \end{bmatrix})^T \begin{bmatrix} 1+m^2 & m \\ m & 1 \end{bmatrix} (U\begin{bmatrix} S_1 \\ S_2 \end{bmatrix}) = \begin{bmatrix} S_1 \\ S_2 \end{bmatrix}^T \begin{bmatrix} 1+m^2 & m \\ m & 1 \end{bmatrix} \begin{bmatrix} S_1 \\ S_2 \end{bmatrix} = (1+m^2)S_1^2 + S_2^2 + 2mS_1S_2$$

For $G^{-1} = \begin{bmatrix} 1+m^2 & m \\ m & 1 \end{bmatrix}^{-1}$

$$(\rho^C)^2 = \begin{bmatrix} S_1{}^C \\ S_2{}^C \end{bmatrix}^T \begin{bmatrix} 1+m^2 & m \\ m & 1 \end{bmatrix}^{-1} \begin{bmatrix} S_1{}^C \\ S_2{}^C \end{bmatrix} = (S_1^C)^2 + (1+m^2)(S_2^C)^2 - 2mS_1^C S_2^C$$

When

$$\begin{bmatrix} S_1{}^C \\ S_2{}^C \end{bmatrix} = \begin{bmatrix} 1 & 0 \\ m & 1 \end{bmatrix}^T \begin{bmatrix} A \\ B \end{bmatrix}$$

we obtain

$$(\rho^C)^2 = (S_1^C)^2 + (1+m^2)(S_2^C)^2 - 2mS_1^C S_2^C = A^2 + B^2$$

We have

$$U^C \begin{bmatrix} S_1{}^C \\ S_2{}^C \end{bmatrix} = U^C \begin{bmatrix} 1 & 0 \\ m & 1 \end{bmatrix}^T \begin{bmatrix} A \\ B \end{bmatrix} = \begin{bmatrix} 1 & 0 \\ m & 1 \end{bmatrix}^T \begin{bmatrix} \cos(\alpha) & \sin(\alpha) \\ -\sin(\alpha) & \cos(\alpha) \end{bmatrix} \begin{bmatrix} A \\ B \end{bmatrix}$$

and we obtain

$$U^C = \begin{bmatrix} 1 & 0 \\ m & 1 \end{bmatrix}^T = \begin{bmatrix} 1 & 0 \\ m & 1 \end{bmatrix}^T \begin{bmatrix} \cos(\alpha) & \sin(\alpha) \\ -\sin(\alpha) & \cos(\alpha) \end{bmatrix} (\begin{bmatrix} 1 & 0 \\ m & 1 \end{bmatrix}^T)^{-1}$$

and

$$(\rho^C)^2 = (U^C \begin{bmatrix} S_1^C \\ S_2 \end{bmatrix})^T \begin{bmatrix} 1+m^2 & m \\ m & 1 \end{bmatrix} (U^C \begin{bmatrix} S_1^C \\ S_2^C \end{bmatrix}) = M = (S_1^C)^2 + (1+m)(S_2^C)^C - 2mS_1^C S_2^C$$

Given the quadratic form

$$\rho^2 = \begin{bmatrix} S_1 \\ S_2 \end{bmatrix}^T \begin{bmatrix} 1 & 0 \\ 0 & -1 \end{bmatrix} \begin{bmatrix} S_1 \\ S_2 \end{bmatrix} = S_1^2 - S_2^2$$

The pseudo metric tensor is

$$G = \begin{bmatrix} 1 & 0 \\ 0 & -1 \end{bmatrix}$$

The metric tensor can be decomposed as follows

$$G = \begin{bmatrix} 1 & 0 \\ 0 & -1 \end{bmatrix} = \begin{bmatrix} 1 & 0 \\ 0 & i \end{bmatrix}^T \begin{bmatrix} 1 & 0 \\ 0 & i \end{bmatrix}$$

The transformation is $T = \begin{bmatrix} 1 & 0 \\ 0 & i \end{bmatrix}$.

The transformation U where the quadratic form is invariant gives

$$U = \begin{bmatrix} 1 & 0 \\ 0 & i \end{bmatrix}^{-1} \begin{bmatrix} \cos(\alpha) & \sin(\alpha) \\ -\sin(\alpha) & \cos(\alpha) \end{bmatrix} \begin{bmatrix} 1 & 0 \\ 0 & i \end{bmatrix} = \begin{bmatrix} \cos(\alpha) & i\sin(\alpha) \\ i\sin(\alpha) & \cos(\alpha) \end{bmatrix} \tag{3.11}$$

The quadratic form $\rho^2 = S_1^2 - S_2^2$ is independent of the angle α and we can assume this angle α = -i A. So we have

$$U = \begin{bmatrix} \cos(-iA) & i\sin(-iA) \\ i\sin(-iA) & \cos(-iA) \end{bmatrix} = \begin{bmatrix} Ch(A) & Sh(A) \\ Sh(A) & Ch(A) \end{bmatrix} \tag{3.12}$$

That is the Lorenz transformation and using this we can obtain the Einstein relativistic effects.

3.3.1 Geometric Psychology and Coherence

Using the work of William C. Hoffman, we obtain a deeper meaning than that found in our previous consideration. He argues about the methods used to process the forms. In particular on how we are able to process visual stimuli at the local level related to global phenomena which is presented on the macroscopic perceptual scale.

How are the line elements detected at the single cell level integrated into the visual figure?

In the first instance visual contours are formed by using the vector fields of the cortical visual manifold V. Given the Lie Derivative

$$D = x_2 \frac{\partial}{\partial x_1} - x_1 \frac{\partial}{\partial x_2} \tag{3.13}$$

for

$$X = \begin{bmatrix} \frac{dx_1}{dt} \\ \frac{dx_2}{dt} \end{bmatrix} = \begin{bmatrix} x_2 \\ -x_1 \end{bmatrix} = \begin{bmatrix} 0 & 1 \\ -1 & 0 \end{bmatrix} \begin{bmatrix} x_1 \\ x_2 \end{bmatrix} \tag{3.14}$$

Observe that (3.13) is the differential equation

$$\begin{cases} \frac{dx_1}{dt} = x_2 \\ \frac{dx_2}{dt} = -x_1 \end{cases} \tag{3.15}$$

The derivatives can be represented in a geometric way is as shown figure 3.19.

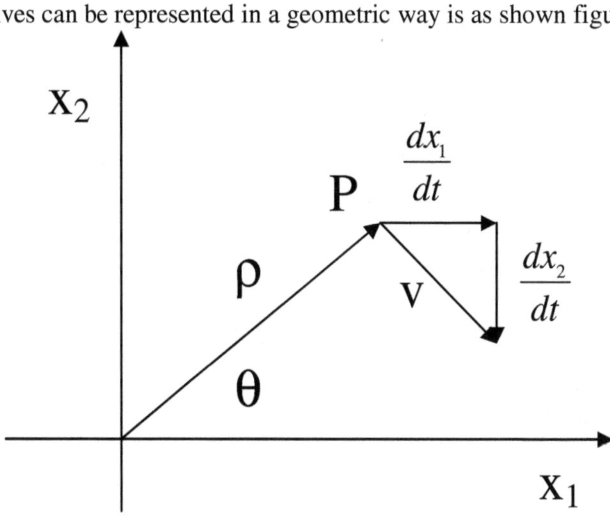

Fig. 3.19 Movement of the point P in the space x_1 and x_2.

We know locally, the components of the velocity of the point P but globally we have no idea of the form the trajectory in which the point P will move. Assuming the trajectory takes the form of a circle (invariant for rotation) where equation is

$$\rho^2 = {x_1}^2 + {x_2}^2 \tag{3.16}$$

Using the form given in equation (3.12) we obtain

$$D\rho^2 = x_2 \frac{\partial({x_1}^2 + {x_2}^2)}{\partial x_1} - x_1 \frac{\partial({x_1}^2 + {x_2}^2)}{\partial x_2} = 2x_2x_1 - 2x_1x_2 = 0 \tag{3.17}$$

So for any point having the form of a "circle", the Lie derivative is zero. In this particular case the square of the radius y of the circle is invariant on the circle. When we move on the ellipse we obtain

$$\begin{aligned} D\rho^2 &= w_1 \frac{\partial\rho^2}{\partial x_1} + w_2 \frac{\partial\rho^2}{\partial x_2} = \\ x_2 &\frac{\partial(\alpha{x_1}^2 + \beta{x_2}^2)}{\partial x_1} - x_1 \frac{\partial(\alpha{x_1}^2 + \beta{x_2}^2)}{\partial x_2} = \\ 2\alpha x_2x_1 &- 2\beta x_1x_2 = \frac{d\rho^2}{dt} \end{aligned} \tag{3.18}$$

this show us the difference between the form of an "ellipse " with respect to the form of a "circle". In the Ellipse form there is *incoherence between* the local definition of the velocity and the computation of the invariance by using the operator D.

In general we can have different forms obtained by the set of differential equations (3.9). Given the quadratic form

$$\rho^2 = a_{1,1}S_1^2 + a_{2,2}S_2^2 + 2a_{1,2}S_1S_2 = \begin{bmatrix} S_1 \\ S_2 \end{bmatrix}^T \begin{bmatrix} a_{1,1} & a_{1,2} \\ a_{2,1} & a_{2,2} \end{bmatrix} \begin{bmatrix} S_1 \\ S_2 \end{bmatrix}$$

We write G = $\begin{bmatrix} a_{1,1} & a_{1,2} \\ a_{2,1} & a_{2,2} \end{bmatrix}$ as G = T^T T and we obtain

$$U = T^{-1}\, B\, T$$

Where B is the unitary matrix.

Now because we have

$$U(t) = \exp(tD)$$

we have

$$\frac{dU(t)}{dt} = D\exp(tD)$$

and

$$D = \frac{dU(t)}{dt}U^{-1}$$

For example we know that the form (3.16) is invariant for the rotation

$$U(t) = \begin{bmatrix} cos(t) & sin(t) \\ -sin(t) & cos(t) \end{bmatrix}$$

The derivative of U(t) is

$$\frac{dU(t)}{dt} = \begin{bmatrix} -sin(t) & -cos(t) \\ cos(t) & -sin(t) \end{bmatrix}$$

and

$$D = \frac{dU(t)}{dt}U(t)^{-1} = \begin{bmatrix} -sin(t) & cos(t) \\ -cos(t) & -sin(t) \end{bmatrix}\begin{bmatrix} cos(t) & sin(t) \\ -sin(t) & cos(t) \end{bmatrix} = \begin{bmatrix} 0 & 1 \\ -1 & 0 \end{bmatrix}$$

For (3.13) the differential form is (3.12).

Given the transformation

$$U = \begin{bmatrix} 1 & 0 \\ m & 1 \end{bmatrix}^{-1}\begin{bmatrix} \cos(\alpha) & \sin(\alpha) \\ -\sin(\alpha) & \cos(\alpha) \end{bmatrix}\begin{bmatrix} 1 & 0 \\ m & 1 \end{bmatrix} =$$
$$\begin{bmatrix} \cos(\alpha) - m\sin(\alpha) & -\sin(\alpha) \\ \sin(\alpha) + m^2\sin(\alpha) & m\sin(\alpha) + \cos(\alpha) \end{bmatrix}$$

and

$$\frac{dU}{d\alpha} = \frac{d}{d\alpha}\begin{bmatrix} \cos(\alpha) - m\sin(\alpha) & -\sin(\alpha) \\ \sin(\alpha) + m^2\sin(\alpha) & m\sin(\alpha) + \cos(\alpha) \end{bmatrix} =$$
$$\begin{bmatrix} -\sin(\alpha) - m\cos(\alpha) & -\cos(\alpha) \\ \cos(\alpha) + m^2\cos(\alpha) & m\cos(\alpha) - \sin(\alpha) \end{bmatrix}$$

We obtain

$$\frac{dU}{d\alpha}U(\alpha)^{-1} = \begin{bmatrix} -m & 1 \\ 1+m^2 & m \end{bmatrix}$$

and

$$\begin{bmatrix} -m & 1 \\ 1+m^2 & m \end{bmatrix}\begin{bmatrix} x_1 \\ x_2 \end{bmatrix} = \begin{bmatrix} -mx_1 + x_2 \\ (1+m)x_1 + mx_2 \end{bmatrix}$$

The differential form is as follows

$$D=(-mx_1+x_2)\frac{\partial}{\partial x_1}+((1+m^2)x_1+mx_2)\frac{\partial}{\partial x_2}$$

for which

$$D\rho^2(x_1,x_2)=D((1+m^2)x_1^2+x_2^2+2mx_1x_2)=0$$

3.4 Local and Global Coherence Principle

Given the space where any point P = (y_1, y_2). We use different methods to calculate the local velocity. The method to compute the velocity is context dependent on P. Where the relation between the movements for which the point goes from P to Q = SP and on the local definition of the velocity. In the set of equations (3.19), we show the positions P and Q and the algorithms needed to compute the velocity at the different points. Given the point $P=(y_1,y_2)$ and the differential equation

$$\begin{cases}\dfrac{dy_1}{dt}=y_2\\[2mm]\dfrac{dy_2}{dt}=-y_1\end{cases}$$

and

$$\frac{dP}{dt}=\left[\frac{dy_1}{dt},\frac{dy_2}{dt}\right]$$

$$\text{and} \tag{3.19}$$

$$MP=\begin{bmatrix}0&1\\-1&0\end{bmatrix}\begin{bmatrix}y_1\\y_2\end{bmatrix}=\begin{bmatrix}y_2\\-y_1\end{bmatrix}$$

$$\begin{gathered}Q=SP=\begin{bmatrix}s_{11}&s_{12}\\s_{21}&s_{22}\end{bmatrix}\begin{bmatrix}y_1\\y_2\end{bmatrix}=\begin{bmatrix}s_{11}y_1+s_{12}y_2\\s_{21}y_1+s_{22}y_2\end{bmatrix},\\SMP=\begin{bmatrix}s_{11}&s_{12}\\s_{21}&s_{22}\end{bmatrix}\begin{bmatrix}y_2\\-y_1\end{bmatrix}=\begin{bmatrix}s_{11}y_2-s_{12}y_1\\s_{21}y_2-s_{22}y_1\end{bmatrix}\end{gathered}\tag{3.20}$$

The set of the relations (3.19) and (3.20) are the algebraic image of the second order agent given in the section 2.1. We note that given Q we obtain

$$MQ=MSP=\begin{bmatrix}s_{21}y_1+s_{22}y_2\\-(s_{11}y_1+s_{12}y_2)\end{bmatrix}$$

this is not equal to

$$SMP = \begin{bmatrix} s_{11} & s_{12} \\ s_{21} & s_{22} \end{bmatrix} \begin{bmatrix} y_2 \\ -y_1 \end{bmatrix} = \begin{bmatrix} s_{11}y_2 - s_{12}y_1 \\ s_{21}y_2 - s_{22}y_1 \end{bmatrix}$$

In general, we have

$$[MS - SM]\,P \neq 0$$

In a particular case $[MS - SM] = 0$ when

$$S = \begin{bmatrix} s_{11} & s_{12} \\ -s_{12} & s_{11} \end{bmatrix}.$$

For the transformation

$$\mathrm{Q} = \mathrm{S\ P}$$

We can compute the quadratic form L

$$\mathrm{L} = \mathrm{Q}^{\mathrm{T}}\,\mathrm{Q} = \mathrm{P}^{\mathrm{T}}\,\mathrm{S}^{\mathrm{T}}\,\mathrm{S\ P}$$

L is equal when we change the reference for the point P. The metric tensor G is then

$$G = S^T S =$$

$$\begin{bmatrix} s_{11} & s_{12} \\ -s_{12} & s_{11} \end{bmatrix}^T \begin{bmatrix} s_{11} & s_{12} \\ -s_{12} & s_{11} \end{bmatrix} =$$

$$(s_{11}{}^2 + s_{12}{}^2)\begin{bmatrix} 1 & 0 \\ 0 & 1 \end{bmatrix}$$

As shown in the first chapter G is a compensatory term and helps to determine the value of L before and after the transformation S

$$\mathrm{L} = \mathrm{P}^{\mathrm{T}}\,\mathrm{S}^{\mathrm{T}}\,\mathrm{S\ P}$$

which becomes

$$\mathrm{L} = \mathrm{Q}^{\mathrm{T}}\,\mathrm{Q}$$

When $s_{11}{}^2 + s_{12}{}^2 = 1$ we have $\mathrm{L} = \rho^2$

In conclusion the coherent transformation S is the rotation of the point P for which the radius ρ is invariant. We have

$$S = \begin{bmatrix} \cos(\alpha) & \sin(\alpha) \\ -\sin(\alpha) & \cos(\alpha) \end{bmatrix}$$

$$S = \begin{bmatrix} \sin(\alpha) & \cos(\alpha) \\ -\cos(\alpha) & \sin(\alpha) \end{bmatrix} =$$

$$\begin{bmatrix} \cos(\frac{\pi}{2}-\alpha) & \sin(\frac{\pi}{2}-\alpha) \\ -\sin(\frac{\pi}{2}-\alpha) & \cos(\frac{\pi}{2}-\alpha) \end{bmatrix}$$

The invariant (3.16) is the expression of the coherence existing between the methods used to compute the velocity of the point P and finally the position Q.

When the agents of the second order change the task; source graph in the space (y_1, y_2) which exist in the task source in the velocity space. Only set of points obtained by the rotation S or Coherent Transformation give the same set of points in the velocity space for different positions. The original rotation rule is also preserved in the velocity space.

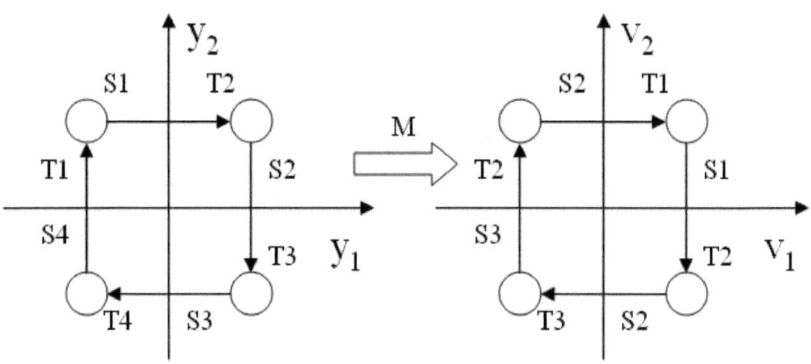

Fig. 3.20 Showing that he second order agent changes the Space of The Position into the Change of the Velocity. For a transformation in the Position Space where ρ = constant, the structure after the action of the operator M remains the same with different positions of sources and tasks.

Constancy Processing for Form Perception may be defined as a very special adaptation of the brain for invariance of form of field) recognition. Human beings can adapt to the distortions produced by lenses, prisms and other devices. Adaptation here means to transform from one form into another more standard or invariant form. An example of adaptation is the transition of “ellipse” form to the “circle” from where Dy = 0 and y is invariant. The task of the visual system to achieve constancy and to encounter *distortion* may be regarded as a means of achieving invariance of a perception and stability of a Perceptional Frame.

The visual system seeks to find a transformation on the distorted form and apply it to restore the original form so as to produce normal visual representation of the world and extend the sensory motor coordination. Using a simple local rule D y = 0 we can generate and control the form in which y is constant. The form is called the Global Entity. Similarly to Hoffman's work we can say the following Geometric Invariance imposed upon a visual manifold by considering the perceptual constancies such as shape, size, colour, motion, are equivalent to the action of the local agents. These include neurons for example where D y = 0. The form obtained by the action of agents may be identified by the object for which y is invariant.

Geometric Psychology studies the action of the neurons that use the local methods obtained by the Lie derivative D. This is used to control the production of special types of forms, which are actively restored or adapted, by the action of the external distortion.

In the visual system similarity refers to the tendency of the visual system to group together those pattern elements of forms or fields that are alike in some respect. These may include factors such as luminance, shape or orientation. The vector fields of the cortical visual manifold V factored by qualities such as brightness, shape for example can be sorted into equivalent classes having similar forms.

Any type of circle with different radius or position in the visual field are similar to one and other because the local property D y = 0 is independent of the position and dimension of the circle form.

a Example of the Coherence in the Ant Minimum Path

A fully connected graph *G* where *n* is the set of points and *e* is the set of connections between the points is called a fully connected graph in the Euclidean space.

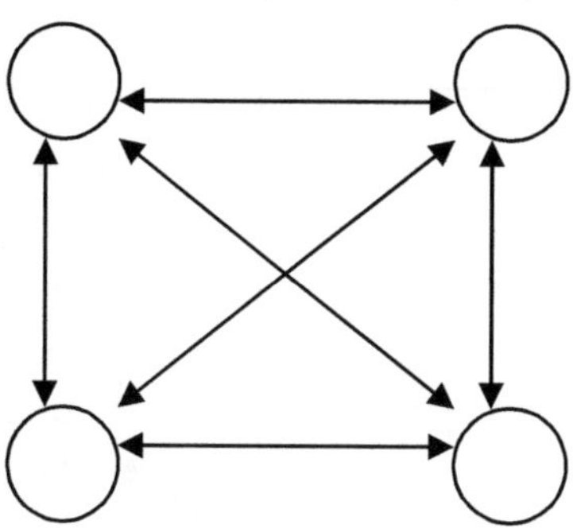

Fig. 3.21 Fully Connected Graph

The problem is to find a minimal length closed tour that visits each point once. The length of the path between points *i* and *j*

$$d_{i,j} = \sqrt{(x_i - x_j)^2 + (y_i - y_j)^2} \, .$$

We assume that there exist m agents which may be Ants that are intitially uniformly distributed at all points of the graph in Figure 3.21. Each Ant is a simple agent having the following characteristics:

1. The Ant chooses a point to reach with a probability that is a function of the distance and of the amount of trail present on the connecting edge;
2. When it completes a tour, it marks the complete traversedsection by a hypothetical substance called here "trail" each edge (*i,j*) visited.

Let τ be the intensity of trail on the edge (*i,j*) at time *t*. Each Ant at time *t* chooses the next point, where it will be at time *t*+1. When the time is *t+n* each ant will have completed a tour. The trail intensity is then updated according to the formula.

$$\tau_{i,j}(t+n)=\rho\tau_{i,j}(t)+\Delta\tau_{i,j} \tag{3.21}$$

Here ρ is a coefficient where (1 - ρ) represents the evaporation of the trail between the times *t* and *t+n*,

$$\Delta\tau_{i,j}=\sum_{k=1}^{m}\Delta\tau_{i,j}^{k} \tag{3.22}$$

Here $\Delta\tau_{i,j}^{k}$ is the quantity which is laid per unit of length of the trail substance laid on edge (*i,j*) by the *k*-th ant between time *t* and *t+n*. The coefficient ρ must be set to a value <1 to avoid an unlimited accumulation of the substance laid by Real Ants known as the pherormone. The initial intensity of the substance will be a small positive constant. The field of trial subastance can be represented in a space of the relationship. Any point in this space is a called a Connection Edge. At any Connection Edge we assign an intensity of the trial substance generated by all the Ants as a result of their movement within the graph.

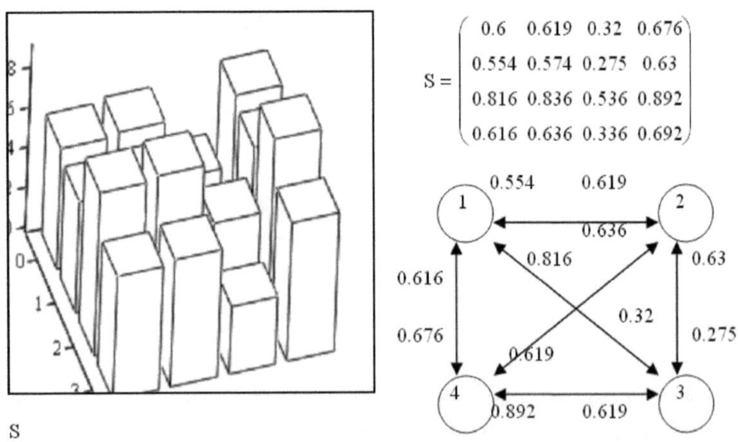

Fig. 3.22 The Field of the Trial Substance within the Fully Connected Graph.

The Source, Task and Actions of the Agents Ants required to generate the Field F is indicated in figure 3.23.

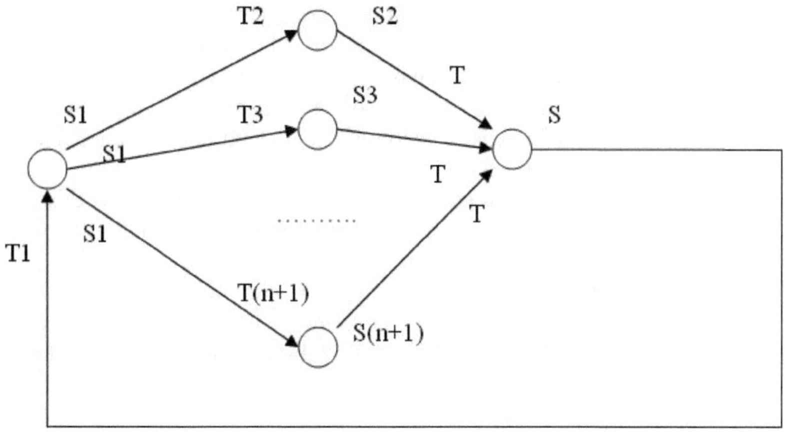

Fig. 3.23 Graph Sources and Tasks to Generate the Field S.

In figure 3.23 the Feedback whereby m Ants generate the field of the trail substance. The graph is repeated for all the edges of the Fully Connected Graph. At the first node in Figure 3.21 the source S1 that is the value of the trial substance that any Ant on a particular edge can measure. With a value of S1 the Ant p has the task Tp to decide whether to move inside the edge. The decision task for the Ants is given as follows. Given the value of visibility $\eta_{i,j} = \frac{1}{d_{i,j}}$. This is a quantity not modified during the runningof the the Ant System. The normalised decision making index from point *i* to point *j* for the *k*-th Ant is

$$D_{i,j}(t) = \tau_{i,j}^{\alpha}(t)\eta_{i.j}^{\beta}$$

Here α and β are the parameters which control the relative importance of the trial versus the visibility. Where the ant decides to generate the Sp trial with the task T to change the field S. From S the task T1 is to mantain the field S in time in the way to be used by the other ants. The coherence of the ant system consists in the invariance of the graph in Figure 3.21 with a different assignment of the sources and task. This was discussed in the chapter 2.1.The decision index is the result of a trade-off between *visibility, visibility requires close points should be chosen with high decision index. This results in a 'greedy costructive heuristic' and trail intensity at time t. this requires that if at an edge(i,j) there has been much traffic. It is then highly desirable.* An Ant Algorithm has the following characteristics:

1. It is a natural algorithm since it is based on the behavior of ants in establishing paths going from their colony to food sources and returning to the colony.
2. It is parallel and distributed as it comprises of a population of agents which are moving simultaneously, independently and without supervision;

3. It is cooperative as each agent chooses a path based on the basis of the information, pheromone trails, laid by the other agents using the same path. This cooperative behavior is autocatalytic. That is it provides positive feedback. This is because the probability of an agent choosing a particular path increases according to the number of agents that have previously chosen that path.
4. It is versatile in that it can be applied to similar versions of this problem. There is an extension of the Traveling Salesman Problem (TSP) to the Asymmetric Traveling Salesman Problem (ATSP).
5. It is robust in that it can be applied with minimal changes to other combinatorial optimization problems. These include the Quadratic Assignment Problem (QAP) and the Job-Shop Scheduling problem (JSP).

Chapter 4
Field Theory for Knowledge

4.1 Introduction

In this chapter we show the spectrum of the possible field of research of the non Euclidean geometry and morphogenetic fields. The Green function extensively used in the solution of linear differential equation assumes a new geometric image and is extended beyond the traditional field of differential equations.

4.2 Field of Conditional Probability and Tensor Image

It is shown that conditional probability has a deeper tensor representation. In fact we know that the join probability can be represented by the matrix form

$$P(Y,X)=B=\begin{bmatrix} p(Y_1 \mid X_1) & p(Y_1 \mid X_2) & \dots & p(Y_1 \mid X_n) \\ p(Y_2 \mid X_1) & p(Y_2 \mid X_2) & \dots & p(Y_2 \mid X_n) \\ \dots & \dots & \dots & \dots \\ p(Y_m \mid X_1) & p(Y_m \mid X_2) & \dots & p(Y_m \mid X_n) \end{bmatrix} \tag{4.1}$$

Now B is the matrix that show in the colons the reference fields one for any colons. Now for

$$S=p(X)=\begin{bmatrix} p(X_1) \\ p(X_2) \\ \dots \\ p(X_n) \end{bmatrix} \tag{4.2}$$

We have

$$p(Y)=Bp(X)=BS=\begin{bmatrix} p(Y_1) \\ p(Y_2) \\ \dots \\ p(Y_m) \end{bmatrix}=\begin{bmatrix} p(Y_1 \mid X_1) & p(Y_1 \mid X_2) & \dots & p(Y_1 \mid X_n) \\ p(Y_2 \mid X_1) & p(Y_2 \mid X_2) & \dots & p(Y_2 \mid X_n) \\ \dots & \dots & \dots & \dots \\ p(Y_m \mid X_1) & p(Y_m \mid X_2) & \dots & p(Y_m \mid X_n) \end{bmatrix}\begin{bmatrix} p(X_1) \\ p(X_2) \\ \dots \\ p(X_n) \end{bmatrix} \tag{4.3}$$

G. Resconi: Geometry of Knowledge for Intelligent Systems, SCI 407, pp. 61–105.
springerlink.com

Now for

$$B^T p(Y) = B^T Bp(X)$$
$$S_B = p(X) = (B^T B)^{-1} B^T p(Y) \tag{4.4}$$

We have

$$p(Y) = \sum_j p(Y \mid X_j) X_j = BS_B = B(B^T B)^{-1} B^T p(Y) = Q_B p(Y) \tag{4.5}$$

Where Q_B is the projection operator of B that project p(Y) into itself. We have also that

$$S_\alpha = B^T p(Y)$$
$$S^\alpha = (B^T B)^{-1} B^T p(Y) = p(X) \tag{4.6}$$

Are the covariant and the contravariant of the vector p(Y) in the space of the reference B. Now with the source interpretation we can say that p(X) is the sources of p(Y). For the previous result we have the conjugate operator

$$B^C = (B^T B)^{-1} B^T = \begin{bmatrix} p(X_1 \mid Y_1) & p(X_1 \mid Y_2) & \dots & p(X_1 \mid Y_m) \\ p(X_2 \mid Y_1) & p(X_2 \mid Y_2) & \dots & p(X_2 \mid Y_m) \\ \dots & \dots & \dots & \dots \\ p(X_n \mid Y_1) & p(X_n \mid Y_1) & \dots & p(X_n \mid Y_m) \end{bmatrix} \tag{4.7}$$

The metric L of the non Euclidean geometry which metric tensor $G = B^T B$ is

$$L = S^\alpha S_\alpha = (S^\alpha)^T (B^T B) S^\alpha = p(X)^T G p(X) \tag{4.8}$$

The two probabilities p(X) and p(Y) are one the dual of the other. Because we have

$$Q = BB^C = B(B^T B)^{-1} B^T =$$
$$\begin{bmatrix} p(Y_1 \mid X_1) & p(Y_1 \mid X_2) & \dots & p(Y_1 \mid X_n) \\ p(Y_2 \mid X_1) & p(Y_2 \mid X_2) & \dots & p(Y_2 \mid X_n) \\ \dots & \dots & \dots & \dots \\ p(Y_m \mid X_1) & p(Y_m \mid X_2) & \dots & p(Y_m \mid X_n) \end{bmatrix} \begin{bmatrix} p(X_1 \mid Y_1) & p(X_1 \mid Y_2) & \dots & p(X_1 \mid Y_m) \\ p(X_2 \mid Y_1) & p(X_2 \mid Y_2) & \dots & p(X_2 \mid Y_m) \\ \dots & \dots & \dots & \dots \\ p(X_n \mid Y_1) & p(X_n \mid Y_1) & \dots & p(X_n \mid Y_m) \end{bmatrix} =$$
$$\begin{bmatrix} 1 & 0 & \dots & 0 \\ 0 & 1 & \dots & 0 \\ \dots & \dots & \dots & \dots \\ 0 & 0 & \dots & 1 \end{bmatrix}$$

So we have

$$\sum_j p(X_S \mid Y_j)p(Y_j) = p(X_S)$$

Now for the normalization of the probability definition the sum of the all the probabilities X conditioned to Y is equal to one so we have

$$\sum_k p(X_k \mid Y_j) = 1$$

With simple computations we obtain

$$\begin{aligned}
&p(X_S) = \sum_j p(X_S \mid Y_j)p(Y_j) = \sum_j p(X_S \mid Y_j)p(Y_j)\sum_k p(X_k \mid Y_j) = \\
&\sum_k \sum_j p(X_S \mid Y_j)p(X_k \mid Y_j)p(Y_j) = \\
&\sum_k \sum_j p(X_S \mid Y_j)\frac{p(X_k \mid Y_j)p(Y_j)}{p(X_k)}p(X_k) = p(X_S) \\
&\textit{but} \\
&B(B^T B)^{-1}B^T p(X) = \sum_k \sum_j p(X_S \mid Y_j)p(Y_j \mid X_k)p(X_k) = p(X_S) \\
&\textit{So we have} \\
&\frac{p(X_k \mid Y_j)p(Y_j)}{p(X_k)} = p(Y_j \mid X_k)
\end{aligned} \tag{4.9}$$

That I the Bayesian law obtained by the property of the projection operator Q. For the Bayesian law we can have also that

$$p(Y \mid X) = \frac{p(X \mid Y)p(Y)}{\sum_j p(X \mid Y_j)p(Y_j)} = \frac{p(X \mid Y)p(Y)}{p(X)}$$

and

$$\sum_j p(Y_j \mid X) = \sum_j \frac{p(X \mid Y_j)p(Y_j)}{\sum_j p(X \mid Y_j)p(Y_j)} = \frac{\sum_j p(X \mid Y_j)p(Y_j)}{\sum_j p(X \mid Y_j)p(Y_j)} = 1$$

The normalization rule and the Bayesian rule are connected one from the other. Now we know that the Green function has this meaning Technically, a Green's function, G(x, s), of a linear operator L acting on distributions over a manifold M, at a point s, is any solution of

$$LG(x,s) = \delta(x-s)$$

where δ is the Dirac delta function. This technique can be used to solve differential equations of the form;

$$Lu(x) = f(x)$$

If the kernel of L is nontrivial, then the Green's function is not unique. However, in practice, some combination of symmetry, boundary conditions and/or other externally imposed criteria would give us a unique Green's function. Also, Green's functions in general are distributions, not necessarily proper functions. Green's functions are also a useful tool in condensed matter theory, where they allow the resolution of the diffusion equation - and in quantum mechanics, where the Green's function of the Hamiltonian is a key concept, with important links to the concept of density of states. The Green's functions used in those two domains are highly similar, due to the analogy in the mathematical structure of the diffusion equation and Schrödinger equation.

Loosely speaking, if such a function G can be found for the operator L, then if we multiply the equation (4.1) for the Green's function by f(s), and then perform an integration in the s variable, we obtain;

$$\int LG(x,s)f(s)ds = \int \delta(x-s)f(s)ds = f(x)$$

The right hand side is now given by the equation (4.2) to be equal to Lu(x), thus:

$$Lu(x) = \int LG(x,s)f(s)ds$$

Because the operator L is linear and acts on the variable x alone (not on the variable of integration s), we can take the operator L outside of the integration on the right hand side obtaining;

$$Lu(x) = L\int G(x,s)f(s)ds$$

And this implies;

$$u(x) = \int G(x,s)f(s)ds$$

Thus, we can obtain the function u(x) through knowledge of the Green's function in equation (4.1), and the source term on the right hand side in equation (4.2). This process has resulted from the linearity of the operator L. In other words, the solution of equation (4.2), u(x), can be determined by the integration given in equation (4.3). Although f(x) is known, this integration cannot be performed unless G is also known. The problem now lies in finding the Green's function G that satisfies equation (4.1).

Not every operator L admits a Green's function. A Green's function can also be thought of as a right inverse of L. Aside from the difficulties of finding a Green's functions for a particular operator, the integral in equation (4.3), may be quite difficult to perform. However the method gives a theoretically exact result.

Convolving with a Green's function gives solutions to inhomogeneous differential-integral equations, most commonly a Sturm-Liouville problem. If G is the Green's function of an operator L, then the solution for u of the equation Lu = f is given by

$$u(x) = \int G(x,s) f(s) ds$$

Such an integral is known as a Fredholm integral equation, the study of which constitutes Fredholm theory. **The integral can be thought of as an expansion of f according to a Dirac delta function basis (projecting f over δ(x − s)) and a superposition of the solution on each projection**. In fact G(x,s) can be considered a set of s fields generate by a set of punctual sources in x = s. The value of the sources are

$$S(x,s) = \delta(x-s)$$

So for the delta of Dirac any sources has density infinite in x = s and zero in all the other points. Now the (4.1) is the superposition of the fields G(x,s) generate by punctual sources where the weights of the superposition are the values of the function f(s). The function f(s) is the function that give us the density of the sources, Now in the conditional probability the function

$$p(x, y) = p(y \mid x) = G(x, y) \tag{4.10}$$

can be considered as a Green function for which the sources are p(x), So we have that

$$\begin{aligned}
p(y) &= \int G(x, y) p(x) dx = \int p(y \mid x) p(x) dx = \\
&\int p(y \mid x) \left(\int p(x \mid s) p(s) ds \right) dx = \\
&\int \left(\int p(y \mid x) p(x \mid s) dx \right) p(s) ds = \\
&\int \delta(y-s) p(s) ds = p(y) = Qp(y)
\end{aligned}$$

Where the projection operator is the identity. From the Green function definition we have that p(x) are the sources of p(y). So the probability p(y) is generate by the sources p(x). In fact given for example

$$B = \begin{bmatrix} p(Red \mid Car) & p(Red \mid Notcar) \\ p(Green \mid Car) & p(Green \mid Notcar) \end{bmatrix}$$

the sources are

$$S = p(X) = \begin{bmatrix} p(Car) \\ p(Notcar) \end{bmatrix}$$

So we have

$$p(Y) = BS = \begin{bmatrix} p(\text{Red} \mid Car) & p(\text{Red} \mid Notcar) \\ p(Green \mid Car) & p(Green \mid Notcar) \end{bmatrix} \begin{bmatrix} p(Car) \\ p(Notcar) \end{bmatrix} =$$

$$\begin{bmatrix} p(\text{Red} \mid car)P(car) + p(\text{Red} \mid Notcar)p(Notcar) \\ p(Green \mid car)P(car) + p(Green \mid Notcar)p(Notcar) \end{bmatrix} =$$

$$\begin{bmatrix} p(Red) \\ p(Green) \end{bmatrix}$$

and

$$(B^T B)^{-1} B^T = \begin{bmatrix} p(Car \mid \text{Red}) & p(Car \mid Green) \\ p(Notcar \mid \text{Red}) & p(Notcar \mid Green) \end{bmatrix} =$$

and

$$(B^T B)^{-1} B^T \begin{bmatrix} p(Red) \\ p(Green) \end{bmatrix} =$$

$$\begin{bmatrix} p(Car \mid \text{Red}) & p(Car \mid Green) \\ p(Notcar \mid \text{Red}) & p(Notcar \mid Green) \end{bmatrix} \begin{bmatrix} p(Red) \\ p(Green) \end{bmatrix} =$$

$$\begin{bmatrix} p(Car) \\ p(NotCar) \end{bmatrix}$$

4.3 Geometry of Invariance and Symmetry in Population of Neurons by QMS (Quantum Morphogenetic System)

4.3.1 Introduction

Basis vectors on a Cartesian grid visualized in Figure 4.1. The pixel base can be defined as a set of Dirac basis vectors on a Cartesian grid. In the QMS language we have that the Dirac basis vectors define the Objective space E and the Cartesian grid is the reference space. The normalized grey level y_i is represented as $y_i = \lambda \cdot e_i$ for $(0 \le \lambda \le 1)$. Any $n \times m$ grey image can be represented by a set of

$$N = n \cdot m$$

basic fields $\{F_1, F_2, \ldots F_N\}$ or objective space in the reference space where any image is a vector which components are the n x m grey values. We show the basic fields and the values of the pixels for one image H

$$H = y_1 \begin{bmatrix} 1 \\ 0 \\ \dots \\ 0 \end{bmatrix} + y_2 \begin{bmatrix} 0 \\ 1 \\ \dots \\ 0 \end{bmatrix} + \dots\dots + y_M \begin{bmatrix} 0 \\ 0 \\ \dots \\ 1 \end{bmatrix} = \begin{bmatrix} y_1 \\ y_2 \\ \dots \\ y_M \end{bmatrix} =$$

$$y_1 e_1 + y_2 e_2 + \dots\dots + y_M e_M = F_1 + F_2 + \dots\dots + F_M$$

$$where\ e_i e_j = \delta_{i,j}\ and\ H(P_1) = y_1,\ H(P_2) = y_2,\ \dots\dots\dots,\ H(P_M) = y_M$$

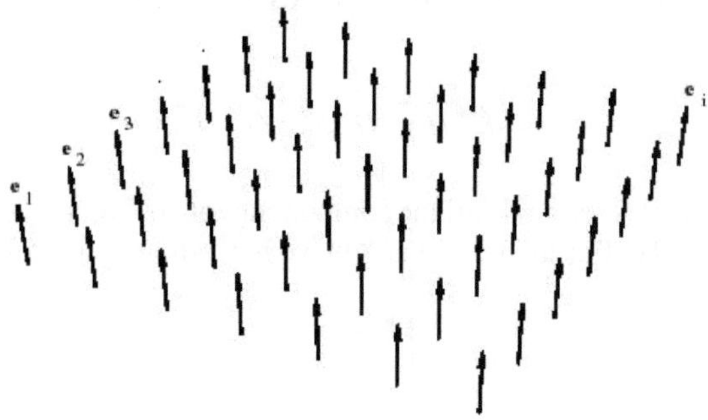

Fig. 4.1 The standard base can be seen as a set of Dirac vectors on Cartesian grid

The pixel description of the grey image can be interpreted as defining a multidimensional signal space. Now after the definition of the object space and the reference space that we consider as the syntactic or technological spaces we define the semantic space of the fields as features. we show the set of fields H_j denoted prototype field (prototype patterns of features) the points P_k in the reference spaces and the input field X.

$$H_j(P) = \begin{bmatrix} H_j(P_1) \\ H_j(P_2) \\ \dots \\ H_j(P_M) \end{bmatrix} = H_1(P_1) \begin{bmatrix} 1 \\ 0 \\ \dots \\ 0 \end{bmatrix} + H_1(P_2) \begin{bmatrix} 0 \\ 1 \\ \dots \\ 0 \end{bmatrix} + \dots H_1(P_M) \begin{bmatrix} 0 \\ 0 \\ \dots \\ 1 \end{bmatrix}$$

So we have

$$H = H_1(P) \oplus H_2(P) \oplus \dots\dots \oplus H_N(P) =$$

$$\begin{bmatrix} H_1(P_1) & H_2(P_1) & \dots & H_N(P_1) \\ H_1(P_2) & H_2(P_2) & \dots & H_N(P_2) \\ \dots & \dots & \dots & \dots \\ H_1(P_M) & H_2(P_M) & \dots & H_N(P_M) \end{bmatrix} \quad (4.11)$$

and the input X is

$$X(P)=\begin{bmatrix} X(P_1) \\ X(P_2) \\ \dots \\ X(P_M) \end{bmatrix} = X(P_1)\begin{bmatrix} 1 \\ 0 \\ \dots \\ 0 \end{bmatrix} + X(P_2)\begin{bmatrix} 0 \\ 1 \\ \dots \\ 0 \end{bmatrix} + \dots X(P_M)\begin{bmatrix} 0 \\ 0 \\ \dots \\ 1 \end{bmatrix}$$

The set of basic fields or prototype fields $H_1, H_2, \dots, H_N$ are vectors in the objective space and form the non-orthogonal space of the fields (features). Given the input filed X and the set of prototype fields H, we can compute the sources S^α in this way

$$S^\alpha = \left(H^T H\right)^{-1} H^T X = G^{-1} H^T X \qquad (4.12)$$

The sources S^α are the contra-variant components of the vector X in the semantic space of the fields H_j. For the relation (4.1) we have the input output relation

$$Y = H\, S^\alpha \qquad (4.13)$$

or

$$Y = S^1\begin{bmatrix} H_1(P_1) \\ H_1(P_2) \\ \cdots \\ H_1(P_M) \end{bmatrix} + S^2\begin{bmatrix} H_2(P_1) \\ H_2(P_2) \\ \cdots \\ H_2(P_M) \end{bmatrix} + \cdots S^N\begin{bmatrix} H_N(P_1) \\ H_N(P_2) \\ \cdots \\ H_N(P_M) \end{bmatrix} = HS^\alpha =$$

$$H(H^T H)^{-1} H^T X = HG^{-1} H^T X = QX$$

Where Q is the projection operator of X into the space of the prototype fields H_j. We remark that Q^2 X= Q X

The sources

$$S_\alpha = G\, S^\alpha = H^T X$$

are the covariant components of X into the space of the prototype fields H_j. Graphic example of the projection operator is shown in Figure 4.2.

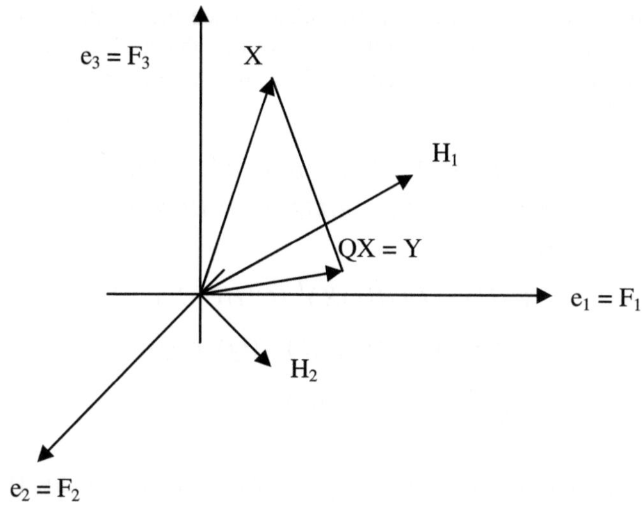

Fig. 4.2 Projection QX of the input vector X (input field) into the space of the prototype fields H_1 and H_2

Input – Output in the QMS (quantum morphogenetic system)

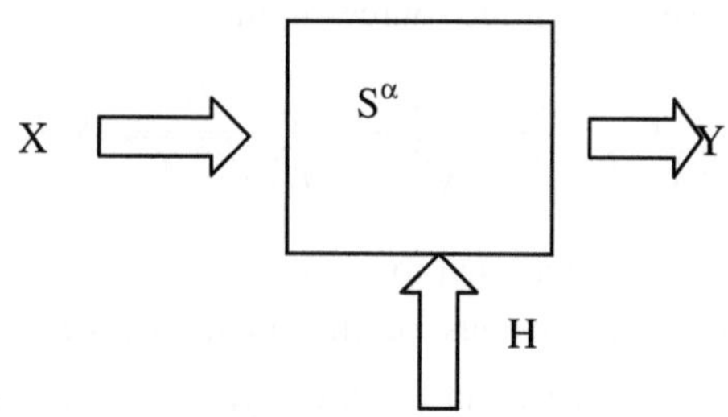

Fig. 4.3 Architecture of the QMS

In Figure 4.3 we show the architecture of the QMS. The important feature of the QMS is that the architecture that from X generates Y is always the same. But with the formal input H we adapt the QMS to any problem. We remark that

$$\begin{aligned} L &= (S^{\alpha})^T (S_{\alpha}) = (S^{\alpha})^T G(S^{\alpha}) = (S^{\alpha})^T H^T H(S^{\alpha}) \\ &= (HS^{\alpha})^T (HS^{\alpha}) = (QX)^T (QX) = \sum_{j,k} G_{j,k} S^j S^k \end{aligned} \tag{4.14}$$

In geometrical sense L is the metric of the space for the prototype fields H_j. In the same time L is the Lagrangian of the QMS. Given a geodetic path in the space of the prototype fields H, for this path L assume the minimum value. Because the space of the prototype field is non Euclidean space the geodetic path is not the straight line but a generic curve in the space of the objects. For H* = H U where U is N x N quadratic matrix we obtain. The projection operator is invariant. We have also that

$$(HU)[(HU)^T HU)]^{-1}(HU)^T X = (HU)[U^T H^T HU)]^{-1}(HU)^T X =$$
$$(HU)[U^T GU)]^{-1}(HU)^T X = (HU)U^{-1}G^{-1}(U^T)^{-1}U^T H^T X =$$
$$HG^{-1}H^T X = QX$$

$$\begin{aligned} L(HU) &= (Q(HU)X)^T (Q(HU)X) \\ &= (Q(H)X)^T (Q(H)X) = (QX)^T (QX) \\ &= \sum_{j,k} G_{j,k} S^j S^k = L(H) \end{aligned} \tag{4.15}$$

For any linear transformation of the space of the prototype fields H by U. The quadratic form $\sum_{j,k} G_{j,k} S^j S^k$ make by the sources is invariant and also the projection operator QX is invariant. The quadratic form is also invariant for any unitary transformation B. Formally we have

$$H^* = BH,\ B^T B = 1$$

With the scale transformation of the sources in this way

$$\sum_{j,k} G_{j,k} \frac{S^J}{\sqrt{G_{j,j}}} \frac{S^k}{\sqrt{G_{k,k}}} = \sum_{j,k} \frac{G_{j,k}}{\sqrt{G_{j,j} G_{k,k}}} S^j S^k$$

we can transform the G into a normal form G*

where $\cos(\alpha_{j,k})$ is the cosinus of the angle $\alpha_{j,k}$ between the vector that represent the prototype field H_j and the prototype field H_k . The quadratic form $\sum_{j,k} G_{j,k} S^j S^k$ can be used to measure the distance between two inputs X_1 and X_2 with the same set of prototype images H. In fact the distance is computed by the expression

$$D(X_1, X_2) = \sqrt{\sum_{j,k} G_{j,k} (S_1^j - S_2^j)(S_1^k - S_2^k)} \quad \textit{where}$$
$$S_1^j = G^{-1} H^T X_1\ ,\ S_2^j = G^{-1} H^T X_2, G = H^T H$$

with the distance $D(X_1, X_2)$ we can define the ranking , the similarity , the cluster operations for inputs ,patterns, or images X.

4.3.2 Invariants and Symmetry

In general for any affine transformation

$$\begin{bmatrix} x' \\ y' \end{bmatrix} = \begin{bmatrix} a_{1,1} & a_{1,2} \\ a_{2,1} & a_{2,2} \end{bmatrix} \begin{bmatrix} x \\ y \end{bmatrix} = H \begin{bmatrix} x \\ y \end{bmatrix} \tag{4.16}$$

we take the QMS

$$H = \begin{bmatrix} a_{11} & a_{12} \\ a_{21} & a_{22} \end{bmatrix} and \; X = \begin{bmatrix} X_1 \\ X_2 \end{bmatrix}$$

So we obtain

$$G = \begin{bmatrix} a_{1,1} & a_{1,2} \\ a_{2,1} & a_{2,2} \end{bmatrix}^T \begin{bmatrix} a_{1,1} & a_{1,2} \\ a_{2,1} & a_{2,2} \end{bmatrix} = \begin{bmatrix} a_{1,1}{}^2 + a_{2,1}{}^2 & a_{1,1}a_{1,2} + a_{2,1}a_{2,2} \\ a_{1,1}a_{1,2} + a_{2,1}a_{2,2} & a_{1,2}{}^2 + a_{2,2}{}^2 \end{bmatrix} \tag{4.17}$$

and

$$S^{\alpha} = \begin{bmatrix} S^1 \\ S^2 \end{bmatrix} = (H^T H)^{-1} H^T X = G^{-1} H^T X = H^{-1} X = \begin{bmatrix} a_{11} & a_{12} \\ a_{21} & a_{22} \end{bmatrix}^{-1} \begin{bmatrix} X_1 \\ X_2 \end{bmatrix}$$

So we have Y = H S^{α} = X. Now the Lagrangian is

$$\begin{aligned} L = \rho^2 &= g_{1,1}S_1{}^2 + g_{2,2}S_2{}^2 + 2g_{1,2}S_1S_2 = \\ &\begin{bmatrix} S_1 \\ S_2 \end{bmatrix}^T \begin{bmatrix} g_{11} & g_{12} \\ g_{21} & g_{22} \end{bmatrix} \begin{bmatrix} S_1 \\ S_2 \end{bmatrix} = \\ &\begin{bmatrix} S_1 \\ S_2 \end{bmatrix}^T \begin{bmatrix} a_{1,1}{}^2 + a_{2,1}{}^2 & a_{1,1}a_{1,2} + a_{2,1}a_{2,2} \\ a_{1,1}a_{1,2} + a_{2,1}a_{2,2} & a_{1,2}{}^2 + a_{2,2}{}^2 \end{bmatrix} \begin{bmatrix} S_1 \\ S_2 \end{bmatrix} = \\ &(a_{1,1}{}^2 + a_{2,1}{}^2)S_1^2 + (a_{1,2}{}^2 + a_{2,2}{}^2)S_2^2 + \\ &2(a_{1,1}a_{1,2} + a_{2,1}a_{2,2})S_1S_2 = (QX)^T (QX) = Y^T Y \end{aligned} \tag{4.18}$$

When we substitute the sources S^α in the Lagrangian we obtain

$$L = \rho^2 = g_{1,1}S_1{}^2 + g_{2,2}S_2{}^2 + 2g_{1,2}S_1S_2 =$$
$$(a_{1,1}{}^2 + a_{2,1}{}^2)S_1^2 + (a_{1,2}{}^2 + a_{2,2}{}^2)S_2^2 +$$
$$2(a_{1,1}a_{1,2} + a_{2,1}a_{2,2})S_1S_2 = (QX)^T(QX) = Y^TY = X_1^2 + X_2^2$$

Now we know that for the rotation of the vector X

$$RX = \begin{bmatrix} \cos(\alpha) & \sin(\alpha) \\ -\sin(\alpha) & \cos(\alpha) \end{bmatrix} \begin{bmatrix} X_1 \\ X_2 \end{bmatrix} \tag{4.19}$$

We have that

$$L = \rho^2 = X_1^2 + X_2^2 \tag{4.20}$$

is invariant, we search the transformation U of S for which

$$L = \rho^2 = (a_{1,1}{}^2 + a_{2,1}{}^2)S_1^2 + (a_{1,2}{}^2 + a_{2,2}{}^2)S_2^2 +$$
$$2(a_{1,1}a_{1,2} + a_{2,1}a_{2,2})S_1S_2$$

is invariant and for which

$$U\,S = U\,H^{-1}\,X. = H^{-1}\,R\,X$$

Now $U = H^{-1}\,R\,H$ is the wanted transformation of S, In fact we obtain

$$H^{-1}\,R\,H\,S = H^{-1}\,R\,H\,H^{-1}\,X = H^{-1}\,R\,X$$

By the rotation of the input vectors X we can compute the transformation U of the sources for which the lagrangian L is invariant. For example given the share transformation

$$\begin{cases} x' = x \\ y' = mx + y \end{cases} \text{ or } \begin{bmatrix} x' \\ y' \end{bmatrix} = \begin{bmatrix} 1 & 0 \\ m & 1 \end{bmatrix} \begin{bmatrix} x \\ y \end{bmatrix} \tag{4.21}$$

We have

$$H = \begin{bmatrix} 1 & 0 \\ m & 1 \end{bmatrix} \textit{and } X = \begin{bmatrix} X_1 \\ X_2 \end{bmatrix}$$

$$G = \begin{bmatrix} 1 & 0 \\ m & 1 \end{bmatrix}^T \begin{bmatrix} 1 & 0 \\ m & 1 \end{bmatrix} = \begin{bmatrix} 1+m^2 & m \\ m & 1 \end{bmatrix} \tag{4.22}$$

and

$$\begin{bmatrix} s^1 \\ s^2 \end{bmatrix} = \begin{bmatrix} 1+m^2 & m \\ m & 1 \end{bmatrix}^{-1} \begin{bmatrix} 1 & 0 \\ m & 1 \end{bmatrix}^T \begin{bmatrix} X_1 \\ X_2 \end{bmatrix} = \begin{bmatrix} 1 & 0 \\ m & 1 \end{bmatrix}^{-1} \begin{bmatrix} X_1 \\ X_2 \end{bmatrix}$$
$$= \begin{bmatrix} 1 & 0 \\ -m & 1 \end{bmatrix} \begin{bmatrix} X_1 \\ X_2 \end{bmatrix} = \begin{bmatrix} X_1 \\ -mX_1 + X_2 \end{bmatrix} \tag{4.23}$$

and

$$L = \rho^2 = g_{1,1} S_1{}^2 + g_{2,2} S_2{}^2 + 2 g_{1,2} S_1 S_2 =$$
$$\begin{bmatrix} S_1 \\ S_2 \end{bmatrix}^T \begin{bmatrix} 1+m^2 & m \\ m & 1 \end{bmatrix} \begin{bmatrix} S_1 \\ S_2 \end{bmatrix} =$$
$$(1+m^2) S_1^2 + S_2^2 + 2m S_1 S_2 = X_1{}^2 + X_2{}^2$$

The transformation U of the sources for which L is invariant is

$$U = \begin{bmatrix} 1 & 0 \\ -m & 1 \end{bmatrix} \begin{bmatrix} \cos(\alpha) & \sin(\alpha) \\ -\sin(\alpha) & \cos(\alpha) \end{bmatrix} \begin{bmatrix} 1 & 0 \\ m & 1 \end{bmatrix}$$
$$= \begin{bmatrix} 1 & 0 \\ m & 1 \end{bmatrix}^{-1} \begin{bmatrix} \cos(\alpha) & \sin(\alpha) \\ -\sin(\alpha) & \cos(\alpha) \end{bmatrix} \begin{bmatrix} 1 & 0 \\ m & 1 \end{bmatrix}$$
$$= \begin{bmatrix} \cos(\alpha) + m\sin(\alpha) & \sin(\alpha) \\ -(1+m^2)\sin(\alpha) & \cos(\alpha) - m\sin(\alpha) \end{bmatrix}$$
$$= \begin{bmatrix} \cos(\alpha) & \sin(\alpha) \\ -\sin(\alpha) & \cos(\alpha) \end{bmatrix} + \begin{bmatrix} 1 & 0 \\ -m & -1 \end{bmatrix} m\sin(\alpha) \tag{4.24}$$

Now for the previous U we can prove the L as function of the sources is really invariant. In fact we have

$$L = (U \begin{bmatrix} S_1 \\ S_2 \end{bmatrix})^T G (U \begin{bmatrix} S_1 \\ S_2 \end{bmatrix}) = \begin{bmatrix} S_1 \\ S_2 \end{bmatrix}^T U^T \begin{bmatrix} 1+m^2 & m \\ m & 1 \end{bmatrix} U \begin{bmatrix} S_1 \\ S_2 \end{bmatrix}$$
$$= \begin{bmatrix} S_1 \\ S_2 \end{bmatrix}^T U^T \begin{bmatrix} 1+m^2 & m \\ m & 1 \end{bmatrix} U \begin{bmatrix} S_1 \\ S_2 \end{bmatrix}$$

But

$$U^T \begin{bmatrix} 1+m^2 & m \\ m & 1 \end{bmatrix} U = (H^{-1}RH)^T (H^T H)(H^{-1}RH)$$
$$= H^T R^T (H^{-1})^T (H^T H)(H^{-1}RH)$$
$$= H^T R^T RH = H^T H \tag{4.25}$$
$$= \begin{bmatrix} 1 & 0 \\ m & 1 \end{bmatrix}^T \begin{bmatrix} 1 & 0 \\ m & 1 \end{bmatrix} = \begin{bmatrix} 1+m^2 & m \\ m & 1 \end{bmatrix} = G$$

So we obtain

$$L = \rho^2 = (U \begin{bmatrix} S_1 \\ S_2 \end{bmatrix})^T \begin{bmatrix} 1+m^2 & m \\ m & 1 \end{bmatrix} (U \begin{bmatrix} S_1 \\ S_2 \end{bmatrix}) =$$
$$\begin{bmatrix} S_1 \\ S_2 \end{bmatrix}^T \begin{bmatrix} 1+m^2 & m \\ m & 1 \end{bmatrix} \begin{bmatrix} S_1 \\ S_2 \end{bmatrix} = (1+m^2)S_1^2 + S_2^2 + 2mS_1S_2 = X_1^2 + X_2^2$$

Given the share transformation (4.4) we have that

$$\text{L}= \begin{bmatrix} x' \\ y' \end{bmatrix}^T \begin{bmatrix} x' \\ y' \end{bmatrix} = (x')^2 + (y')^2 = (\begin{bmatrix} 1 & 0 \\ m & 1 \end{bmatrix} \begin{bmatrix} x \\ y \end{bmatrix})^T \begin{bmatrix} 1 & 0 \\ m & 1 \end{bmatrix} \begin{bmatrix} x \\ y \end{bmatrix} =$$
$$(1+m^2)x^2 + y^2 + 2mxy = (1+m^2)S_1{}^2 + S_2{}^2 + 2mS_1S_2$$

Now when the sources S are equal to the initial values (x, y) with the QMS we can compute the transformation U for which L is invariant in the space (x,y). Now we compute the compensation transformation on (x, y) by which the transformation U become the simple rotation R. The transformation of the sources is

$$S = \begin{bmatrix} x \\ y \end{bmatrix} = H^{-1} \begin{bmatrix} X_1 \\ X_2 \end{bmatrix} = H^{-1} \begin{bmatrix} x' \\ y' \end{bmatrix}$$

Because any affine transformation is the composition of the Share transformation, rotation and scale, we can find also invariants for affine transformations in general.

We can also reverse the process. Now we begin with the Lagrangian L and compute the transformation U of the sources by which L is invariant. In fact given

$$L = \rho^2 = \begin{bmatrix} S_1 \\ S_2 \end{bmatrix}^T \begin{bmatrix} 1 & 0 \\ 0 & -1 \end{bmatrix} \begin{bmatrix} S_1 \\ S_2 \end{bmatrix} = S_1^2 - S_2^2 = x^2 - c^2 t^2$$

with

$$G = \begin{bmatrix} 1 & 0 \\ 0 & -1 \end{bmatrix} = \begin{bmatrix} 1 & 0 \\ 0 & i \end{bmatrix}^T \begin{bmatrix} 1 & 0 \\ 0 & i \end{bmatrix} = H^T H$$

The transformation U for which the quadratic form is invariant is

$$U = \begin{bmatrix} 1 & 0 \\ 0 & i \end{bmatrix}^{-1} \begin{bmatrix} \cos(\alpha) & \sin(\alpha) \\ -\sin(\alpha) & \cos(\alpha) \end{bmatrix} \begin{bmatrix} 1 & 0 \\ 0 & i \end{bmatrix} = \begin{bmatrix} \cos(\alpha) & i\sin(\alpha) \\ i\sin(\alpha) & \cos(\alpha) \end{bmatrix}$$

because the quadratic form

$$L = \rho^2 = S_1^2 - S_2^2$$

is independent of the angle α, we can take this angle $\alpha = -i\, A$. So we have

$$U = \begin{bmatrix} \cos(-iA) & i\sin(-iA) \\ i\sin(-iA) & \cos(-iA) \end{bmatrix} = \begin{bmatrix} Ch(A) & Sh(A) \\ Sh(A) & Ch(A) \end{bmatrix} \tag{4.26}$$

That is the Lorenz transformation. With QMS we can study more complex transformations as

$$\begin{bmatrix} y_1 \\ y_2 \\ \dots \\ y_n \\ y_{n+1} \\ y_{n+2} \\ \dots \\ y_m \end{bmatrix} = \begin{bmatrix} a_{1,1} & a_{1,2} & \dots & a_{1,n} \\ a_{2,1} & a_{2,2} & \dots & a_{2,n} \\ \dots & \dots & \dots & \dots \\ a_{n,1} & a_{n,2} & \dots & a_{n,n} \\ b_{1,1} & b_{1,2} & \dots & b_{1,n} \\ b_{2,1} & b_{2,2} & \dots & b_{2,n} \\ \dots & \dots & \dots & \dots \\ b_{m-n,1} & b_{m-n,2} & \dots & b_{m-n,n} \end{bmatrix} \begin{bmatrix} x_1 \\ x_2 \\ \dots \\ x_n \end{bmatrix} = HS \tag{4.27}$$

For which we have

$$G = \begin{bmatrix} a_{1,1} & a_{1,2} & \dots & a_{1,n} \\ a_{2,1} & a_{2,2} & \dots & a_{2,n} \\ \dots & \dots & \dots & \dots \\ a_{n,1} & a_{n,2} & \dots & a_{n,n} \\ b_{1,1} & b_{1,2} & \dots & b_{1,n} \\ b_{2,1} & b_{2,2} & \dots & b_{2,n} \\ \dots & \dots & \dots & \dots \\ b_{m-n,1} & b_{m-n,2} & \dots & b_{m-n,n} \end{bmatrix}^T \begin{bmatrix} a_{1,1} & a_{1,2} & \dots & a_{1,n} \\ a_{2,1} & a_{2,2} & \dots & a_{2,n} \\ \dots & \dots & \dots & \dots \\ a_{n,1} & a_{n,2} & \dots & a_{n,n} \\ b_{1,1} & b_{1,2} & \dots & b_{1,n} \\ b_{2,1} & b_{2,2} & \dots & b_{2,n} \\ \dots & \dots & \dots & \dots \\ b_{m-n,1} & b_{m-n,2} & \dots & b_{m-n,n} \end{bmatrix} \tag{4.28}$$

With the condition

$$Y = \begin{bmatrix} y_1 \\ y_2 \\ \cdots \\ y_n \\ y_{n+1} \\ y_{n+2} \\ \cdots \\ y_m \end{bmatrix} = \begin{bmatrix} Y_1 \\ Y_2 \end{bmatrix}$$

$$= \begin{bmatrix} a_{1,1} & a_{1,2} & \cdots & a_{1,n} \\ a_{2,1} & a_{2,2} & \cdots & a_{2,n} \\ \cdots & \cdots & \cdots & \cdots \\ a_{n,1} & a_{n,2} & \cdots & a_{n,n} \\ b_{1,1} & b_{1,2} & \cdots & b_{1,n} \\ b_{2,1} & b_{2,2} & \cdots & b_{2,n} \\ \cdots & \cdots & \cdots & \cdots \\ b_{m-n,1} & b_{m-n,2} & \cdots & b_{m-n,n} \end{bmatrix} \begin{bmatrix} x_1 \\ x_2 \\ \cdots \\ x_n \end{bmatrix} = \begin{bmatrix} A \\ B \end{bmatrix} X = \begin{bmatrix} A \\ MA \end{bmatrix} X = \begin{bmatrix} AX \\ M(AX) \end{bmatrix} \tag{4.29}$$

so we have the invariant form or constrain

$$Y_2 = MY_1, M = BA^{-1} \tag{4.30}$$

4.4 Retina Model and Rotation Symmetry

The optical image received from pupillary input is convoluted with activities of retinal receptors consist of retinal ganglion cells (RGC) in terms of center-surround receptive fields. A good model for the RF sensitivity profile can be the Laplacian of the Gaussian kernel known as 'Mexican Hat' function. In the present model, the circular symmetrical function [2] is used:

$$\nabla^2 G(x,y) = -\frac{1}{\pi\sigma^4}\left(1 - \frac{x^2+y^2}{2\sigma^2}\right) e^{-\frac{x^2+y^2}{2\sigma^2}} \tag{4.31}$$

With the Gaussian function

$$G(x,y) = -\frac{1}{2\pi\sigma^2} e^{-\frac{x^2+y^2}{2\sigma^2}} \tag{4.32}$$

And the sources

$$S(x,y) = -\frac{1}{\pi\sigma^4}\left(1-\frac{x^2+y^2}{2\sigma^2}\right)e^{-\frac{x^2+y^2}{2\sigma^2}}$$

Are the 'Mexican Hat'. The equations (4.5) and (4.6) equation become the potential equation or Poisson equation

$$\nabla^2 G(x,y) = S(x,y) \tag{4.33}$$

where S(x,y) is the volume density sources of the potential.. When S(x,y) = 0 we have the Laplace equation

$$\nabla^2 G(x,y) = \frac{\partial^2 G(x,y)}{\partial x^2} + \frac{\partial^2 G(x,y)}{\partial y^2} = 0$$

The solution G of the Laplace equation as we know is invariant fro the Rotation of the reference. Now we know that the Green function for the two dimensional potential theory is

$$G(x,y) = \log(\sqrt{x^2+y^2})$$

That is invariant for the rotation of the reference. The solution of the non homogeneous equation

$$\nabla^2 G = \left(\frac{\partial^2 G}{\partial x'^2} + \frac{\partial^2 G}{\partial y'^2}\right) = S(x,y)$$

Where S(x,y) are sources, is given by the integral form

$$\begin{aligned} G(u,v) &= \iint_G S(x,y)\ln\frac{1}{\rho(x,y,u,v)}dxdy = \\ &-\iint_G S(x,y)\ln(\rho(x,y,u,v))dxdy \end{aligned} \tag{4.34}$$

Where

$$\rho(x,y,u,v) = \sqrt{(x-u)^2+(y-v)^2}$$

where ln ρ (x,y) is the Green function for the equation of the potential or two dimensional potential equation. Now for the invariance of the equation of potential by Laplacian respect to the rotation and scale we have the logmap

$$w = \ln\frac{1}{z} = \ln\frac{1}{\rho} - j(\theta+2k\pi) = -(\ln\rho + j(\theta+2k\pi))$$

Now the Laplacian can be extended to functions defined on surface or more general on Riemannian and pseudo –Riemannian manifolds value. So we have the Laplace Beltrami operator

$$\nabla^2 G = \frac{1}{\sqrt{|g|}} \frac{\partial}{\partial x_i} (\sqrt{|g|} g^{i,j} \frac{\partial}{\partial x_j})$$

where the metric of the surface is

$$L = D^2 = g_{1,1}x^2 + g_{2,2}y^2 + 2g_{1,2}xy$$

That is a non Euclidean metric as in QMS. With the Laplace Beltrami operators we can simulate by the potential theory in the retina complex transformations U as we show in (4.2).

4.4.1 Sources and Transformations in Diffusion Reaction Equation

We know that in a more general case we have that the sources in the diffusion reaction equation is given by a chemical network. In fact for example given the chemical network in fig. 4.4

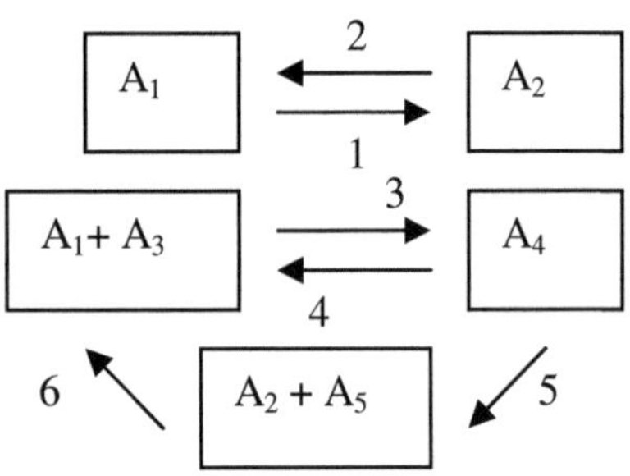

Fig. 4.4 Set of chemical reactions

The diffusion-reaction equation is

$$\begin{aligned}
\frac{dG_1(x,y)}{dt} &= D_1\left(\frac{\partial^2 G_1}{\partial x^2}+\frac{\partial^2 G_1}{\partial y^2}\right)-k_1G_1+k_2G_2^2-k_3G_1G_3+k_4G_4+k_6G_2G_5\\
\frac{dG_2(x,y)}{dt} &= D_2\left(\frac{\partial^2 G_2}{\partial x^2}+\frac{\partial^2 G_2}{\partial y^2}\right)+2k_1G_1-2k_2G_2^2+k_5G_5-k_6G_2G_5\\
\frac{dG_3(x,y)}{dt} &= D_3\left(\frac{\partial^2 G_3}{\partial x^2}+\frac{\partial^2 G_3}{\partial y^2}\right)-k_3G_1G_3+k_6G_2G_5+k_4G_4\\
\frac{dG_4(x,y)}{dt} &= D_4\left(\frac{\partial^2 G_4}{\partial x^2}+\frac{\partial^2 G_4}{\partial y^2}\right)+k_3G_1G_3-k_4G_4-k_5G_4\\
\frac{dG_5(x,y)}{dt} &= D_5\left(\frac{\partial^2 G_5}{\partial x^2}+\frac{\partial^2 G_5}{\partial y^2}\right)+k_5G_5-k_6G_2G_5
\end{aligned} \tag{4.35}$$

Given the set of equations

$$\frac{\partial G_j(x_1,x_2,\ldots,x_n)}{\partial t}-D_j\nabla^2 G_j(x_1,x_2,\ldots,x_n)=S_j(x_1,x_2,\ldots,x_n) \tag{4.36}$$

With the Green function as H , and S_j the sources we can write

$$G_j(x_1,x_2,\ldots,x_n)=H\ S_j(x_1,x_2,\ldots,x_n)$$

And the for the transformation T of G we have

$$\begin{aligned}
&\frac{\partial TG_j}{\partial t}-D_j\nabla^2 TG_j\\
&=T\left(\frac{\partial G_j}{\partial t}-D_j\nabla^2 G_j\right)+G_j\left(\frac{\partial T}{\partial t}-D_j\nabla^2 T\right)\\
&=TS_j+G_j\left[\frac{\partial T}{\partial t}-D_j\nabla^2 T\right]
\end{aligned} \tag{4.37}$$

With the new sources

$$S(T)=TS_j+G_j\left[\frac{\partial T}{\partial t}-D_j\nabla^2 T\right] \tag{4.38}$$

we generate the new field G' = TG. When we know G' we can compute the sourcers S(T) with the inverse function. The connection between sources and field that is fundamental in the morphogenetic system is used to control fields. So we have

$$T^{-1}\left(S(T)-G_j\left(\frac{\partial T}{\partial t}-D_j\nabla^2 T\right)\right)=S_j \tag{4.39}$$

We can compute the original sources.

4.4.2 *Computational Experiments by Loo Chu Kiong [ ]*

The experiment to check assertion of Quantum Morphology System (QMS) was done on a Pentium-4 1.3-GHz PC using the following algorithm programmed in Matlab with Image Processing Toolbox:

- P images of front face from 10 different individual with index i were coded into $H_1, H_2, \ldots, H_i, \ldots, H_P$ prototype field where a pixel grey level is described by basic field value F_{ij}. For example, $H_1 = \left[F_{11}, F_{12,} F_{13,} \ldots, F_{1N}\right]$. N represents the dimension of the encoded prototype image vectors.
- Images were preprocessed according to: $\overline{F}_{ij} = F_{ij} - \frac{1}{2N}\sum_{j=1}^{N} F_{ij}$. The resulting basic field values $\overline{F}_{ij}$ was then normalized according to $\vec{F}_j = \frac{\vec{F}_j}{\left\|\vec{F}_j\right\|}$. Such normalization was assumed to be quantum-implemented into plane-wave.
- The QMS projection operator Q is generated according to $H\left(H^T H\right)^{-1} H^T$.
- Later, in the "selective reconstructive stage", a partially occluded or noise contaminated query image, X_{query} that was inserted. The QMS source was calculated according to

$$S = \left(H^T H\right)^{-1} H^T X_{query}.$$

The QMS reacted according to

$$Y = H\left(H^T H\right)^{-1} H^T X_{query} = QX_{query} = HS.$$

The "query" input field was reconstructed or corrected based on memorized model in QMS.

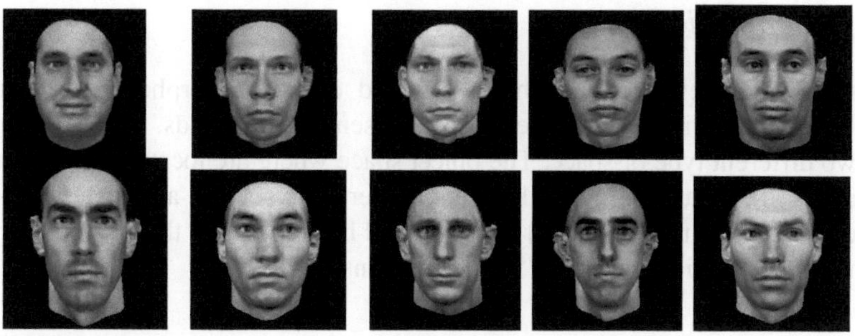

Fig. 4.5 Training images for face reconstruction experiments

In the experiment, all images are stored simultaneously in the same morphogenetic field, Q^*. One of the images is rotated or scaled and reconstructed out of Q^* as triggered by a "query"-image X^{query}. Fig. 4.10(b) and 4.11(b) shows the image-recognition results. The results show that the cortical-model and quantum morphogenetic system (QMS) can achieve the rotation and scale invariant face recognition. The invariant property of the QMS on rotation angle and scale factor of the query image is shown in Fig. 4.12. The invariant D^* varies slightly for both the rotation and scale invariant experiment.

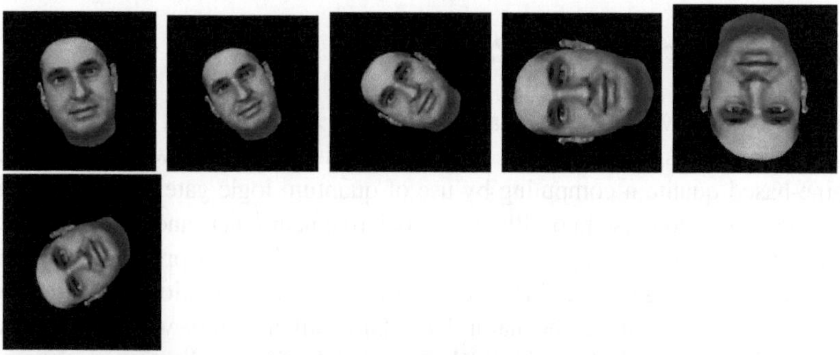

(a)

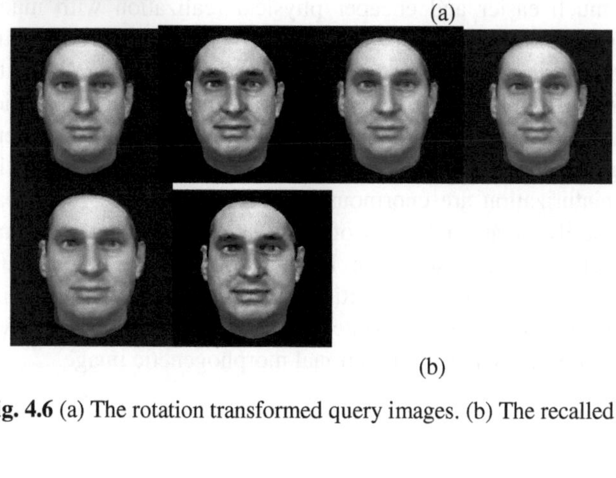

(b)

Fig. 4.6 (a) The rotation transformed query images. (b) The recalled images from QMS

4.4.3 Conclusion

In this part we present a new theory denoted quantum morphogenetic theory (QMS). The theory is based on the space representation of fields. In the QMS we have two different type of space. The object space where are locate the points and the space of the prototype fields. Projection operators, sources and invariant with symmetry are study to give a solid mathematical background to the image recognition and to the invariant for image transformations.

4.5 Quantum Mechanics and Non Euclidean Geometry

4.5.1 Introduction

We show that by a non orthogonal prototype image we can built Green function method as in the quantum mechanics but embedded into a non Euclidean geometry. Non Euclidean geometry allow to use information included in the correlation among prototype images. The new information improve the reconstruction process for input noised image as we can show in this paper. When we reshape the space of the prototypes with the inclusion of the correlations, we introduce morphogenetic element inside the original Euclidean space.

4.5.2 Introduction to the Problem

There is growing evidence that quantum-physical systems could be harnessed for information processing, including image recognition in two ways: by Turing-machine-based quantum computing by use of quantum logic gates and by quantum processing similar to those in oscillatory associative neural nets and holography.

This interdisciplinary paper reports how it is possible, in principle, to implement successful image recognition, as verified by our simulations, in a quantum holographic process. Since the natural fundamental quantum-wave dynamics is harnessed, it allows much easier and cheaper physical realization with much-bigger sizes and resolutions of images than the mainstream quantum-computing approaches The main contribution of this paper is to propose a generally better image-recognition method, and present its potential alternative implementation into a quantum-wave medium and morphogenetic system. Like quantum-net's capacities in morphogenetic system of connectivity, parallelism, storage, associativity, speed, and miniaturization are enormous. The Hopfield model with real-valued thus not necessarily binary activities of units neurons, having linear not sigmoid or signum activation function, can be transformed into a quantum-holographic procedure and in its morphogenetic system extension. The Hebbian memory storage is replaced by multiple self-interferences of quantum plane waves and in more complex non Euclidean and operational morphogenetic image.

This translation succeeded by the simplest variable exchange of the Hopfield's real-valued variables with the complex-valued variables changing as sinusoids waves. Thereby, all input-to-output transformations are preserved. Thus quantum-wave image recognition functions equivalently to Hopfield's one; only the implementation is much miniaturized, enabling almost infinitely large associative memory. The same for the successive extensions by morphogenetic systems. Since the opposite translation, i.e., digitalization of holography, was done in the 1960s to get the first computational associative memories, one might wonder what is new in the present proposal. Additionally to the big experimental success of classical optical, acoustic, microwave, x-ray, atom, and electron holography,the recent fast development of quantum optics gave birth to quantum holography and successive extensions by morphogenetic systems. The latter might implement the well known Hopfield model and its generalizations in a completely new framework in which the former obstacles memory-capacity limitations, problems with nonorthogonality producing cross are included in the extensions by morphogenetic systems.

4.5.3 Hopfield Net and Quantum Holography by Morphogenetic System in Euclidean Geometry

Using neuroquantum analogies we transform the Hopfield-like associative neural net into quantum formalism. Quantum wave-function acts as net's state vector q. Quantum eigenwave functions

$$H = \psi_k(r_j) = \begin{bmatrix} \psi_1(r_1) & \psi_2(r_1) & \dots & \psi_N(r_1) \\ \psi_1(r_2) & \psi_2(r_2) & \dots & \psi_N(r_2) \\ \dots & \dots & \dots & \dots \\ \psi_1(r_M) & \psi_2(r_M) & \dots & \psi_N(r_M) \end{bmatrix} \tag{4.40}$$

With

$$\begin{bmatrix} \psi_1(r_1) & \psi_2(r_1) & \dots & \psi_N(r_1) \\ \psi_1(r_2) & \psi_2(r_2) & \dots & \psi_N(r_2) \\ \dots & \dots & \dots & \dots \\ \psi_1(r_M) & \psi_2(r_M) & \dots & \psi_N(r_M) \end{bmatrix}^T \begin{bmatrix} \psi_1(r_1) & \psi_2(r_1) & \dots & \psi_N(r_1) \\ \psi_1(r_2) & \psi_2(r_2) & \dots & \psi_N(r_2) \\ \dots & \dots & \dots & \dots \\ \psi_1(r_M) & \psi_2(r_M) & \dots & \psi_N(r_M) \end{bmatrix}$$

$$= \begin{bmatrix} 1 & 0 & \dots & 0 \\ 0 & 1 & \dots & 0 \\ \dots & \dots & \dots & 0 \\ 0 & 0 & \dots & 1 \end{bmatrix} \tag{4.41}$$

The quantum Green-function propagator G (The hologram) is

$$G = Q = H(H^T H)^{-1} H^T = \begin{bmatrix} \psi_1(r_1) & \psi_2(r_1) & \cdots & \psi_N(r_1) \\ \psi_1(r_2) & \psi_2(r_2) & \cdots & \psi_N(r_2) \\ \cdots & \cdots & \cdots & \cdots \\ \psi_1(r_M) & \psi_2(r_M) & \cdots & \psi_N(r_M) \end{bmatrix} \begin{bmatrix} \psi_1(r_1) & \psi_2(r_1) & \cdots & \psi_N(r_1) \\ \psi_1(r_2) & \psi_2(r_2) & \cdots & \psi_N(r_2) \\ \cdots & \cdots & \cdots & \cdots \\ \psi_1(r_M) & \psi_2(r_M) & \cdots & \psi_N(r_M) \end{bmatrix}^T = \sum_j \psi_j(r_h)\psi_j(r_k)^* = \psi_j \otimes \psi_j \tag{4.42}$$

replaces the Hebb memory matrix J. We remark that Q is a projection operator for which we have

$$Q^2 = (H(H^T H)^{-1} H^T)(H(H^T H)^{-1} H^T) = H(H^T H)^{-1} H^T = Q \tag{4.44}$$

Now we can reconstruct one image by presenting the input $\Psi^{\text{Input}}(r_j)$ and project into the set of wave function in (4.1) in this way

$$\Psi^{Output} = Q\,\Psi^{Input} = H(H^T H)^{-1} H^T \Psi^{Input} = HH^T \Psi^{Input} = \Psi_h^{Output} = \sum_j G_{h,j} \Psi_j^{Input} \tag{4.43}$$

When we write $G = H\,H^T$ in this way

$$G = \begin{bmatrix} w_{1,1} & w_{1,2} & \cdots & w_{1,N} \\ w_{2,1} & w_{2,2} & \cdots & w_{2,N} \\ \cdots & \cdots & \cdots & \cdots \\ w_{N,1} & w_{N,2} & \cdots & w_{N,N} \end{bmatrix} \tag{4.44}$$

We can write

$$\begin{bmatrix} \Psi^{Output}(r_1) \\ \Psi^{Output}(r_2) \\ \cdots \\ \Psi^{Output}(r_N) \end{bmatrix} = \begin{bmatrix} w_{1,1} & w_{1,2} & \cdots & w_{1,N} \\ w_{2,1} & w_{2,2} & \cdots & w_{2,N} \\ \cdots & \cdots & \cdots & \cdots \\ w_{N,1} & w_{N,2} & \cdots & w_{N,N} \end{bmatrix} \begin{bmatrix} \Psi^{Intput}(r_1) \\ \Psi^{Intput}(r_2) \\ \cdots \\ \Psi^{Intput}(r_N) \end{bmatrix} \tag{4.45}$$

That is connected with the MIMO (multi input, multi output) system

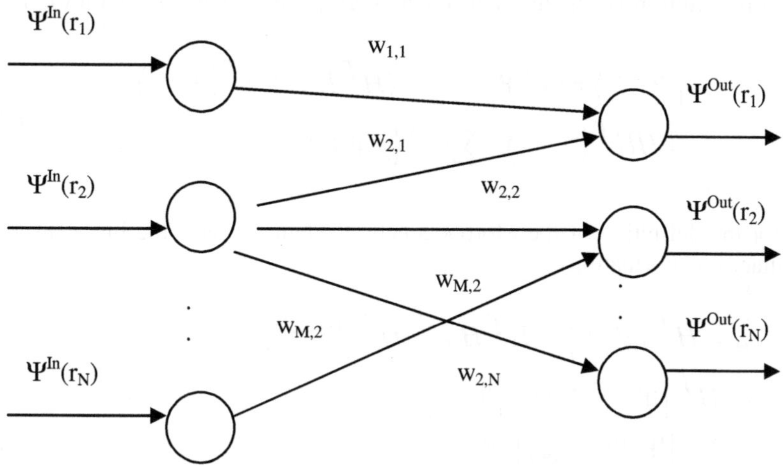

Fig. 4.7 Multi Input and Multi Output system

When $\Psi^{Input} = \psi_{K0}$ where ψ_{K0} is one of the functions in (4.1) we have that

$$
\begin{aligned}
&G\psi_{k0} = Q\psi_{k0} \\
&\begin{bmatrix} \psi_1(r_1) & \psi_2(r_1) & \dots & \psi_N(r_1) \\ \psi_1(r_2) & \psi_2(r_2) & \dots & \psi_N(r_2) \\ \dots & \dots & \dots & \dots \\ \psi_1(r_M) & \psi_2(r_M) & \dots & \psi_N(r_M) \end{bmatrix} \\
&\begin{bmatrix} \psi_1(r_1) & \psi_2(r_1) & \dots & \psi_N(r_1) \\ \psi_1(r_2) & \psi_2(r_2) & \dots & \psi_N(r_2) \\ \dots & \dots & \dots & \dots \\ \psi_1(r_M) & \psi_2(r_M) & \dots & \psi_N(r_M) \end{bmatrix}^T \\
&\begin{bmatrix} \psi_{k0}(r_1) \\ \psi_{k0}(r_2) \\ \dots \\ \psi_{k0}(r_M) \end{bmatrix} = \sum_k (\psi_k(r_i)\psi_k(r_j)^*)\psi_{k0}(r_j) \\
&= \sum_k (\psi_k(r_i)\psi_{k0}(r_j)\psi_k(r_j)^*) = \psi_{k0}(r_j)
\end{aligned}
\tag{4.46}
$$

The projection operator Q = G describes the resulting collapse-like selective retrieval (recognition) of image A encoded in ψ_{K0}. In the Dirac notation we have

$$\begin{aligned}\left|\Psi^{Output}\right\rangle &= Q\left|\Psi^{Input}\right\rangle = H(H^T H)^{-1} H^T \left|\Psi^{Input}\right\rangle \\ &= HH^T \left|\Psi^{Input}\right\rangle = \sum_j \left|\psi_j\right\rangle \left\langle\psi_j\right| \left|\Psi^{Input}\right\rangle\end{aligned} \tag{4.47}$$

Now for the definition of the Morphogenetic System we have the invariant form for unitary transformations.

$$\begin{aligned}L &= (H^T \left|\Psi^{Input}\right\rangle)^T (H^T H)^{-1} (H^T \left|\Psi^{Input}\right\rangle) \\ &= (H^T \left|\Psi^{Input}\right\rangle)^T (H^T \left|\Psi^{Input}\right\rangle) \\ &= (\left\langle\psi_k\right| \left|\Psi^{Input}\right\rangle)^T (\left\langle\psi_k\right| \left|\Psi^{Input}\right\rangle) \\ &= (E^T)^{Input} E^{Input} = \sum_j \left\langle E_j^{Input}\right| \left|E_j^{Input}\right\rangle = \sum_j (E_j^{Input})(E_j^{Input})^* = \\ &= \left\langle\Psi^{Input}\right| HH^T \left|\Psi^{Input}\right\rangle) = \left\langle\Psi^{Input}\right| Q \left|\Psi^{Input}\right\rangle = \left\langle\Psi^{Input}\right| G \left|\Psi^{Input}\right\rangle\end{aligned} \tag{4.48}$$

In fact for unitary U transformation in the N dimensional space of the positions r_j, where $U^T U = 1$, we have for $E^{Input} = E$

$$\begin{aligned}E' &= UE \\ L' &= E^T E = (UE)^T (UE) = E^T U^T U E = E^T E = L\end{aligned} \tag{4.49}$$

We remark that any physical phenomena in quantum mechanics is an unitary transformation for which quantum probability is an invariant. When wave functions are substitute by images we can found the suitable image transformation like quantum transformations. We have also that

$$Q' = G' = (UH)(UH)^T = UHH^T U^T = UGU^T = UQU^T \tag{4.50}$$

For the unitary transformation the new Green function G' is congruent or similar to the Green function G.

When $\Psi^{Input} = \psi_{K0}$ we have

$$L = \left\langle\Psi^{Input}\right| G \left|\Psi^{Input}\right\rangle = \left\langle\psi_{k0}\right| \left|\psi_{k0}\right\rangle = \sum_j \psi_{k0}(r_j)\psi_{k0}(r_j)^* = 1 \tag{4.51}$$

4.5.4 Hopfield Net and Quantum Holography by Morphogenetic System in Non Euclidean Geometry

Given a set of basis non orthonormal functions

$$H = \eta_k(r_j) = \begin{bmatrix} \eta_1(r_1) & \eta_2(r_1) & \dots & \eta_N(r_1) \\ \eta_1(r_2) & \eta_2(r_2) & \dots & \eta_N(r_2) \\ \dots & \dots & \dots & \dots \\ \eta_1(r_M) & \eta_2(r_M) & \dots & \eta_N(r_M) \end{bmatrix} \quad (4.52)$$

For which

$$g = \begin{bmatrix} \eta_1(r_1) & \eta_2(r_1) & \dots & \eta_N(r_1) \\ \eta_1(r_2) & \eta_2(r_2) & \dots & \eta_N(r_2) \\ \dots & \dots & \dots & \dots \\ \eta_1(r_M) & \eta_2(r_M) & \dots & \eta_N(r_M) \end{bmatrix}^T \begin{bmatrix} \eta_1(r_1) & \eta_2(r_1) & \dots & \eta_N(r_1) \\ \eta_1(r_2) & \eta_2(r_2) & \dots & \eta_N(r_2) \\ \dots & \dots & \dots & \dots \\ \eta_1(r_M) & \eta_2(r_M) & \dots & \eta_N(r_M) \end{bmatrix} = \quad (4.53)$$

$$\begin{bmatrix} g_{1,1} & g_{1,2} & \dots & g_{1,N} \\ g_{2,1} & g_{2,2} & \dots & g_{2,N} \\ \dots & \dots & \dots & \dots \\ g_{N,1} & g_{N,2} & \dots & g_{N,N} \end{bmatrix}$$

The non Euclidean like quantum Green-function propagator G (The hologram) is

$$G = Q = H(H^T H)^{-1} H^T = \sum_{i,j} \eta_i(r_h)(g_{i,j})^{-1} \eta_j(r_k)^* = \eta_i \otimes \eta_j \quad (4.54)$$

replaces the generalisation of Hebb memory matrix J. We remark that Q is a projection operator for which we have

$$Q^2 = (H(H^T H)^{-1} H^T)(H(H^T H)^{-1} H^T) = H(H^T H)^{-1} H^T = Q \quad (4.55)$$

The projection operator can write in this way

$$\begin{aligned} &G\Psi = Q\Psi = H\ J \\ &where\ J_i = \sum_j (g_{i,j})^{-1} \eta_j(r_k)^* \ \Psi(r_k) \end{aligned} \quad (4.56)$$

So we have

$$G\Psi = Q\Psi = H\ J = \sum_i J_i \eta_i(r_j) \quad (4.57)$$

The projection operator Q or the Green function G project Ψ into the basis H. So $Q\Psi$ is a linear combination of the basis in H. Now we can reconstruct one image by presenting the input $\Psi^{\text{Input}}(r_j)$ and project into the set of wave function in (4.2) in this way

$$\Psi^{Output} = Q\,\Psi^{Input} = H(H^T H)^{-1} H^T \Psi^{Input} = \sum_j G_{h,j} \Psi_j^{Input} \tag{4.58}$$

When we write $G = Q = H(H^T H)^{-1} H^T$ in this way

$$G = \begin{bmatrix} w_{1,1} & w_{1,2} & \dots & w_{1,N} \\ w_{2,1} & w_{2,2} & \dots & w_{2,N} \\ \dots & \dots & \dots & \dots \\ w_{N,1} & w_{N,2} & \dots & w_{N,N} \end{bmatrix} \tag{4.59}$$

We can write the extension of MIMO in this way

$$\begin{bmatrix} \Psi^{Output}(r_1) \\ \Psi^{Output}(r_2) \\ \dots \\ \Psi^{Output}(r_N) \end{bmatrix} = \begin{bmatrix} w_{1,1} & w_{1,2} & \dots & w_{1,N} \\ w_{2,1} & w_{2,2} & \dots & w_{2,N} \\ \dots & \dots & \dots & \dots \\ w_{N,1} & w_{N,2} & \dots & w_{N,N} \end{bmatrix} \begin{bmatrix} \Psi^{Intput}(r_1) \\ \Psi^{Intput}(r_2) \\ \dots \\ \Psi^{Intput}(r_N) \end{bmatrix} \tag{4.60}$$

That is connected with the MIMO (multi input, multi output) system When $\Psi^{Input} = \psi_{K0}$ where ψ_{K0} is one of the functions in (4.2) we have for (4.3) and the projection operator property that

$$G\psi_{k0} = Q\psi_{k0} = \psi_{k0} \tag{4.61}$$

The projection operator Q = G describes the resulting collapse-like selective retrieval (recognition) of image A encoded in ψ_{K0} also in the general case.

In the Dirac notation for

$$\begin{aligned} \left|\Psi^{Output}\right\rangle &= Q\left|\Psi^{Input}\right\rangle = H(H^T H)^{-1} H^T \left|\Psi^{Input}\right\rangle \\ &= \sum_{i,j} \left|\eta_i\right\rangle (\left\langle\eta_i\right\|\left.\eta_j\right\rangle)^{-1} \left\langle\eta_j\right\|\left.\Psi^{Input}\right\rangle \\ &= \sum_{i,j} \left|\eta_i\right\rangle g_{i,j}^{-1} \left\langle\eta_j\right\|\left.\Psi^{Input}\right\rangle \end{aligned} \tag{4.62}$$

where $g_{i,j}$ is the metric tensor. The previous expression can be extended to operator quantum formalism (Tilmann Hickel , Theory of Many – Body effects in the Kondo- Lattice model : Projection operator method, Phd thesis Berlin 1974) in this way

$$G = Q = \left|A_i\right\rangle g_{i,j}^{-1} \left\langle A_j\right| = \left|A_i\right\rangle \chi_{i,j}^{-1} \left\langle A_j\right|$$
$$g_{i,j} = \left\langle A_i \mid A_j\right\rangle \tag{4.63}$$

Where $\{|A_1\rangle, |A_2\rangle, \ldots\ldots, |A_N\rangle\}$ are the operator Basis that in the matrix form is given by the usual representation

$$\left|A_i\right\rangle = \left\langle \psi_h \mid A_i \mid \psi_k\right\rangle \tag{4.64}$$

χ = g is denoted the susceptible matrix (Tilmann Hickel 1974). In the space of the operators, that substitute the space of the functions (Hilbert Space) , we define a subspace of the basis operators where we project any general operator $|B\rangle$.

Now in the morphogenetic the projection operator is more complex that in the traditional projection operator methods, In fact the projection operator is separate in different parts which meaning has deeper geometric consequences.

Now Given

$$\left|\Psi^{Output}\right\rangle = Q \left|\Psi^{Input}\right\rangle = \sum_{i,j} |\eta_i\rangle (\langle \eta_i \| \eta_j\rangle)^{-1} \left\langle \eta_j \| \Psi^{Input}\right\rangle$$

where (4.65)

$$E_j = \left\langle \eta_j \| \Psi^{Input}\right\rangle$$

form the covariant basis of $\left|\Psi^{Input}\right\rangle$

$$\text{and } J^i = \sum_j \langle \eta_i \| \eta_j\rangle)^{-1} \left\langle \eta_j \| \Psi^{Input}\right\rangle \tag{4.66}$$

form the controvariant basis for $\left|\Psi^{Input}\right\rangle$. Given the unitary transformation U, we have that

$$L = (UJ)^T (UE) = J^T U^T UE = J^T E \tag{4.67}$$

is invariant for the unitary transformation U. We remark that for the unitary transformation H' = U H, we have

$$\begin{aligned} Q' &= UH((UH)^T (UH))^{-1} (UH)^T = UH(H^T U^T UH))^{-1} H^T U^T = \\ & U(H(H^T H))^{-1} H^T) U^T = UQU^T \end{aligned} \tag{4.68}$$

For the unitary transformation the new projection operator Q' is congruent with the previous projection operator Q. Now the invariant form L because E = g J. The relation between J and E is the cross relation between E and J. We have also that E and J are sources of the morphogenetic system.

$$L = J^T g\, J = E^T g^{-1} E \tag{4.69}$$

For the unitary transformation the new Green function G' is congruent or similar to the Green function G.

For the basis of operators we have

$$\left| {}_B Output \right\rangle = Q \left| {}_B Input \right\rangle = \sum_{i,j} \left| A_i \right\rangle \left(\left\langle A_i \middle\| A_j \right\rangle \right)^{-1} \left\langle A_j \middle\| {}_B Input \right\rangle$$

where (4.70)

$$E_j = \left\langle A_j \middle\| {}_B Input \right\rangle$$

form the covariant basis of $\left| {}_B Input \right\rangle$ and

$$J^i = \sum_j \left\langle A_i \middle\| A_j \right\rangle^{-1} \left\langle A_j \middle\| {}_B Input \right\rangle \tag{4.71}$$

form the controvariant basis for $\left| \Psi^{Input} \right\rangle$. Given the unitary transformation U, we have that

$$L = (UJ)^T (UE) = J^T U^T U E = J^T E \tag{4.72}$$

is invariant for the unitary transformation U. We remark that for the unitary transformation H' = U H, we have

$$\begin{aligned} Q' &= UA_i((UAj)^T (UA_i))^{-1} (UA_j)^T = UA_i (A_j^T U^T UA_i))^{-1} A_j^T U^T = \\ &U(A_i (A_j^T A_i))^{-1} A_j^T) U^T = UQU^T \end{aligned} \tag{4.73}$$

4.5.5 Computational Experiments by Professor Loo Chu Kiong

All experiments were performed on a Pentium 4, 1.3-GHz PC in Matlab with the Image Processing Toolbox.

- P images of front face from 10 different individual with index i were coded into $H_1, H_2, \ldots, H_i, \ldots, H_P$ prototype field where a pixel grey level is described by basic field value F_{ij} . For example, $H_1 = \left[F_{11}, F_{12,} F_{13,} \ldots, F_{1N} \right]$. N represents the dimension of the encoded prototype image vectors.

- Images were preprocessed according to: $\overline{F}_{ij} = F_{ij} - \frac{1}{2N} \sum_{j=1}^{N} F_{ij}$. The resulting basic field values $\overline{F}_{ij}$ was then normalized according to $\vec{F}_j = \frac{\vec{F}_j}{\left\| \vec{F}_j \right\|}$. Such normalization was assumed to be quantum-implemented into plane-wave.
- The QMS projection operator Q is generated according to $H\left(H^T H\right)^{-1} H^T$.
- Later, in the "selective reconstructive stage", a partially occluded or noise contaminated query image, X_{query} that was inserted. The QMS source was calculated according to $S = \left(H^T H\right)^{-1} H^T X_{query}$. The QMS reacted according to $Y = H\left(H^T H\right)^{-1} H^T X_{query} = QX_{query} = HS$. The "query" input field was reconstructed or corrected based on memorized model in QMS.

Thirty-five images (five faces, each one taken from seven different viewpoints, as in Figure 4.2) were preprocessed and stored in the "quantum hologram" as in Eq. 4.16. Retrieval, as in Eq. (4.24) followed. The quality of a reconstructed image **v** was measured with peak signal-to-noise ratio (in dB; for 255 gray values):

$$PSNR = 20\log\left(255/RMSE\right) \text{ where } RMSE = \sqrt{\frac{1}{N}\sum_{j=1}^{N}\left(v_j^{original} - v_j^{recontructed}\right)^2}$$

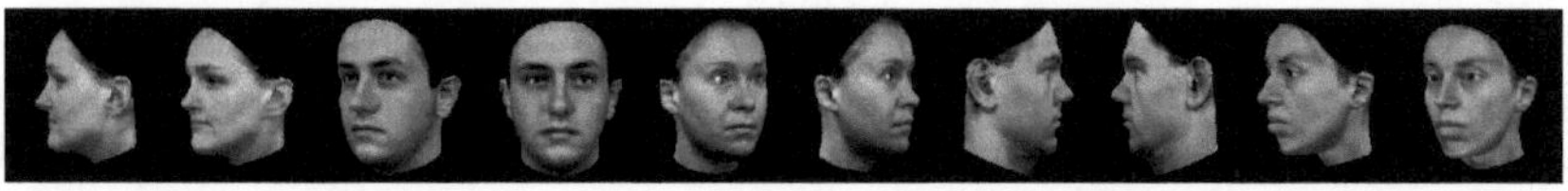

Fig. 4.8 Some examples of training images to be stored in the "quantum hologram" G.

The plot in Figure 4.3 demonstrates that the reconstruction quality (y axis) only slightly decreases with increasing number of images stored simultaneously (*x* axis) and falls with increased noising of the query image. As is evident from the reconstruction results in Figure 4.4, the image that shared the most pixels with the query image was selected from the memory matrix and reconstructed ("recognized"). Up to about 50% noise, the results are almost perfect. On the other hand, the reconstructions from quantum hologram by morphogenetic field in Euclidean geometry are worst (Figure 4.4(i) and (j). We assume that the cause of bad results with objects is correlations of quantum morphogenetic system in Euclidean geometry.

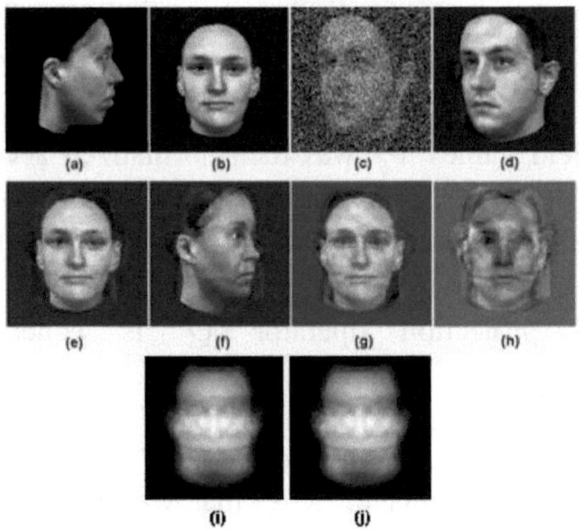

Fig. 4.9 Recontructions from 35 simultaneously stored images by quantum morphogenetic system in non-Eucludean geometry: (a) retrieval when query equals the stored original.; (b) reconstruction triggered by 30% noised query face 1; (c) query image of 50% noised face 2; (d) reconstruction from (c); (e-h) reconstructions from query images with very high salt and pepper noise rates of (e) 60%, (f) 70%, (g) 80% and (h) 95%. (i)-(j) Reconstructions from quantum associative network with 30% and 60% noise face 2.

4.5.6 Conclusions

Simulations demonstrate that "quantum hologram" by morphogenetic system in non-Euclidian geometry is crucial for clear selective reconstruction of an object. Even with an increase in the number of stored images, the reconstruction quality persists if no noise is added and decreases gradually if the query is noised. The novelty of our simulations is a reconsideration of the extension of quantum associative network [ref] to quantum morphogenetic system in non-Euclidean geometry. The problem of crosstalk in [ref] is almost eliminated in our non-Euclidean quantum morphogenetic system. Moreover, instead of plane waves, images could be encoded into Gabor wavelets [ref], which are similar to quantum wave packets. This is a natural (future) alternative to the present plane wave encoding method.

4.6 Superposition of Basis Field and Morphogenetic Field

In this chapter, we apply the general concept of morphogenetic neuron (MN) introduced in order to present an innovative way of mining semantic data. We give the neuron a formal representation as a non-Euclidian representation in order to express correlations The machine learning, statistical learning and support vector machine have witnessed a resurgence of interest over the last few years. Data is no longer a scarce resource, it is abundant and its analysis needs "intelligent" methods to extract relevant information from it. We can find general rules that explain data given only one set of samples, as requested in the learning process. If the objects under consideration are associated with target values, the learning problem is called supervised; otherwise, if the data is a sample of objects without target values, the learning problem is called unsupervised. Statistical learning takes care of risk and conditional probability in the supervised and unsupervised learning. One of the devices mostly used to describe the learning process is the Support Vector Machine theory, where a kernel operator and features define an Euclidean vector space, in which data are located and separated, according to their class values. Beside that, we know that current theory of multi-sensory representation is inconsistent with the existence of a large proportion of multimodal neurons with gain fields and partially shifting receptive fields and do not fully resolve the recoding problem and statistical issues involved in the multisensory integration. As published by Lynn C. Robertson different areas of the cortex receive sensory information through different receptors, corresponding to different features of the sensory mosaic. This modular organization of the brain led to the question of how features that are registered separately are bound together to produce our unified experience of the world (binding problem). The field theory used in the morphogenetic neuron and the linear or non linear model of the prediction is used in this paper.

4.6.1 Objectives and Principles

We have developed the concept of the morphogenetic neuron as a generalization of the classical neurons constituting neural networks. A morphogenetic neuron implements a particular input-output function whose form is *shaped* (hence the attribute "morphogenetic") by suitable *sources*. The sources can be derived from desired constraints imposed upon the input space. We proposed a way to solve the synthesis problem of a morphogenetic neuron by realizing a given input-output functional association, without the necessity to resort to training with examples.

Recent research on neural network architectures indicated the need for the introduction of a concept wider than the one of *neural unit*, or *node*, dating back to the pioneering work by McCulloch and Pitts [8]. Today we deal with neural units which can be considered as input-output filters of a much more general form Moreover, research on the so-called *integrated* neural architectures, in which many different neural networks cooperate together needs a descriptive language in which a single network can be considered as a single entity. A fundamental problem in computational neuroscience concerns how information is encoded by neural architecture of the brain. What are the units of computation and how is information represented at the neural level. An important part of the answers to these questions is that individual elements of information are encoded not by a single neuron, but rather by populations or clusters of neurons that we denote as population code. At each neuron we associate a basis function or conceptual reference. The three tasks which are computation, learning, and short-term memory can be efficiently handled using the basis function approach that represents the space of the attributes or morphogenetic basis fields. These morphogenetic fields correspond to attributes, they can be seen as basis functions too.

On the basis of these considerations we introduce a novel generalization of the concept of neural unit, which has been named *morphogenetic neuron.*

From an abstract point of view, the latter is a generic analog input-output device through which two elementary operations are possible: the "*Write*" and "*Read*" operations :

- the *Write* operation: starting from a suitable *basis function* and from the *samples of morphogenetic field*, we compute the sources or weights that can generate the model of the *morphogenetic field*
- the *Read* operation: starting from the sources, we build a model of the *morphogenetic field* or a shape.

The attribute *morphogenetic* that means *form generator* was chosen because, according to our point of view, the existence of samples of morphogenetic fields on a reference space, induces a "shape" of the morphogenetic neuron within this space, The name "neuron" was adopted because the activation function of such a device is characterized, in the same way as usual neural units, by a *weighted sum* of suitable nonlinear functions (basis functions) or *elementary morphogenetic fields.*

Because elementary morphogenetic fields are functions on independent variables (Reference Space), these variables are the input. The output is computed by the independent variables of the morphogenetic field which constraint is determined by its samples in a finite number of input values.

The advantages of this new concept can be listed as follows:

- Once a suitable way to implement *Write* and *Read* operations has been found, we have obtained an analog device able to satisfy or to minimise the error with given constraints (samples of the morphogenetic field) on input data *without any need for introducing either logical operators in an explicit way, or training procedures based on examples,*
- The concept of morphogenetic neuron allows to unify, within one single theory, concepts belonging to very different research domains: namely, all different types of neural units are particular kinds of morphogenetic neurons. Moreover, even holography can be seen as equivalent to the operation of a particular morphogenetic neuron. And an *interference* process giving rise to the formation of the hologram corresponds to the *Write* operation, whereas the *diffraction* process (giving rise to object's image) corresponds to the *Read* operation,
- Since the morphogenetic neuron is essentially analog, its operations depend on suitable continuous quantities (the sources and the morphogenetic fields). We are allowed to apply standard methods of

dynamical systems theory or methods from the framework of field theory, as formulated within theoretical physics. The classical representation of neural network models, on the contrary, precludes a direct application of these methods,

- By introducing a suitable approximation of the *Read* operation, it becomes possible to obtain automatically the design of a suitable network architecture consisting of conventional neural units implementing the input-output transfer function of a given morphogenetic neuron, without the need to resort to laborious methods of synthesis, such as training.

All these advantages depend on a careful definition of *Write* and *Read* operations.

In the following, we give a more formal definition of the morphogenetic neuron. A so called morphogenetic neuron is defined as a 7-uple

$$(R, S, H, D, F, Write, Read) \tag{4.74}$$

where

R is the Reference Space (context in input),

S is the set of *Sources* or weights in the elementary *Morphogenetic Fields superposition*

H is the set of the elementary Morphogenetic Fields or Basis Functions (Filter Space) within the Reference Space F is the morphogenetic field (context in output).

D is the set of Constraints within the Reference Space (samples of morphogenetic fields),

Write denotes a function Hx D → S which builds the Sources from the knowledge of H(Filter Space) and D,

Read denotes a function H x S → F which builds the morphogenetic field F from the knowledge of X and S. We write

$$\mathbf{MN \equiv (R, S, H , D, F, Write, Read} \tag{4.75}$$

It should be noted that a MN is endowed with a much greater flexibility than classical neural networks, regarding to constraint satisfaction.

Most sensory motor transformations are nonlinear mappings V of the sensory and posture signals into motor coordinates. Hence given a set of elementary morphogenetic field or basis functions (filter space H), the motor command can be obtained by a linear combination of these functions. Each neuron generates one of the basis functions and the nonlinear mapping is the weighted sum of the activity of every neuron that generates one elementary or basis function. So we have the following equation:

$$F = H\,S = w_1\,H_1 + w_2\,H_2 + \ldots.. + w_n\,H_n$$

with

$$H = \begin{bmatrix} X_{1,1} & X_{1,2} & \cdots & X_{1,N} \\ X_{2,1} & X_{2,2} & \cdots & X_{2,N} \\ \cdots & \cdots & \cdots & \cdots \\ X_{M,1} & X_{M,2} & \cdots & X_{M,N} \end{bmatrix}, \quad S = \begin{bmatrix} S_1 \\ S_2 \\ \cdots \\ S_N \end{bmatrix}$$

and $H_k = \begin{bmatrix} X_{1,k} \\ X_{2,k} \\ \cdots \\ X_{n,k} \end{bmatrix}$

Table 4.1 gives an overview of the data H and constrain D in the reference space R of the objects.

Table 4.1 DATA and morphogenetic fields H_k

	X_1	X_2	…	X_N	Mapping D
Ob_1	$X_{1,1}$	$X_{1,2}$	…	$X_{1,N}$	D_1
Ob_2	$X_{2,1}$	$X_{2,2}$	…	$X_{2,N}$	D_2
…	…	…	…	…	…
Ob_M	$X_{M,1}$	$X_{M,2}$	…	$X_{M,N}$	D_M

In the next chapters we explain how to connect the constrain D , the reference space, the filter reference, the sources and the field F in the morphogenetic neuron.

4.6.2 *Example of Elementary Morphogenetic Field and Sources*

In figure 4.10 we show an example of two different elementary *Morphogenetic Fields or Basis Functions*

$$F = Ke^{-h[(x-x_0)^2+(y-y_0)^2]} \tag{4.76}$$

the field F_1 has K=1, h =2 and x_0 = -0.5 and y_0 = -0.5, the second field F_2 has K=1, h =2 and x_0 = 0.5 and y_0 = 0.5.

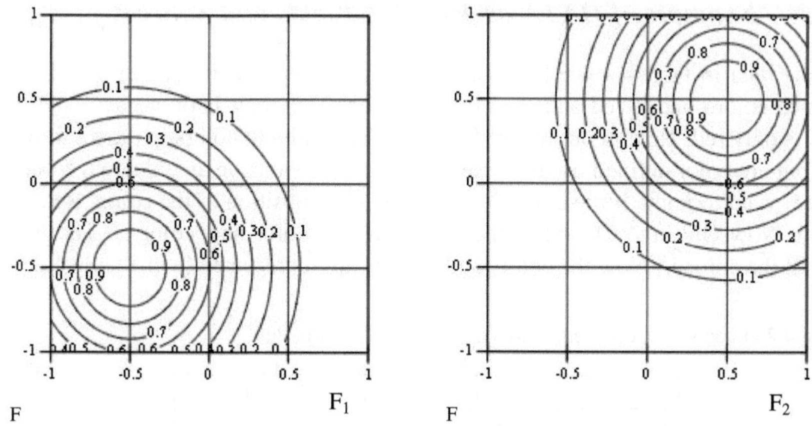

Fig. 4.10 Two different elementary *Morphogenetic Fields or Basis Functions* (Filter Space) In the two dimensional reference space R

In figure 4.11 we show two field obtained by the superposition of the elementary fields in figure 4.10. The sources in the first field in figure 4.1 are $S_1 = 1$ and $S_2 = 1$. The sources of the elementary fields in figure 1 for the second field in figure 4.11 are $S_1 = 1$ and $S_2 = 2$.

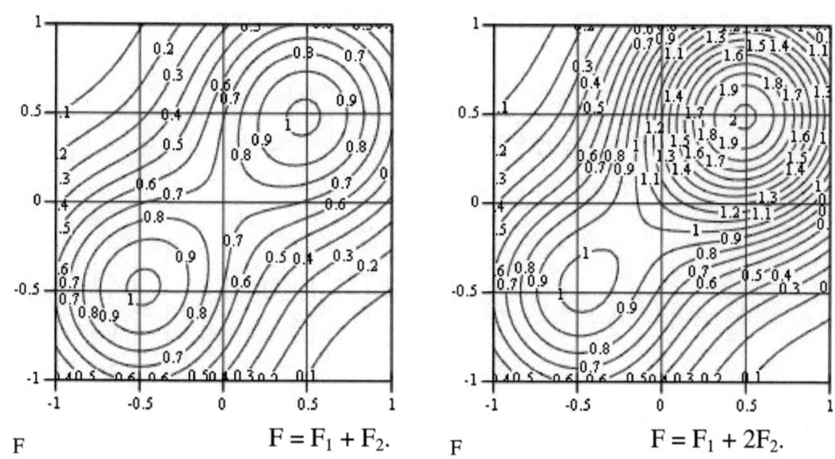

Fig. 4.11 At the left we have the field F obtained by the superposition of the two fields in figure 1 $F = F_1 + F_2$. At the right we have another field $F = F_1 + 2\ F_2$

4.6.3 *Example of DATA as Morphic Fields and Sources*

For the data in table 4.1 we have three objects (Lamp, Chair, Table) now Lamp has two colors one is red and the other is green. The chair is only green and the table is only red. Now in our model Red and Green are not colors in the traditional way but are color defined by the three objects.. Now we have an unknown color D and we want to know if the color D can be the superposition of the colors Red and Green. Now Only the Chair and the Table has the color new color D. The lamp has not the color D, Now the colors and objects are represented by the table 4.2

Table 4.2 Objects with different colors and sample D of the field

	Red	Green	D
Lamp	1	1	0
Chair	0	1	1
Table	1	0	1

We know that H is given by

$$H = \begin{bmatrix} 1 & 1 \\ 0 & 1 \\ 1 & 0 \end{bmatrix} \quad D = \begin{bmatrix} 0 \\ 1 \\ 1 \end{bmatrix}$$

And we have the sources

$$S_k = H^T D = \begin{bmatrix} 1 & 0 & 1 \\ 1 & 1 & 0 \end{bmatrix} \begin{bmatrix} 0 \\ 1 \\ 1 \end{bmatrix} = \begin{bmatrix} 1 \\ 1 \end{bmatrix}$$

$$S^k = (X^T X)^{-1} X^T D = \begin{bmatrix} 2 & 1 \\ 1 & 2 \end{bmatrix}^{-1} \begin{bmatrix} 1 & 0 & 1 \\ 1 & 1 & 0 \end{bmatrix} \begin{bmatrix} 0 \\ 1 \\ 1 \end{bmatrix} = \begin{bmatrix} \frac{1}{3} \\ \frac{1}{3} \end{bmatrix}$$

We remark that for the metric tensor

$$G = H^T H = \begin{bmatrix} 2 & 1 \\ 1 & 2 \end{bmatrix}$$

$$S_k = GS^k = \begin{bmatrix} 2 & 1 \\ 1 & 2 \end{bmatrix} \begin{bmatrix} \frac{1}{3} \\ \frac{1}{3} \end{bmatrix} = \begin{bmatrix} 1 \\ 1 \end{bmatrix}$$

Now we remark that the simple table 4.1 has a very interesting geometric image by the attribute space H and the vector D in the objects space. The best approximation of the constrains D is the field obtained by the (4.45)

$$F = HS^k = \begin{bmatrix} 1 & 1 \\ 0 & 1 \\ 1 & 0 \end{bmatrix} \begin{bmatrix} \frac{1}{3} \\ \frac{1}{3} \end{bmatrix} = \frac{1}{3}\begin{bmatrix} 1 \\ 0 \\ 1 \end{bmatrix} + \frac{1}{3}\begin{bmatrix} 1 \\ 1 \\ 0 \end{bmatrix} = \begin{bmatrix} \frac{2}{3} \\ \frac{1}{3} \\ \frac{1}{3} \end{bmatrix}$$

Among the possible superposition of the elementary fields

$$\alpha \begin{bmatrix} 1 \\ 0 \\ 1 \end{bmatrix} + \beta \begin{bmatrix} 1 \\ 1 \\ 0 \end{bmatrix} = \begin{bmatrix} \alpha+\beta \\ \beta \\ \alpha \end{bmatrix}$$

with the error E

$$\Delta = \begin{bmatrix} \alpha+\beta \\ \beta \\ \alpha \end{bmatrix} - \begin{bmatrix} 0 \\ 1 \\ 1 \end{bmatrix} = \begin{bmatrix} \alpha+\beta \\ \beta-1 \\ \alpha-1 \end{bmatrix}, \ E = (\alpha+\beta)^2 + (\beta-1)^2 + (\alpha-1)^2$$

The minimum value of E is given by the condition

$$\begin{cases} \dfrac{\partial E}{\partial \alpha} = 2(\alpha+\beta) + 2(\alpha-1) = 0 \\ \dfrac{\partial E}{\partial \beta} = 2(\alpha+\beta) + 2(\beta-1) = 0 \end{cases}$$

which solution is $\alpha = \beta = \frac{1}{3}$ that are the values of the compute sources S^k .In conclusion among all possible fields obtained by the superposition of the elementary fields the best approximation is obtained by the solution of the previous system.

4.6.4 What Is Data Mining?

In this section, we present the data mining domain and its issues. Data Mining [1, 14, 15, 18, 19, 32] uses the mining metaphor for representing the discovery of unexpected relationships by analyzing such large volumes of data that automated processes are necessary. In [15] it was defined as "*a new generation of computational theories and tools to assist humans in extracting useful information (knowledge) from the rapidly growing volumes of digital data*". Generally the extracted knowledge is expressed in global models or local patterns like sets of rules, neural networks or clusters. Numerous applications are concerned with the search of patterns and models: banking, insurance, medicine, biology, genomics provides success stories in this domain. One can note that data to be mined are observation data which were collected for other purpose than analysis , but they are exploited in order to identify specific and hidden trends.

We generally use the expression of *data mining process* which involves combination of different techniques coming mainly from data analysis, artificial intelligence, machine learning and databases areas. A typical mining process may be described in three steps : first, a data preparation step, followed by a step of search for frequent phenomena by mining and finally a step for model presentation to final users.

In this paper, we are only interested in the central step of pattern discovery among data. Mining tasks are essentially two kinds: description or prediction. In each of them, mining techniques are chosen according to the type and the structure of the expected result. Since no result at this point has proved that one technique is the best in any case, analysts proceed by checking different available techniques. For instance, the *Weather* data set given in Table 3.5 is characterized by five descriptive attributes on *outlook*, *temperature*, *humidity* and *windy* conditions of a match which was played or not. This data set is a typical example for prediction

Table 4.3 *Weather* data set

outlook	temperature	humidity	windy	play = D
sunny	hot	high	false	No
sunny	hot	high	True	No
overcast	hot	high	false	Yes
rainy	mild	high	false	Yes
rainy	cool	normal	false	Yes
rainy	cool	normal	True	No
overcast	cool	normal	True	Yes
sunny	mild	high	false	No
sunny	cool	normal	false	Yes
rainy	mild	normal	false	Yes
sunny	mild	normal	True	Yes
overcast	mild	high	True	Yes
overcast	hot	normal	false	Yes
rainy	mild	high	True	No

task since the *play* attribute is considered as the goal attribute to be predicted. The task consists in discovering a predictive model in order to classify a new record in the class *plaly=yes* or in the class *play=no*.

4.6.5 Second Order Data Mining

For the table 4.3 we assume that any word as sunny, overcast, rainy and so on are elementary morphogenetic fields or basis functions in the reference space R.

The reference space R is the space of the parameters that describe the position of any agent (world) that gives a logic value (YES, NO) or (TRUE, FALSE).

So for the example the agent "Charles" that live in New York " that has the age of 51 years old say that

" is true that the weather is sunny"

The reference space consist of two coordinates. The first is the age and the other the geographic position of the city.

We can repeat the same process for all the worlds in one row of the table 3.5. For example given the first row of the table 3.5 we have

Table 4.4 First row of the table

	outlook	temperature	humidity	windy	Play = D
Agents	sunny	hot	high	false	No

So we have the filter context

Table 4.5 Agents and data logic value in data mining

Agents	SUNNY	HOT	HIGH	FALSE	NO = D
Charles	true	true	true	true	true
Antony	true	false	true	false	true
George	false	true	false	false	fasle
Peter	true	false	false	true	false
Diana	false	false	true	true	true

We remark that FALSE is a fuzzy variable and the same for NO. The five agent have different opinions of the four fuzzy attributes

(SUNNY, HOT, HIGH, FALSE)

and different opinions on the reply NO. When true = 1 and false = 0 we can study the structure of the fuzzy sets by the agents opinion. So we have

$$X = \begin{bmatrix} 1 & 1 & 1 & 1 \\ 1 & 0 & 1 & 0 \\ 0 & 1 & 0 & 0 \\ 1 & 1 & 1 & 1 \\ 0 & 0 & 1 & 1 \end{bmatrix}, \quad D = \begin{bmatrix} 1 \\ 1 \\ 0 \\ 0 \\ 1 \end{bmatrix}$$

So we can compute the sources

$$S^k = (X^T X)^{-1} X^T D = \begin{bmatrix} -\frac{1}{2} \\ 0 \\ \frac{3}{2} \\ -\frac{1}{2} \end{bmatrix}$$

The best approximation of the reply NO is

$$F = XS = -\frac{1}{2}\begin{bmatrix} 1 \\ 1 \\ 0 \\ 1 \\ 0 \end{bmatrix} + \frac{3}{2}\begin{bmatrix} 1 \\ 1 \\ 0 \\ 1 \\ 1 \end{bmatrix} - \frac{1}{2}\begin{bmatrix} 1 \\ 0 \\ 0 \\ 1 \\ 1 \end{bmatrix} = -\frac{1}{2}(SUNY) + \frac{3}{2}(HIGH) - \frac{1}{2}(FALSE) = \begin{bmatrix} \frac{1}{2} \\ 1 \\ 0 \\ \frac{1}{2} \\ 1 \end{bmatrix}$$

The fuzzy play NO is not a linear superposition of the four fuzzy attributes (SUNNY, HOT, HIGH, FALSE).

4.7 Example of Computation of Sources of Fields

Give the elementary fields in the reference space R = (X, Y)

$$F_k(X,Y) = Ke^{-h[(X-X_k)^2+(Y-Y_k)^2]} \tag{4.77}$$

and with the constrain field

$$D = D(X,Y) = Ke^{-hX^2} \tag{4.78}$$

For (4.6) and $X_{h,k} = F_k(X_h, Y_h)$ we have the write function

$$S_k = X^T D = \sum_X \sum_Y F_k(X,Y) D(X,Y) \tag{4.79}$$

and for

$$G_{h,k} = \sum_X \sum_Y F_h(X,Y) F_k(X,Y) \tag{4.80}$$

So we have

$$S^k = G^{-1} S_k \tag{4.81}$$

and the best approximation of D is obtained by the read function

$$F(X,Y) = \sum_k S^k F_k(X,Y) \tag{4.82}$$

For four morphogenetic fields of the type (4.46) with the (X_k, Y_k) in the points

$(X_k, Y_k) = (0.5, 0.5), (-0.5, -0.5), (0.5, -0.5), (-0.5, -0.5)$,

as we can see in figure 4.12

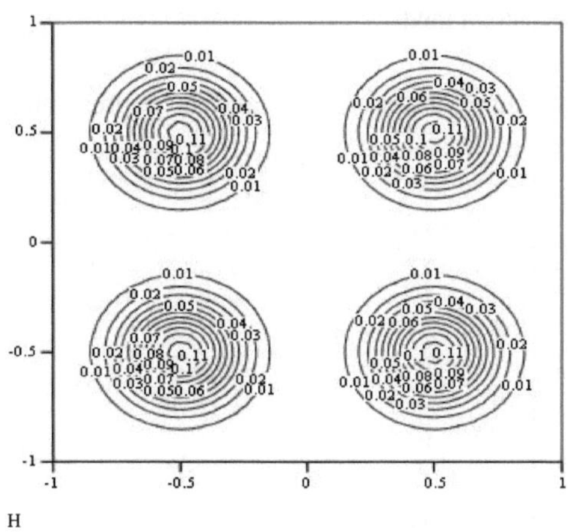

Fig. 4.12 Four elementary morphogenetic fields

Constrain field D

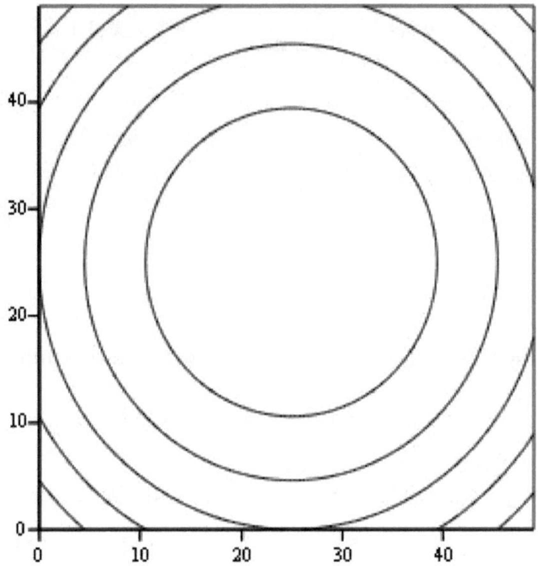

D

Fig. 4.13 Field D or constrain field

best approximation of D by the sources 50 and the expression 51

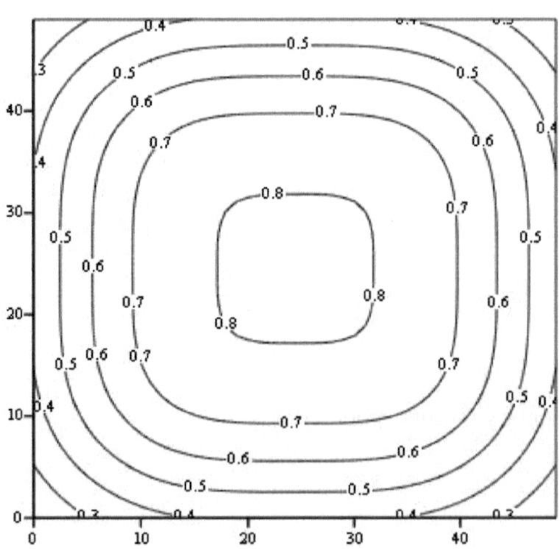

F

Fig. 4.14 Best approximation of the constrain field D

References

1. Feng, N., Yang, Y., Zheng, X.: Spatiotemporal Neuronal Signaling by Nitric Oxide: Diffusion-Reaction Modeling and Analysis. In: 27th Annual International Conference of the Engineering in Medicine and Biology Society, IEEE-EMBS 2005, Shanghai, pp. 6085–6088 (2005)
2. Marr, D., Hildreth, E.: Theory of edge detection. Proc. R. Soc. Lond. B 207, 187–217 (1980)
3. Bolduc, M., Levine, M.: A review of biologically motivated space-variant data reduction models for robotic vision. CVIU 69(2), 170–184 (1998)
4. Wilson, S.W.: On the Retino-Cortical Mapping. International Journal of Man-Machine Studies 18(4), 361–389 (1983)
5. Hubel, D., Wiesel, T.: Receptive fields, binocular interaction and functional architecture in the cat's visual cortex. Journal of Physiology of London 160, 106–154
6. Marcelja, S.: Mathematical description of the response of simple cortical cells. Journal of Optics Society America 70, 1297–1300 (1980)
7. Pribram, K.H.: Brain and Preception (Holonomy and Structure in Figural Processing). Lawrence Erlbaum Associates, Hillsdale
8. Lee, T.S.: Image representation using 2-dimensional Gabor wavelets. IEEE Transactions on Pattern Analysis and Machine Intelligence 18(10), 1–13
9. Jones, J.P., Palmer, L.A.: An evaluation of the two-dimensional gabor filter model of simple receptive fields in cat striate cortex. Journal of Neurophysiology 58, 1233–1258 (1987)
10. Grossberg, S., Mingolla, E.: Neural dynamics of perceptual grouping: texture, boundaries and emergent segmentations. Perception and Psychophysics 36, 428–456 (1984)
11. Daugman, J.G.: Complete discrete 2-D Gabor transform by neural networks for image analysis and compression. IEEE Transactions on Acoustics, Speech, and Signal Processing 36, 1169–1179 (1988)
12. Pollen, D.A., Ronner, S.F.: Visual cortical neurons as localized spatial frequency filters. IEEE Transactions on System, Man and Cybernetics 13, 907–916 (1983)
13. Kruizinga, P., Petkov, N., Grigorescu, S.E.: Comparison of texture features based on Gabor filters. In: Proceedings of the 10th International Conference on Image Analysis and Processing (ICIAP 1999), Venice, Italy, September 27-29, pp. 142–147. IEEE Computer Society, Los Alamitos (1999)
14. Morrone, M., Burr, D.: Feature detection in human vision: A phase dependent energy model. Proc. Royal Society of London B 235(3), 221–245 (1988)
15. Schwartz, E.L.: Computational anatomy and functional architecture of striate cortex: a spatial mapping approach to perceptual coding. Vision Research 20, 645–669 (1980)
16. Resconi, G., Jain, L.C.: Intelligent Agents. Springer, Heidelberg (2004)

Chapter 5
Brain Neurodynamic and Tensor Calculus

This chapter presents the neurodynamic of brain as a part of a geometric space. Logic is not only true and false but new topics in logic as fuzzy set and many valued logic extend the simple idea of true and false to values or coordinates in geometric space where the true and false are positions in this space. Thus, the holistic approach to Fuzzy and many value logic of the agent can be well represented by geometry of agent knowledge. We assume that Logic is included into the neurodynamic of the Brain

5.1 Properties and Geometry of Transformations Using Tensor Calculus

The connection between logic and tensor calculus is now explained. Given the generic transformation

$$y^j = y^j(x^i) \tag{5.1}$$

Where y^j are components of vectors in the Euclidean space x^i. They are the components of the same vectors in the non Euclidean Space. Now considering the differential property we obtain

$$dy_j = \frac{\partial y_j}{\partial x^1} dx^1 + \frac{\partial y_j}{\partial x^2} dx^2 + ... + \frac{\partial y_j}{\partial x^n} dx^n =$$
$$= \begin{bmatrix} \frac{\partial y_1}{\partial x^1} & \frac{\partial y_1}{\partial x^2} & \cdots & \frac{\partial y_1}{\partial x^n} \\ \frac{\partial y_2}{\partial x^1} & \frac{\partial y_2}{\partial x^2} & \cdots & \frac{\partial y_2}{\partial x^n} \\ \cdots & \cdots & \cdots & \cdots \\ \frac{\partial y_n}{\partial x^1} & \frac{\partial y_n}{\partial x^2} & \cdots & \frac{\partial y_n}{\partial x^n} \end{bmatrix} \begin{bmatrix} dx^1 \\ dx^2 \\ \cdots \\ dx^n \end{bmatrix} = \begin{bmatrix} dy_1 \\ dy_2 \\ \cdots \\ dy_n \end{bmatrix} = Hdx^i = dy_i \tag{5.2}$$

G. Resconi: Geometry of Knowledge for Intelligent Systems, SCI 407, pp. 107–158.
springerlink.com

When the determinant of H is not zero we write

$$\begin{bmatrix} dx^1 \\ dx^2 \\ \dots \\ dx^n \end{bmatrix} = \begin{bmatrix} \frac{\partial y_1}{\partial x^1} & \frac{\partial y_1}{\partial x^2} & \dots & \frac{\partial y_1}{\partial x^n} \\ \frac{\partial y_2}{\partial x^1} & \frac{\partial y_2}{\partial x^2} & \dots & \frac{\partial y_2}{\partial x^n} \\ \dots & \dots & \dots & \dots \\ \frac{\partial y_n}{\partial x^1} & \frac{\partial y_n}{\partial x^2} & \dots & \frac{\partial y_n}{\partial x^n} \end{bmatrix}^{-1} \begin{bmatrix} dy_1 \\ dy_2 \\ \dots \\ dy_n \end{bmatrix} \tag{5.3}$$

The previous expression may be written as follows

$$\begin{bmatrix} dx^1 \\ dx^2 \\ \dots \\ dx^n \end{bmatrix} = \left(\begin{bmatrix} \frac{\partial y_1}{\partial x^1} & \frac{\partial y_1}{\partial x^2} & \dots & \frac{\partial y_1}{\partial x^n} \\ \frac{\partial y_2}{\partial x^1} & \frac{\partial y_2}{\partial x^2} & \dots & \frac{\partial y_2}{\partial x^n} \\ \dots & \dots & \dots & \dots \\ \frac{\partial y_n}{\partial x^1} & \frac{\partial y_n}{\partial x^2} & \dots & \frac{\partial y_n}{\partial x^n} \end{bmatrix}^T \begin{bmatrix} \frac{\partial y_1}{\partial x^1} & \frac{\partial y_1}{\partial x^2} & \dots & \frac{\partial y_1}{\partial x^n} \\ \frac{\partial y_2}{\partial x^1} & \frac{\partial y_2}{\partial x^2} & \dots & \frac{\partial y_2}{\partial x^n} \\ \dots & \dots & \dots & \dots \\ \frac{\partial y_n}{\partial x^1} & \frac{\partial y_n}{\partial x^2} & \dots & \frac{\partial y_n}{\partial x^n} \end{bmatrix} \right)^{-1} \begin{bmatrix} \frac{\partial y_1}{\partial x^1} & \frac{\partial y_1}{\partial x^2} & \dots & \frac{\partial y_1}{\partial x^n} \\ \frac{\partial y_2}{\partial x^1} & \frac{\partial y_2}{\partial x^2} & \dots & \frac{\partial y_2}{\partial x^n} \\ \dots & \dots & \dots & \dots \\ \frac{\partial y_n}{\partial x^1} & \frac{\partial y_n}{\partial x^2} & \dots & \frac{\partial y_n}{\partial x^n} \end{bmatrix}^T \begin{bmatrix} dy_1 \\ dy_2 \\ \dots \\ dy_n \end{bmatrix}$$

Where

$$g_{i,k} = g_{k,i} = \begin{bmatrix} \frac{\partial y_1}{\partial x^1} & \frac{\partial y_1}{\partial x^2} & \dots & \frac{\partial y_1}{\partial x^n} \\ \frac{\partial y_2}{\partial x^1} & \frac{\partial y_2}{\partial x^2} & \dots & \frac{\partial y_2}{\partial x^n} \\ \dots & \dots & \dots & \dots \\ \frac{\partial y_n}{\partial x^1} & \frac{\partial y_n}{\partial x^2} & \dots & \frac{\partial y_n}{\partial x^n} \end{bmatrix}^T \begin{bmatrix} \frac{\partial y_1}{\partial x^1} & \frac{\partial y_1}{\partial x^2} & \dots & \frac{\partial y_1}{\partial x^n} \\ \frac{\partial y_2}{\partial x^1} & \frac{\partial y_2}{\partial x^2} & \dots & \frac{\partial y_2}{\partial x^n} \\ \dots & \dots & \dots & \dots \\ \frac{\partial y_n}{\partial x^1} & \frac{\partial y_n}{\partial x^2} & \dots & \frac{\partial y_n}{\partial x^n} \end{bmatrix} =$$

$$\begin{bmatrix} (\frac{\partial y_1}{\partial x^1}\frac{\partial y_1}{\partial x^1} + \dots + \frac{\partial y_n}{\partial x^1}\frac{\partial y_n}{\partial x^1}) & (\frac{\partial y_1}{\partial x^1}\frac{\partial y_1}{\partial x^2} + \dots + \frac{\partial y_n}{\partial x^1}\frac{\partial y_n}{\partial x^2}) & \dots & (\frac{\partial y_1}{\partial x^1}\frac{\partial y_1}{\partial x^n} + \dots + \frac{\partial y_n}{\partial x^1}\frac{\partial y_n}{\partial x^n}) \\ (\frac{\partial y_1}{\partial x^2}\frac{\partial y_1}{\partial x^1} + \dots + \frac{\partial y_n}{\partial x^2}\frac{\partial y_n}{\partial x^1}) & (\frac{\partial y_1}{\partial x^2}\frac{\partial y_1}{\partial x^2} + \dots + \frac{\partial y_n}{\partial x^2}\frac{\partial y_n}{\partial x^2}) & \dots & (\frac{\partial y_1}{\partial x^2}\frac{\partial y_1}{\partial x^n} + \dots + \frac{\partial y_n}{\partial x^2}\frac{\partial y_n}{\partial x^n}) \\ \dots & \dots & \dots & \dots \\ (\frac{\partial y_1}{\partial x^n}\frac{\partial y_1}{\partial x^1} + \dots + \frac{\partial y_n}{\partial x^n}\frac{\partial y_n}{\partial x^1}) & (\frac{\partial y_1}{\partial x^n}\frac{\partial y_1}{\partial x^2} + \dots + \frac{\partial y_n}{\partial x^n}\frac{\partial y_n}{\partial x^2}) & \dots & (\frac{\partial y_1}{\partial x^n}\frac{\partial y_1}{\partial x^n} + \dots + \frac{\partial y_n}{\partial x^n}\frac{\partial y_n}{\partial x^n}) \end{bmatrix} =$$

$$\sum_j \frac{\partial y_j}{\partial x^i}\frac{\partial y_j}{\partial x^k}$$

Where g is the metric tensor that gives the distance in the non Euclidean space

$$ds^2 = \sum_{i,k} g_{i,k} dx^i dx^k \tag{5.4}$$

Given

$$H = \begin{bmatrix} \frac{\partial y_1}{\partial x^1} & \frac{\partial y_1}{\partial x^2} & \cdots & \frac{\partial y_1}{\partial x^n} \\ \frac{\partial y_2}{\partial x^1} & \frac{\partial y_2}{\partial x^2} & \cdots & \frac{\partial y_2}{\partial x^n} \\ \cdots & \cdots & \cdots & \cdots \\ \frac{\partial y_n}{\partial x^1} & \frac{\partial y_n}{\partial x^2} & \cdots & \frac{\partial y_n}{\partial x^n} \end{bmatrix} \tag{5.5}$$

For our purposes we can write H by the object / attribute structure

$$H = \begin{bmatrix} & Attribute\ 1 & Attribute\ 2 & \ldots & Attribute\ n \\ Object\ 1 & \frac{\partial y_1}{\partial x^1} & \frac{\partial y_1}{\partial x^2} & \ldots & \frac{\partial y_1}{\partial x^n} \\ Object\ 2 & \frac{\partial y_2}{\partial x^1} & \frac{\partial y_2}{\partial x^2} & \ldots & \frac{\partial y_2}{\partial x^n} \\ \ldots & \ldots & \ldots & \ldots & \ldots \\ Object\ n & \frac{\partial y_n}{\partial x^1} & \frac{\partial y_n}{\partial x^2} & \ldots & \frac{\partial y_n}{\partial x^n} \end{bmatrix}$$

Any general coordinates x in a non Euclidean space are attribute and any coordinate y in the Euclidean space is an object. Using the previous considerations we denote the space x as the space of the attributes and the space y as the space of the objects. Here the dimension of the space of the objects is equal to the dimension of the space of the attributes. The previous expressions (5.3) can be written

$$\begin{bmatrix} dx^1 \\ dx^2 \\ \ldots \\ dx^n \end{bmatrix} = \begin{bmatrix} \frac{\partial y_1}{\partial x^1} & \frac{\partial y_1}{\partial x^2} & \cdots & \frac{\partial y_1}{\partial x^n} \\ \frac{\partial y_2}{\partial x^1} & \frac{\partial y_2}{\partial x^2} & \cdots & \frac{\partial y_2}{\partial x^n} \\ \cdots & \cdots & \cdots & \cdots \\ \frac{\partial y_n}{\partial x^1} & \frac{\partial y_n}{\partial x^2} & \cdots & \frac{\partial y_n}{\partial x^n} \end{bmatrix}^{-1} \begin{bmatrix} dy_1 \\ dy_2 \\ \ldots \\ dy_n \end{bmatrix} = dx^i = H^{-1} dy_j = (H^T H)^{-1} H^T dy_j$$

We note that

$$\begin{bmatrix} \frac{\partial y_1}{\partial x^1} & \frac{\partial y_1}{\partial x^2} & \cdots & \frac{\partial y_1}{\partial x^n} \\ \frac{\partial y_2}{\partial x^1} & \frac{\partial y_2}{\partial x^2} & \cdots & \frac{\partial y_2}{\partial x^n} \\ \cdots & \cdots & \cdots & \cdots \\ \frac{\partial y_n}{\partial x^1} & \frac{\partial y_n}{\partial x^2} & \cdots & \frac{\partial y_n}{\partial x^n} \end{bmatrix} \begin{bmatrix} dx^1 \\ dx^2 \\ \cdots \\ dx^n \end{bmatrix} = H(H^T H)^{-1} H^T dy_j = dy_j$$

Now because in the general coordinates we have two basis for the same vector in the Euclidean Space. The covariant basis and the contravariant basis are related as follows

$$\begin{bmatrix} dx^1 \\ dx^2 \\ \cdots \\ dx^n \end{bmatrix} = \begin{bmatrix} \frac{\partial y_1}{\partial x^1} & \frac{\partial y_1}{\partial x^2} & \cdots & \frac{\partial y_1}{\partial x^n} \\ \frac{\partial y_2}{\partial x^1} & \frac{\partial y_2}{\partial x^2} & \cdots & \frac{\partial y_2}{\partial x^n} \\ \cdots & \cdots & \cdots & \cdots \\ \frac{\partial y_n}{\partial x^1} & \frac{\partial y_n}{\partial x^2} & \cdots & \frac{\partial y_n}{\partial x^n} \end{bmatrix}^{-1} \begin{bmatrix} dy_1 \\ dy_2 \\ \cdots \\ dy_n \end{bmatrix} = dx^i = (H^T H)^{-1} dx_i = g^{-1} dx_i = g^{-1} H^T dy_j$$

Where $dx_i = H^T dy_j$ are the covariant components of dy. At any point P we define the local general coordinates inside the matrix H. Consequently we have the following graphical representation.

Note that the metric can be written as follows

$$ds^2 = \sum_{i,k} dx^i g_{i,k} dx^k = \sum_{i,k} dx^i dx_k$$

All the transformations of dx^i and dx_k where ds^2 does not change forms part of the possible behavior in the specific non Euclidean reference, where the metric is always the same. We create this family of transformations where the metric and any tensor invariant are unchanging.

Using the transformations

$$\bar{x}^k = \bar{x}^k(x^i) \textit{ and the inverse transformation } x^i = x^i(\bar{x}^k)$$

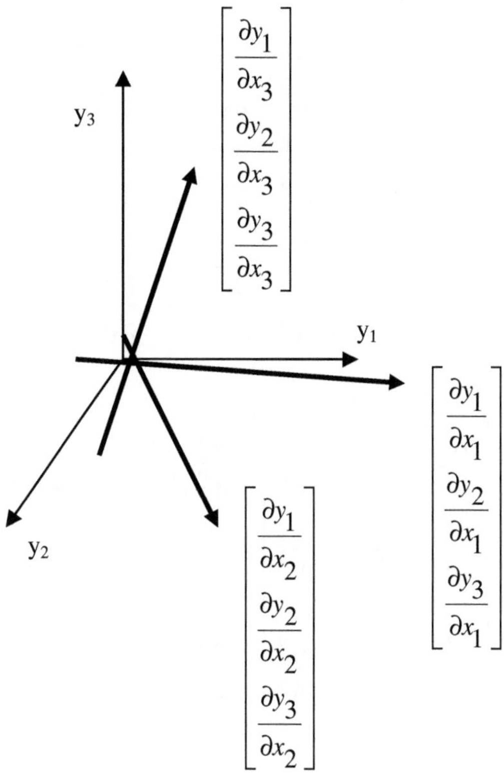

Fig. 5.1 The Geometric image of the general coordinates in H

It is possible to use the differential form

$$A = \frac{\partial \bar{x}^k}{\partial x^i} = \begin{bmatrix} \frac{\partial \bar{x}^1}{\partial x^1} & \frac{\partial \bar{x}^1}{\partial x^2} & \cdots & \frac{\partial \bar{x}^1}{\partial x^n} \\ \frac{\partial \bar{x}^2}{\partial x^1} & \frac{\partial \bar{x}^2}{\partial x^2} & \cdots & \frac{\partial \bar{x}^2}{\partial x^n} \\ \cdots & \cdots & \cdots & \cdots \\ \frac{\partial \bar{x}^n}{\partial x^1} & \frac{\partial \bar{x}^n}{\partial x^2} & \cdots & \frac{\partial \bar{x}^n}{\partial x^n} \end{bmatrix}$$

The inverse transformation is

$$B=\frac{\partial x^i}{\partial \bar{x}^k}=\begin{bmatrix} \frac{\partial x^1}{\partial \bar{x}^1} & \frac{\partial x^2}{\partial \bar{x}^1} & \cdots & \frac{\partial x^n}{\partial \bar{x}^1} \\ \frac{\partial x^1}{\partial \bar{x}^2} & \frac{\partial x^2}{\partial \bar{x}^2} & \cdots & \frac{\partial x^n}{\partial \bar{x}^2} \\ \cdots & \cdots & \cdots & \cdots \\ \frac{\partial x^1}{\partial \bar{x}^n} & \frac{\partial x^2}{\partial \bar{x}^n} & \cdots & \frac{\partial x^n}{\partial \bar{x}^n} \end{bmatrix}$$

We have

$$A^T B=\sum_k \frac{\partial \bar{x}^k}{\partial x^i}\frac{\partial x^i}{\partial \bar{x}^k}=\begin{bmatrix} \frac{\partial \bar{x}^1}{\partial x^1} & \frac{\partial \bar{x}^1}{\partial x^2} & \cdots & \frac{\partial \bar{x}^1}{\partial x^n} \\ \frac{\partial \bar{x}^2}{\partial x^1} & \frac{\partial \bar{x}^2}{\partial x^2} & \cdots & \frac{\partial \bar{x}^2}{\partial x^n} \\ \cdots & \cdots & \cdots & \cdots \\ \frac{\partial \bar{x}^n}{\partial x^1} & \frac{\partial \bar{x}^n}{\partial x^2} & \cdots & \frac{\partial \bar{x}^n}{\partial x^n} \end{bmatrix}^T \begin{bmatrix} \frac{\partial x^1}{\partial \bar{x}^1} & \frac{\partial x^2}{\partial \bar{x}^1} & \cdots & \frac{\partial x^n}{\partial \bar{x}^1} \\ \frac{\partial x^1}{\partial \bar{x}^2} & \frac{\partial x^2}{\partial \bar{x}^2} & \cdots & \frac{\partial x^n}{\partial \bar{x}^2} \\ \cdots & \cdots & \cdots & \cdots \\ \frac{\partial x^1}{\partial \bar{x}^n} & \frac{\partial x^2}{\partial \bar{x}^n} & \cdots & \frac{\partial x^n}{\partial \bar{x}^n} \end{bmatrix}=\begin{bmatrix} 1 & 0 & \cdots & 0 \\ 0 & 1 & \cdots & 0 \\ \cdots & \cdots & \cdots & \cdots \\ 0 & 0 & \cdots & 1 \end{bmatrix}$$

We have the relation $A^T = B^{-1}$.

As

$$A^T B = 1\,,\ (A^T B)^T = 1$$

and

$$B^T A = 1$$

So we obtain

$$B^T = A^{-1}\,.$$

Because we have

$$\bar{S}_i=\begin{bmatrix} \bar{S}_1 \\ \bar{S}_2 \\ \cdots \\ \bar{S}_n \end{bmatrix}=\begin{bmatrix} \frac{\partial x^1}{\partial \bar{x}^1} & \frac{\partial x^2}{\partial \bar{x}^1} & \cdots & \frac{\partial x^n}{\partial \bar{x}^1} \\ \frac{\partial x^1}{\partial \bar{x}^2} & \frac{\partial x^2}{\partial \bar{x}^2} & \cdots & \frac{\partial x^n}{\partial \bar{x}^2} \\ \cdots & \cdots & \cdots & \cdots \\ \frac{\partial x^1}{\partial \bar{x}^n} & \frac{\partial x^2}{\partial \bar{x}^n} & \cdots & \frac{\partial x^n}{\partial \bar{x}^n} \end{bmatrix}\begin{bmatrix} S_1 \\ S_2 \\ \cdots \\ S_n \end{bmatrix}$$

And

$$\bar{S}^i = \begin{bmatrix} \bar{S}^1 \\ \bar{S}^2 \\ \dots \\ \bar{S}^n \end{bmatrix} = \begin{bmatrix} \frac{\partial \bar{x}^1}{\partial x^1} & \frac{\partial \bar{x}^1}{\partial x^2} & \dots & \frac{\partial \bar{x}^1}{\partial x^n} \\ \frac{\partial \bar{x}^2}{\partial x^1} & \frac{\partial \bar{x}^2}{\partial x^2} & \dots & \frac{\partial \bar{x}^2}{\partial x^n} \\ \dots & \dots & \dots & \dots \\ \frac{\partial \bar{x}^n}{\partial x^1} & \frac{\partial \bar{x}^n}{\partial x^2} & \dots & \frac{\partial \bar{x}^n}{\partial x^n} \end{bmatrix} \begin{bmatrix} S^1 \\ S^2 \\ \dots \\ S^n \end{bmatrix}$$

We now are able to obtain

$$\bar{S}_i \bar{S}^i = \begin{bmatrix} \bar{S}_1 \\ \bar{S}_2 \\ \dots \\ \bar{S}_n \end{bmatrix}^T \begin{bmatrix} \bar{S}^1 \\ \bar{S}^2 \\ \dots \\ \bar{S}^n \end{bmatrix} = (BS_i)^T (AS^i) = (\begin{bmatrix} \frac{\partial x^1}{\partial \bar{x}^1} & \frac{\partial x^2}{\partial \bar{x}^1} & \dots & \frac{\partial x^n}{\partial \bar{x}^1} \\ \frac{\partial x^1}{\partial \bar{x}^2} & \frac{\partial x^2}{\partial \bar{x}^2} & \dots & \frac{\partial x^n}{\partial \bar{x}^2} \\ \dots & \dots & \dots & \dots \\ \frac{\partial x^1}{\partial \bar{x}^n} & \frac{\partial x^2}{\partial \bar{x}^n} & \dots & \frac{\partial x^n}{\partial \bar{x}^n} \end{bmatrix} \begin{bmatrix} S_1 \\ S_2 \\ \dots \\ S_n \end{bmatrix})^T \begin{bmatrix} \frac{\partial \bar{x}^1}{\partial x^1} & \frac{\partial \bar{x}^1}{\partial x^2} & \dots & \frac{\partial \bar{x}^1}{\partial x^n} \\ \frac{\partial \bar{x}^2}{\partial x^1} & \frac{\partial \bar{x}^2}{\partial x^2} & \dots & \frac{\partial \bar{x}^2}{\partial x^n} \\ \dots & \dots & \dots & \dots \\ \frac{\partial \bar{x}^n}{\partial x^1} & \frac{\partial \bar{x}^n}{\partial x^2} & \dots & \frac{\partial \bar{x}^n}{\partial x^n} \end{bmatrix} \begin{bmatrix} S^1 \\ S^2 \\ \dots \\ S^n \end{bmatrix} =$$

$$\begin{bmatrix} S_1 \\ S_2 \\ \dots \\ S_n \end{bmatrix}^T \begin{bmatrix} \frac{\partial x^1}{\partial \bar{x}^1} & \frac{\partial x^2}{\partial \bar{x}^1} & \dots & \frac{\partial x^n}{\partial \bar{x}^1} \\ \frac{\partial x^1}{\partial \bar{x}^2} & \frac{\partial x^2}{\partial \bar{x}^2} & \dots & \frac{\partial x^n}{\partial \bar{x}^2} \\ \dots & \dots & \dots & \dots \\ \frac{\partial x^1}{\partial \bar{x}^n} & \frac{\partial x^2}{\partial \bar{x}^n} & \dots & \frac{\partial x^n}{\partial \bar{x}^n} \end{bmatrix}^T \begin{bmatrix} \frac{\partial \bar{x}^1}{\partial x^1} & \frac{\partial \bar{x}^1}{\partial x^2} & \dots & \frac{\partial \bar{x}^1}{\partial x^n} \\ \frac{\partial \bar{x}^2}{\partial x^1} & \frac{\partial \bar{x}^2}{\partial x^2} & \dots & \frac{\partial \bar{x}^2}{\partial x^n} \\ \dots & \dots & \dots & \dots \\ \frac{\partial \bar{x}^n}{\partial x^1} & \frac{\partial \bar{x}^n}{\partial x^2} & \dots & \frac{\partial \bar{x}^n}{\partial x^n} \end{bmatrix} \begin{bmatrix} S^1 \\ S^2 \\ \dots \\ S^n \end{bmatrix} = S_i B^T A S^i = S_i S^i$$

Example
Given the transformations

$$\begin{cases} y_1 = R\cos(\alpha) \\ y_2 = R\sin(\alpha) \end{cases}$$

We have

$$H = \begin{bmatrix} \frac{\partial y_1}{\partial x^1} & \frac{\partial y_1}{\partial x^2} \\ \frac{\partial y_2}{\partial x^1} & \frac{\partial y_2}{\partial x^2} \end{bmatrix} = \begin{bmatrix} \frac{\partial y_1}{\partial R} & \frac{\partial y_1}{\partial \alpha} \\ \frac{\partial y_2}{\partial R} & \frac{\partial y_2}{\partial \alpha} \end{bmatrix} = \begin{bmatrix} \cos(\alpha) & -R\sin(\alpha) \\ \sin(\alpha) & R\cos(\alpha) \end{bmatrix}$$

Then

$$g = \begin{bmatrix} \cos(\alpha) & -R\sin(\alpha) \\ \sin(\alpha) & R\cos(\alpha) \end{bmatrix}^T \begin{bmatrix} \cos(\alpha) & -R\sin(\alpha) \\ \sin(\alpha) & R\cos(\alpha) \end{bmatrix} = \begin{bmatrix} 1 & 0 \\ 0 & R^2 \end{bmatrix}$$

The metric is

$$ds^2 = dx^1 + R^2 dx^2$$

Given the transformation

$$\begin{cases} \overset{-1}{x} = \lambda_1 x^1 \\ \overset{-2}{x} = \lambda_2 x^2 \end{cases}$$

$$A = \frac{\partial \overset{-k}{x}}{\partial x^i} = \begin{bmatrix} \frac{\partial \overset{-1}{x}}{\partial x^1} & \frac{\partial \overset{-1}{x}}{\partial x^2} \\ \frac{\partial \overset{-2}{x}}{\partial x^1} & \frac{\partial \overset{-2}{x}}{\partial x^2} \end{bmatrix} = \begin{bmatrix} \lambda_1 & 0 \\ 0 & \lambda_2 \end{bmatrix}$$

and

$$B = \frac{\partial x^i}{\partial \overset{-k}{x}} = \begin{bmatrix} \frac{\partial x^1}{\partial \overset{-1}{x}} & \frac{\partial x^2}{\partial \overset{-1}{x}} \\ \frac{\partial x^1}{\partial \overset{-2}{x}} & \frac{\partial x^2}{\partial \overset{-2}{x}} \end{bmatrix} = \begin{bmatrix} \frac{1}{\lambda_1} & 0 \\ 0 & \frac{1}{\lambda_2} \end{bmatrix}$$

We obtain

$$\overline{S}_i = \begin{bmatrix} \overline{S}_1 \\ \overline{S}_2 \end{bmatrix} = \begin{bmatrix} \frac{\partial x^1}{\partial \overset{-1}{x}} & \frac{\partial x^2}{\partial \overset{-1}{x}} \\ \frac{\partial x^1}{\partial \overset{-2}{x}} & \frac{\partial x^2}{\partial \overset{-2}{x}} \end{bmatrix} \begin{bmatrix} S_1 \\ S_2 \end{bmatrix} = \begin{bmatrix} \frac{1}{\lambda_1} & 0 \\ 0 & \frac{1}{\lambda_2} \end{bmatrix} \begin{bmatrix} S_1 \\ S_2 \end{bmatrix}$$

and

$$\overline{S}^i = \begin{bmatrix} \overline{S}^1 \\ \overline{S}^2 \end{bmatrix} = \begin{bmatrix} \frac{\partial \overset{-1}{x}}{\partial x^1} & \frac{\partial \overset{-1}{x}}{\partial x^2} \\ \frac{\partial \overset{-2}{x}}{\partial x^1} & \frac{\partial \overset{-2}{x}}{\partial x^2} \end{bmatrix} \begin{bmatrix} S^1 \\ S^2 \end{bmatrix} = \begin{bmatrix} \lambda_1 & 0 \\ 0 & \lambda_2 \end{bmatrix} \begin{bmatrix} S^1 \\ S^2 \end{bmatrix}$$

So we have

$$\bar{S}_i\bar{S}^i = \begin{bmatrix}\bar{S}_1\\ \bar{S}_2\end{bmatrix}^T \begin{bmatrix}\bar{S}^1\\ \bar{S}^2\end{bmatrix} = \left(\begin{bmatrix}\frac{\partial x^1}{\partial \bar{x}^1} & \frac{\partial x^2}{\partial \bar{x}^1}\\ \frac{\partial x^1}{\partial \bar{x}^2} & \frac{\partial x^2}{\partial \bar{x}^2}\end{bmatrix}\begin{bmatrix}S_1\\ S_2\end{bmatrix}\right)^T \begin{bmatrix}\frac{\partial \bar{x}^1}{\partial x^1} & \frac{\partial \bar{x}^1}{\partial x^2}\\ \frac{\partial \bar{x}^2}{\partial x^1} & \frac{\partial \bar{x}^2}{\partial x^2}\end{bmatrix}\begin{bmatrix}S^1\\ S^2\end{bmatrix} =$$

$$\left(\begin{bmatrix}\frac{1}{\lambda_1} & 0\\ 0 & \frac{1}{\lambda_2}\end{bmatrix}\begin{bmatrix}S_1\\ S_2\end{bmatrix}\right)^T \begin{bmatrix}\lambda_1 & 0\\ 0 & \lambda_2\end{bmatrix}\begin{bmatrix}S^1\\ S^2\end{bmatrix} = \begin{bmatrix}S_1\\ S_2\end{bmatrix}^T \begin{bmatrix}\frac{1}{\lambda_1} & 0\\ 0 & \frac{1}{\lambda_2}\end{bmatrix}\begin{bmatrix}\lambda_1 & 0\\ 0 & \lambda_2\end{bmatrix}\begin{bmatrix}S^1\\ S^2\end{bmatrix} = S_i S^i$$

The form $V = S_i S^i$ is invariant for the transformations A and B.

5.2 Derivative Operator in Tensor Calculus and Commutators

We know that the two basis contravariant S^i and covariant S_i for the vector V are

$$\begin{bmatrix}S^1\\ S^2\\ \dots\\ S^n\end{bmatrix} = S^i = (H^T H)^{-1} H^T V_j\,, \quad \begin{bmatrix}S_1\\ S_2\\ \dots\\ S_n\end{bmatrix} = S_i = H^T V_j$$

Given the derivative of the vector V we obtain the two basis

$$\begin{bmatrix}S^1{}_{/k}\\ S^2{}_{/k}\\ \dots\\ S^n{}_{/k}\end{bmatrix} = S^i{}_{/k} = (H^T H)^{-1} H^T \frac{\partial V_j}{\partial x_k}\,, \quad \begin{bmatrix}S_{1/k}\\ S_{2/k}\\ \dots\\ S_{n/k}\end{bmatrix} = S_{i/k} = H^T \frac{\partial V_j}{\partial x_k}$$

We can now compute the derivative directly

$$\begin{bmatrix}\frac{\partial S^1}{\partial x_k}\\ \frac{\partial S^2}{\partial x_k}\\ \dots\\ \frac{\partial S^n}{\partial x_k}\end{bmatrix} = \frac{\partial S^i}{\partial x_k} = \frac{\partial[(H^T H)^{-1} H^T V_j]}{\partial x_k} = (H^T H)^{-1} H^T \frac{\partial V_j}{\partial x_k} + \frac{\partial[(H^T H)^{-1} H^T]}{\partial x_k} V_j =$$

$$S^i{}_{/k} + \frac{\partial[(H^T H)^{-1} H^T]}{\partial x_k} V_j$$

Where $S^i_{/k}$ is the controvariant derivatives. For the covariant basis we have

$$\begin{bmatrix} \frac{\partial s_1}{\partial x_k} \\ \frac{\partial s_2}{\partial x_k} \\ \dots \\ \frac{\partial s_i}{\partial x_k} \end{bmatrix} = \frac{\partial s_i}{\partial x_k} = \frac{\partial [H^T V_j]}{\partial x_k} = H^T \frac{\partial V_j}{\partial x_k} + \frac{\partial H^T}{\partial x_k} V_j = S_{i/k} + \frac{\partial H^T}{\partial x_k} V_j$$

For the previous equations

$$S^i_{/k} = \frac{\partial S^i}{\partial x_k} - \frac{\partial [(H^T H)^{-1} H^T]}{\partial x_k} V_j$$

And also

$$S_{i/k} = \frac{\partial s_i}{\partial x_k} - \frac{\partial H^T}{\partial x_k} V_j$$

Given

$$\mathrm{M} = H^T H)^{-1} H^T$$

The previous expression for the contravariant basis is

$$\frac{\partial s^i}{\partial x_k} - S^i_{/k} = \frac{\partial s^i}{\partial x_k} - M \frac{\partial V_j}{\partial x_k} = \frac{\partial M V_j}{\partial x_k} - M \frac{\partial V_j}{\partial x_k} =$$

$$(\frac{\partial}{\partial x_k} M - M \frac{\partial}{\partial x_k}) V_j = [\frac{\partial}{\partial x_k}, M] V_j =$$

$$\frac{\partial [(H^T H)^{-1} H^T]}{\partial x_k} V_j$$

L defined as

$$L = [\frac{\partial}{\partial x_k}, M] V_j$$

Is the commutator between the derivative and the operators

$$M = H^T \, , \, M = (H^T H)^{-1} H^T$$

$$E_k = S_{i/k} S^i{}_{/k} = g_{i,j} S^i{}_{/k} S^j{}_{/k}$$

and

$$\overline{E}_k = \overline{S}_{i/k} \overline{S}^i{}_{/k} = (AS_{i/k})(BS^i{}_{/k}) = S_{i/k} A^T BS^i{}_{/k} = S_{i/k} S^i{}_{/k} = g_{i,j} S^i{}_{/k} S^j{}_{/k} = E_k$$

So covariant and the controvariant derivatives are the basis with the invariant form E_k. Graphically we have

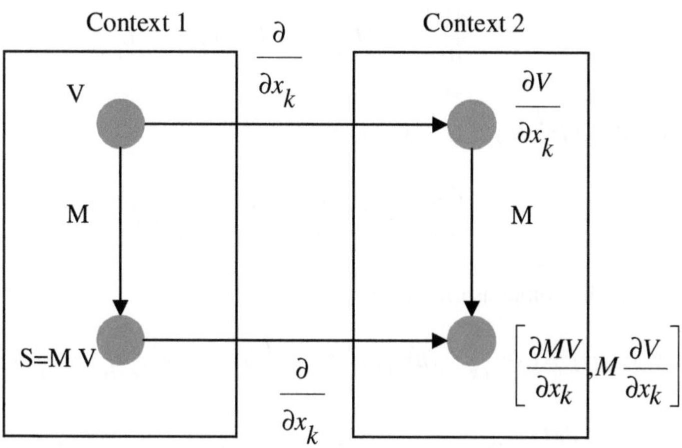

Fig. 5.2 Derivatives and the operator M for the contro -variants and covariants basis

In figure 5.2 we show the commutative diagram with the conflict situation represented by the couple

$$\left[\frac{\partial MV}{\partial x_k}, M \frac{\partial V}{\partial x_k} \right]$$

Obtained by the two paths which initial point is the vector V. The diagram in figure 5.2 solve the conflict situation only when

$$\frac{\partial MV}{\partial x_k} = M\frac{\partial V}{\partial x_k}$$

and

$$L = (\frac{\partial}{\partial x_k}M - M\frac{\partial}{\partial x_k})V = 0$$

We can use the *compensation to solve the conflict situation.* When L is different from zero, we can write

$$\frac{\partial MV}{\partial x_k} = M\frac{\partial V}{\partial x_k} + (\frac{\partial}{\partial x_k}M - M\frac{\partial}{\partial x_k})V = M\frac{\partial V}{\partial x_k} + L$$

When $M = H^T$ we have

$$\frac{\partial MV}{\partial x_k} = M\frac{\partial V}{\partial x_k} + L = V_{i/k} + L_i$$

When $M = (H^T H)^{-1} H^T$ we have

$$\frac{\partial MV}{\partial x_k} = M\frac{\partial V}{\partial x_k} + L = V^i_{/k} + L^i$$

Where L is the compensation term

$$\overline{E}_k = \overline{V}^i{}_{/k}\overline{V}_{i/k} = (AV^i{}_{/k})^T(BV_{i/k}) = V^i{}_{/k}A^T BV_{i/k} = V^i{}_{/k}V_{i/k} = g_{i,j}V^i{}_{/k}V^j{}_{/k} = E_k$$

We separate $\frac{\partial MV}{\partial x_k}$ into two parts. For the first part we have the invariant E_k and for the second part we have a change of the invariant in different points of the space x_k. When $L^i = 0$ *or* $L_i = 0$ we return to the invaraint form.

We can also compute the derivative for which we have the invariant

$$\frac{\partial MV}{\partial x_k} - L_i = V_{i/k} \text{ or } \frac{\partial MV}{\partial x_k} - L^i = V^i{}_{/k}$$

We adapt the derivatives of the covariant or contravariant components of the vector V so as to always have the symmetry property of the invariant form E_k. Given the derivative of the transformation.

$$
\frac{\partial \bar{V}_i}{\partial \bar{x}_k}=\begin{bmatrix}\frac{\partial \bar{V}_1}{\partial \bar{x}_k}\\ \frac{\partial \bar{V}_2}{\partial \bar{x}_k}\\ \cdots\\ \frac{\partial \bar{V}_n}{\partial \bar{x}_k}\end{bmatrix}=\frac{\partial\begin{bmatrix}\frac{\partial x^1}{\partial \bar{x}^1} & \frac{\partial x^2}{\partial \bar{x}^1} & \cdots & \frac{\partial x^n}{\partial \bar{x}^1}\\ \frac{\partial x^1}{\partial \bar{x}^2} & \frac{\partial x^2}{\partial \bar{x}^2} & \cdots & \frac{\partial x^n}{\partial \bar{x}^2}\\ \cdots & \cdots & \cdots & \cdots\\ \frac{\partial x^1}{\partial \bar{x}^n} & \frac{\partial x^2}{\partial \bar{x}^n} & \cdots & \frac{\partial x^n}{\partial \bar{x}^n}\end{bmatrix}\begin{bmatrix}V_1\\ V_2\\ \cdots\\ V_n\end{bmatrix}}{\partial \bar{x}_k}=
$$

$$
\frac{\partial\begin{bmatrix}\frac{\partial x^1}{\partial \bar{x}^1} & \frac{\partial x^2}{\partial \bar{x}^1} & \cdots & \frac{\partial x^n}{\partial \bar{x}^1}\\ \frac{\partial x^1}{\partial \bar{x}^2} & \frac{\partial x^2}{\partial \bar{x}^2} & \cdots & \frac{\partial x^n}{\partial \bar{x}^2}\\ \cdots & \cdots & \cdots & \cdots\\ \frac{\partial x^1}{\partial \bar{x}^n} & \frac{\partial x^2}{\partial \bar{x}^n} & \cdots & \frac{\partial x^n}{\partial \bar{x}^n}\end{bmatrix}}{\partial \bar{x}_k}\begin{bmatrix}V_1\\ V_2\\ \cdots\\ V_n\end{bmatrix}+\begin{bmatrix}\frac{\partial x^1}{\partial \bar{x}^1} & \frac{\partial x^2}{\partial \bar{x}^1} & \cdots & \frac{\partial x^n}{\partial \bar{x}^1}\\ \frac{\partial x^1}{\partial \bar{x}^2} & \frac{\partial x^2}{\partial \bar{x}^2} & \cdots & \frac{\partial x^n}{\partial \bar{x}^2}\\ \cdots & \cdots & \cdots & \cdots\\ \frac{\partial x^1}{\partial \bar{x}^n} & \frac{\partial x^2}{\partial \bar{x}^n} & \cdots & \frac{\partial x^n}{\partial \bar{x}^n}\end{bmatrix}\begin{bmatrix}\frac{\partial V_1}{\partial \bar{x}k}\\ \frac{\partial V_2}{\partial \bar{x}k}\\ \cdots\\ \frac{\partial V_n}{\partial \bar{x}k}\end{bmatrix}=
$$

$$
\frac{\partial\begin{bmatrix}\frac{\partial x^1}{\partial \bar{x}^1} & \frac{\partial x^2}{\partial \bar{x}^1} & \cdots & \frac{\partial x^n}{\partial \bar{x}^1}\\ \frac{\partial x^1}{\partial \bar{x}^2} & \frac{\partial x^2}{\partial \bar{x}^2} & \cdots & \frac{\partial x^n}{\partial \bar{x}^2}\\ \cdots & \cdots & \cdots & \cdots\\ \frac{\partial x^1}{\partial \bar{x}^n} & \frac{\partial x^2}{\partial \bar{x}^n} & \cdots & \frac{\partial x^n}{\partial \bar{x}^n}\end{bmatrix}}{\partial \bar{x}_k}\begin{bmatrix}V_1\\ V_2\\ \cdots\\ V_n\end{bmatrix}+\begin{bmatrix}\frac{\partial x^1}{\partial \bar{x}^1} & \frac{\partial x^2}{\partial \bar{x}^1} & \cdots & \frac{\partial x^n}{\partial \bar{x}^1}\\ \frac{\partial x^1}{\partial \bar{x}^2} & \frac{\partial x^2}{\partial \bar{x}^2} & \cdots & \frac{\partial x^n}{\partial \bar{x}^2}\\ \cdots & \cdots & \cdots & \cdots\\ \frac{\partial x^1}{\partial \bar{x}^n} & \frac{\partial x^2}{\partial \bar{x}^n} & \cdots & \frac{\partial x^n}{\partial \bar{x}^n}\end{bmatrix}\begin{bmatrix}\frac{\partial x^1}{\partial \bar{x}^1} & \frac{\partial x^2}{\partial \bar{x}^1} & \cdots & \frac{\partial x^n}{\partial \bar{x}^1}\\ \frac{\partial x^1}{\partial \bar{x}^2} & \frac{\partial x^2}{\partial \bar{x}^2} & \cdots & \frac{\partial x^n}{\partial \bar{x}^2}\\ \cdots & \cdots & \cdots & \cdots\\ \frac{\partial x^1}{\partial \bar{x}^n} & \frac{\partial x^2}{\partial \bar{x}^n} & \cdots & \frac{\partial x^n}{\partial \bar{x}^n}\end{bmatrix}\begin{bmatrix}\frac{\partial V_1}{\partial x^1}\\ \frac{\partial V_2}{\partial x^2}\\ \cdots\\ \frac{\partial V_n}{\partial x^n}\end{bmatrix}
$$

So the first derivatives of the covariant or controvariant components are not a tensor and cannot used to obtain invariants. The previous expression can be written in the following more simple way

$$\frac{\partial \bar{V}_i}{\partial \bar{x}^k} = \frac{\partial}{\partial \bar{x}^k} \frac{\partial x^j}{\partial \bar{x}^i} V_j + \frac{\partial x^j}{\partial \bar{x}^i} \frac{\partial x^h}{\partial \bar{x}^k} \frac{\partial V_j}{\partial x^h}$$

$$S_{j,h} = \frac{\partial V_j}{\partial x^h}$$

For the second order tensor $S_{j,h}$ we have the invariant

$$E = S_{j,h} S^{j,h} = \sum_{j,h} S_{j,h} S^{j,h}$$

Where

$$S_{j,h} = (H \otimes H)^T V_{i,k},$$

$$S^{j,h} = ((H \otimes H)^T (H \otimes H))^{-1} (H \otimes H)^T V_{i,k} = G^{-1} (H \otimes H)^T V_{i,k} = G^{-1} S_{i.k}$$

So we have

$$E = \sum_{j,h} G_{i,k,j,h} S^{i,k} S^{j,h}$$

Where ⊗ is the tensor product. Using the tensor product we can represent the general transformations for the second tensor order as

$$\bar{S} = \overline{S_{j,h}} = \sum_i \frac{\partial x^i}{\partial \bar{x}^j} \sum_k \frac{\partial x^k}{\partial \bar{x}^h} S_{k,i} = (B \otimes B) S$$

where

$$B^T = A^{-1},$$

$$(B \otimes B)^T = (B^T \otimes B^T) = (A^{-1} \otimes A^{-1}) = (A \otimes A)^{-1}$$

This is an example of the tensor product

$$B \otimes B = \begin{bmatrix} \frac{\partial x^1}{\partial \bar{x}^1} \begin{bmatrix} \frac{\partial x^1}{\partial \bar{x}^1} & \frac{\partial x^2}{\partial \bar{x}^1} \\ \frac{\partial x^1}{\partial \bar{x}^2} & \frac{\partial x^2}{\partial \bar{x}^2} \end{bmatrix} & \frac{\partial x^2}{\partial \bar{x}^1} \begin{bmatrix} \frac{\partial x^1}{\partial \bar{x}^1} & \frac{\partial x^2}{\partial \bar{x}^1} \\ \frac{\partial x^1}{\partial \bar{x}^2} & \frac{\partial x^2}{\partial \bar{x}^2} \end{bmatrix} \\ \frac{\partial x^1}{\partial \bar{x}^2} \begin{bmatrix} \frac{\partial x^1}{\partial \bar{x}^1} & \frac{\partial x^2}{\partial \bar{x}^1} \\ \frac{\partial x^1}{\partial \bar{x}^2} & \frac{\partial x^2}{\partial \bar{x}^2} \end{bmatrix} & \frac{\partial x^2}{\partial \bar{x}^2} \begin{bmatrix} \frac{\partial x^1}{\partial \bar{x}^1} & \frac{\partial x^2}{\partial \bar{x}^1} \\ \frac{\partial x^1}{\partial \bar{x}^2} & \frac{\partial x^2}{\partial \bar{x}^2} \end{bmatrix} \end{bmatrix}$$

We now have

$$(B \otimes B)S = \begin{bmatrix} \dfrac{\partial x^1}{\partial \bar{x}^1}\begin{bmatrix} \dfrac{\partial x^1}{\partial \bar{x}^1} & \dfrac{\partial x^2}{\partial \bar{x}^1} \\ \dfrac{\partial x^1}{\partial \bar{x}^2} & \dfrac{\partial x^2}{\partial \bar{x}^2} \end{bmatrix} & \dfrac{\partial x^2}{\partial \bar{x}^1}\begin{bmatrix} \dfrac{\partial x^1}{\partial \bar{x}^1} & \dfrac{\partial x^2}{\partial \bar{x}^1} \\ \dfrac{\partial x^1}{\partial \bar{x}^2} & \dfrac{\partial x^2}{\partial \bar{x}^2} \end{bmatrix} \\ \dfrac{\partial x^1}{\partial \bar{x}^2}\begin{bmatrix} \dfrac{\partial x^1}{\partial \bar{x}^1} & \dfrac{\partial x^2}{\partial \bar{x}^1} \\ \dfrac{\partial x^1}{\partial \bar{x}^2} & \dfrac{\partial x^2}{\partial \bar{x}^2} \end{bmatrix} & \dfrac{\partial x^2}{\partial \bar{x}^2}\begin{bmatrix} \dfrac{\partial x^1}{\partial \bar{x}^1} & \dfrac{\partial x^2}{\partial \bar{x}^1} \\ \dfrac{\partial x^1}{\partial \bar{x}^2} & \dfrac{\partial x^2}{\partial \bar{x}^2} \end{bmatrix} \end{bmatrix} \begin{bmatrix} S_{1,1} \\ S_{1,2} \\ S_{2,1} \\ S_{2,2} \end{bmatrix} =$$

$$\begin{bmatrix} \dfrac{\partial x^1}{\partial \bar{x}^1}\begin{bmatrix} \dfrac{\partial x^1}{\partial \bar{x}^1} & \dfrac{\partial x^2}{\partial \bar{x}^1} \\ \dfrac{\partial x^1}{\partial \bar{x}^2} & \dfrac{\partial x^2}{\partial \bar{x}^2} \end{bmatrix} & \dfrac{\partial x^2}{\partial \bar{x}^1}\begin{bmatrix} \dfrac{\partial x^1}{\partial \bar{x}^1} & \dfrac{\partial x^2}{\partial \bar{x}^1} \\ \dfrac{\partial x^1}{\partial \bar{x}^2} & \dfrac{\partial x^2}{\partial \bar{x}^2} \end{bmatrix} \\ \dfrac{\partial x^1}{\partial \bar{x}^2}\begin{bmatrix} \dfrac{\partial x^1}{\partial \bar{x}^1} & \dfrac{\partial x^2}{\partial \bar{x}^1} \\ \dfrac{\partial x^1}{\partial \bar{x}^2} & \dfrac{\partial x^2}{\partial \bar{x}^2} \end{bmatrix} & \dfrac{\partial x^2}{\partial \bar{x}^2}\begin{bmatrix} \dfrac{\partial x^1}{\partial \bar{x}^1} & \dfrac{\partial x^2}{\partial \bar{x}^1} \\ \dfrac{\partial x^1}{\partial \bar{x}^2} & \dfrac{\partial x^2}{\partial \bar{x}^2} \end{bmatrix} \end{bmatrix} \begin{bmatrix} S_{1,1} \\ S_{1,2} \\ S_{2,1} \\ S_{2,2} \end{bmatrix} =$$

$$\begin{bmatrix} \sum\limits_{i,k} \dfrac{\partial x^i}{\partial \bar{x}^1}\dfrac{\partial x^k}{\partial \bar{x}^1} S_{i,k} & \sum\limits_{i,k} \dfrac{\partial x^i}{\partial \bar{x}^1}\dfrac{\partial x^k}{\partial \bar{x}^2} S_{i,k} & \sum\limits_{i,k} \dfrac{\partial x^i}{\partial \bar{x}^1}\dfrac{\partial x^k}{\partial \bar{x}^3} S_{i,k} & \sum\limits_{i,k} \dfrac{\partial x^i}{\partial \bar{x}^1}\dfrac{\partial x^k}{\partial \bar{x}^4} S_{i,k} \\ \sum\limits_{i,k} \dfrac{\partial x^i}{\partial \bar{x}^2}\dfrac{\partial x^k}{\partial \bar{x}^1} S_{i,k} & \sum\limits_{i,k} \dfrac{\partial x^i}{\partial \bar{x}^2}\dfrac{\partial x^k}{\partial \bar{x}^2} S_{i,k} & \sum\limits_{i,k} \dfrac{\partial x^i}{\partial \bar{x}^2}\dfrac{\partial x^k}{\partial \bar{x}^3} S_{i,k} & \sum\limits_{i,k} \dfrac{\partial x^i}{\partial \bar{x}^2}\dfrac{\partial x^k}{\partial \bar{x}^4} S_{i,k} \\ \sum\limits_{i,k} \dfrac{\partial x^i}{\partial \bar{x}^3}\dfrac{\partial x^k}{\partial \bar{x}^1} S_{i,k} & \sum\limits_{i,k} \dfrac{\partial x^i}{\partial \bar{x}^3}\dfrac{\partial x^k}{\partial \bar{x}^2} S_{i,k} & \sum\limits_{i,k} \dfrac{\partial x^i}{\partial \bar{x}^3}\dfrac{\partial x^k}{\partial \bar{x}^3} S_{i,k} & \sum\limits_{i,k} \dfrac{\partial x^i}{\partial \bar{x}^3}\dfrac{\partial x^k}{\partial \bar{x}^4} S_{i,k} \\ \sum\limits_{i,k} \dfrac{\partial x^i}{\partial \bar{x}^4}\dfrac{\partial x^k}{\partial \bar{x}^1} S_{i,k} & \sum\limits_{i,k} \dfrac{\partial x^i}{\partial \bar{x}^4}\dfrac{\partial x^k}{\partial \bar{x}^2} S_{i,k} & \sum\limits_{i,k} \dfrac{\partial x^i}{\partial \bar{x}^4}\dfrac{\partial x^k}{\partial \bar{x}^3} S_{i,k} & \sum\limits_{i,k} \dfrac{\partial x^i}{\partial \bar{x}^4}\dfrac{\partial x^k}{\partial \bar{x}^4} S_{i,k} \end{bmatrix}$$

$$(B \otimes B) = ((A^{-1})^T \otimes (A^{-1})^T = ((A \otimes A)^T)^{-1} S$$

where for example we have

$$\sum_{i,k}\frac{\partial x^i}{\partial \overline{x}^1}\frac{\partial x^k}{\partial \overline{x}^1}S_{i,k} = \frac{\partial x^1}{\partial \overline{x}^1}\frac{\partial x^1}{\partial \overline{x}^1}S_{1,1} + \frac{\partial x^1}{\partial \overline{x}^1}\frac{\partial x^2}{\partial \overline{x}^1}S_{1,2} + \frac{\partial x^2}{\partial \overline{x}^1}\frac{\partial x^1}{\partial \overline{x}^1}S_{2,1} + \frac{\partial x^2}{\partial \overline{x}^1}\frac{\partial x^2}{\partial \overline{x}^1}S_{2,2}$$

Now for the invariant

$$E = \overline{S}_{j,h}\overline{S}^{j,h} = \sum_{i,k}\frac{\partial x^i}{\partial \overline{x}^j}\frac{\partial x^k}{\partial \overline{x}^h}S_{i,k}\frac{\partial \overline{x}^i}{\partial x^i}\frac{\partial \overline{x}^k}{\partial x^k}S^{i,k}$$

we have

$$E = \overline{S}_{j,h}\overline{S}^{j,h} = \overline{S}^T\overline{S} = [(A \otimes A)S]^T (B \otimes B)S = S^T (A \otimes A)^T (B \otimes B)S = S^T S$$

$$E = S_{j,h}S^{j,h} = G_{j,h,i,k}S^{i,k}S^{j,h}$$

For

$$A^{-1} = B^T$$

and

$$(B \otimes B)^T = (B^T \otimes B^T) = (A^{-1} \otimes A^{-1}) = (A \otimes A)^{-1}$$

So

$$(B \otimes B)^T (A \otimes A) = (A \otimes A)^{-1}(A \otimes A) = 1$$

and

$$E = \overline{S}_{j,h}\overline{S}^{j,h} = \overline{S}^T\overline{S} = [(B \otimes B)S]^T (A \otimes A)S = S^T (B \otimes B)^T (A \otimes A)S = S^T S$$

Now we remark that

$$\overline{D}_{i,k} = \frac{\partial \overline{V}_i}{\partial \overline{x}^k} = \frac{\partial}{\partial \overline{x}^k}\frac{\partial x^j}{\partial \overline{x}^i}V_j + \frac{\partial x^j}{\partial \overline{x}^i}\frac{\partial x^h}{\partial \overline{x}^k}\frac{\partial V_j}{\partial x^h} = \frac{\partial}{\partial \overline{x}^k}BV_j + (B \otimes B)\frac{\partial V_j}{\partial x^h} = \frac{\partial}{\partial \overline{x}^k}BV_j + (B \otimes B)D_{j,h}$$

So we cannot built the invariant form E , in fact we have

$$E = \overline{D}_{i,k}\overline{D}^{i,k} \neq D_{j,h}D^{j,h}$$

Given a set of local non Euclidean reference H , we can built the covariant and the controvariants components of any vectors defined originally in the Euclidean space. Covaraint and controvariant components can be changes by special transformations for which linear compositions of the basis are invariants. Operators that change original vector in the Euclidean space in general does not commute with the operator M that generate the dual basis covariant and controvariant.

5.3 Neurodynamic and Tensor Space Image

Intelligent sensor network is designed to implement wanted constrains and create best process that begin with sensors as sources to wanted goal (neurocomputation). Constrains define the frame where we locate the possible states. So from constrains we built the non orthogonal space of all possible states. Now this frame is a complex non orthogonal multidimensional space of the states. Straight lines with minimum distance are deformed into geodesic which parameters are defined by the metric tensor generate by constrains. Now the movements in the geodesic lines are solution of a set of ordinary differential equations (ODE) which form is obtained by a principle of minimum action. To found the solution of the ODE nature people built special neural network, connection map, connectome or connectomics. Because we can design electrical circuit with memristor that simulate movements on the geodesic and for the well known connection between electrical circuit and neural network, is possible to built neural network to solve ODE that satisfy at specific constrains and compute in a best way wanted function. Sensors are part of the neural network to solve the ODE. For evident random phenomena in the space of states we built geodesic in the perturbed by random process of the state space and we study the importance of the Fisher information and information space in the solution of ODE.

5.4 Introduction

Walter J. Freeman discusses the question of consciousness [23] in the CNS in this way:

> *Given that current efforts to deal with the problem of consciousness have led either to its dismissal (Dennett, 1991), trivialization (Crick, 1994), or transplantation to fields outside of biology, my view is that an alternate approach may be called for. Rather than pursuing forthrightly the elusive concept of consciousness, I suggest that an alternative target be formulated. For me that target lies in the area of goal-directed behaviors, overlapping largely but not completely with what are commonly called*
> *'voluntary' actions (Smith, 1994), whether or not they are conscious These behaviors emerge from within CNSs, in contrast to evoked or reflex actions, and their flexibility and adaptiveness in the face of unexpected obstacles belies the possibility of genetic or environmental programming. The experimental, mechanistic question is: How can populations of neurons in CNSs generate the neuroactivity patterns directing these movements? A useful set of theoretical tools with which to seek answers is to be found in the self-organizing properties of nonlinear dynamic systems. The theoretical philosophical question is: What principle or organizing concept can be adopted to supplant the notion of consciousness? My choice of focus is the term 'intentionality'.*

In 1980 an artificial neural network was built that work but has high precision components, slow unstable learning, non adaptive (Train and ship) and external control. Now we want Low precision components, fast stable learning, Adapt to environment and autonomous. How we can get this? We can get make components dynamical, add feedback (positive & negative) and close the loop with the outside world. The ordinary differential equations or ODE to control the neural dynamic are a stiff and nonlinear system. Why not just program this on a computer? We know that stiff and nonlinear dynamical systems are inefficient on digital computer. An example is the IBM Blue Gene project with 4096 CPUs and 1000 Terabytes RAM. To simulate the Mouse cortex use 8 x 10^6 neurons, 2 x 10^{10} synapses 10^9 Hz, 40 Kilowatts and digital. The CNS uses 10^{10} neurons, 10^{14} synapses 10 Hz and 20 watts with analog system. Analog is more efficient by several order of magnitude. Greg S.Snider suggests to use analog electrical circuit denoted CrossNet with memristor to solve the problem of the neural computation.

Recall that for Turing the physical devise is not computable by a Turing machine, which is the theoretical version of the digital computer. Snider suggests that the physics or analog computer is more efficient to solve the neural network problem by ordinary differential equation (ODE). In fact, for analog system we do not have algorithms to program the neurons. Rather, we substitute the digital program with the minimum action and geodesic in a non orthogonal space of states. The minimum path in this space of the states is the intention, the dynamics of which is in the geodesic trajectory and the computational process. In conclusion geometric and physical description of the Freeman intentionality is beyond any algorithmic or digital computation.

To clarify better the new computation paradigm, we can refer the following principle:

"Animals and humans use their finite brain's to comprehend and adapt to infinitely complex environment." (Freeman & Kozma, 2003)

5.5 Constrains Description by States Manifold. [18]

We represent the neural network by a set of variables q that form the state of the neural network.

Now we argue that neural network is a devise that implement wanted relations (constrains) that we represent by the system of equations (5.1).

$$\begin{cases} f_1(q_1,q_2,\dots,q_n)=0 \\ f_2(q_1,q_2,\dots,q_n)=0 \\ \dots \\ f_{n-m}(q_1,q_2,\dots,q_n)=0 \end{cases} \tag{5.6}$$

This system define a subspace of the state of the space. Inside this subspace all the relations (5.1) are true. The subspace is not an orthogonal space but is a manifold which definition is

Definition

A manifold is a mathematical space which can everywhere be described locally by a Cartesian coordinate system. A manifold can always be regarded as a surface embedded in a higher dimensional space

For example given the two functions

$$\begin{cases} a_{1,1}q_1 + a_{1,2}q_2 + a_{1,3}q_3 = 0 \\ a_{2,1}q_1 + a_{2,2}q_2 + a_{2,3}q_3 = 0 \end{cases}$$

$$where\ n = 3 \quad m = 1$$

We have

$$\begin{cases} a_{1,1}q_1 + a_{1,2}q_2 = -a_{1,3}q_3 \\ a_{2,1}q_1 + a_{2,2}q_2 = -a_{2,3}q_3 \end{cases}$$

So $$\begin{bmatrix} q_1 \\ q_2 \end{bmatrix} = \begin{bmatrix} a_{1,1} & a_{1,2} \\ a_{2,1} & a_{2,2} \end{bmatrix}^{-1} \begin{bmatrix} -a_{1,3}q_3 \\ -a_{2,3}q_3 \end{bmatrix} \quad and \quad \begin{cases} q_1 = q_1(x_3) \\ q_2 = q_2(x_3) \\ q_3 = x_3 \end{cases}$$

The dimension of the state space is three where only one components x_3 is independent. So the one dimensional subspace define the possible movement. Now the general relation between independent states and dependent state can write in this general form

$$\begin{cases} q_1 = q_1(x_1, x_2, \ldots, x_m, t) \\ q_2 = q_2(x_1, x_2, \ldots, x_m, t) \\ \ldots \\ q_n = q_n(x_1, x_2, \ldots, x_m, t) \end{cases} \tag{5.7}$$

We built the ODE for the sensor network in the space of the independent variables x.

5.6 Metric Tensor and Geodesic in the Space of the States x

For the relation (5.6) the derivative of the functions with time give

$$\begin{aligned} &\frac{df_j}{dt} = \frac{\partial f_j}{\partial t} + \sum_i \frac{\partial f_j}{\partial q_i}\frac{dq_i}{dt} = 0 \\ &for \frac{\partial f_j}{\partial t} = 0 \end{aligned} \tag{5.8}$$

$$\frac{df_j}{dt} = \sum_i \frac{\partial f_j}{\partial q_i}\frac{dq_i}{dt} = \left[\frac{\partial f_j}{\partial q}\right]^T \left[\frac{dq}{dt}\right] = 0 \tag{5.9}$$

So the vectors

$$\left[\frac{\partial f_j}{\partial q}\right] = \begin{bmatrix} \frac{\partial f_1}{\partial q_1} & \frac{\partial f_2}{\partial q_1} & \dots & \frac{\partial f_{n-m}}{\partial q_1} \\ \frac{\partial f_1}{\partial q_2} & \frac{\partial f_2}{\partial q_2} & \dots & \frac{\partial f_{n-m}}{\partial q_2} \\ \dots & \dots & \dots & \dots \\ \frac{\partial f_1}{\partial q_n} & \frac{\partial f_2}{\partial q_n} & \dots & \frac{\partial f_{n-m}}{\partial q_n} \end{bmatrix}, \left[\frac{\partial q}{\partial x_j}\right] = \begin{bmatrix} \frac{\partial q_1}{\partial x_1} & \frac{\partial q_1}{\partial x_2} & \dots & \frac{\partial q_1}{\partial x_m} \\ \frac{\partial q_2}{\partial x_1} & \frac{\partial q_2}{\partial x_2} & \dots & \frac{\partial q_2}{\partial x_m} \\ \dots & \dots & \dots & \dots \\ \frac{\partial q_n}{\partial x_1} & \frac{\partial q_n}{\partial x_2} & \dots & \frac{\partial q_n}{\partial x_m} \end{bmatrix} \tag{5.10}$$

The first set of vectors are orthogonal to the manifold x the second are tangent to the manifold. To prove the orthogonal property we define this system of differential equation

$$\frac{dq_i}{dt} = \sum_k \frac{\partial q_i}{\partial x_k}\frac{dx_k}{dt} \tag{5.11}$$

Now for

$$\frac{df_j}{dt} = \sum_i \frac{\partial f_j}{\partial q_i}\frac{dq_i}{dt} = \sum_i \frac{\partial f_j}{\partial q_i}\sum_k \frac{\partial q_i}{\partial x_k}\frac{dx_k}{dt} = \sum_{i,k} \frac{\partial f_j}{\partial q_i}\frac{\partial q_i}{\partial x_k}\frac{dx_k}{dt} = 0 \tag{5.12}$$

and

$$\sum_{i,k} \frac{\partial f_j}{\partial q_i}\frac{\partial q_i}{\partial x_k} = \left[\frac{\partial q}{\partial x_k}\right]\left[\frac{\partial f_j}{\partial q}\right]^T = 0 \tag{5.13}$$

The two vectors are orthogonal. Now when

$$\left[\frac{\partial q}{\partial x_j}\right]^T = (I - \left[\frac{\partial f}{\partial x_j}\right](\left[\frac{\partial f}{\partial x_j}\right]^T \left[\frac{\partial f}{\partial x_j}\right])^{-1} \left[\frac{\partial f}{\partial x_j}\right]^T \tag{5.14}$$

When we know the set of functions

$$x(t) = (x_1(t), x_2(t), \ldots\ldots, x_m(t))$$

That are solution of the ODE that we will define in chapter 5.8 we can solve the (5.11) by the (5.14) that give us the explicit form of the system (5.10). So from constrains we can compute the tangent vector. The geodesic in the independent variable x is

$$ds^2 = \sum_j dq_j^2 = \sum_j (\sum_k \frac{\partial q_j}{\partial x_k} dx_k)^2 \tag{5.15}$$

$$= \sum_{i,j} (\sum_k \frac{\partial q_k}{\partial x_i} \frac{\partial q_k}{\partial x_j}) dx_i dx_j = \sum_{i,j} g_{i,j} dx_i dx_j$$

5.7 Ordinary Differential Equation (ODE) in the Independent Variables by Geodesic

From the geodesic we can compute the action by the usual method

$$A = \int \sqrt{\sum_{i,j} g_{i,j} \frac{dx_i}{dt} \frac{dx_j}{dt}} \tag{5.16}$$

The variation of A on the geodesic line is equal to zero $\delta A = 0$. We show the Action in the z coordinates the two variables in the x , y coordinates and the minimum path or geodesic.

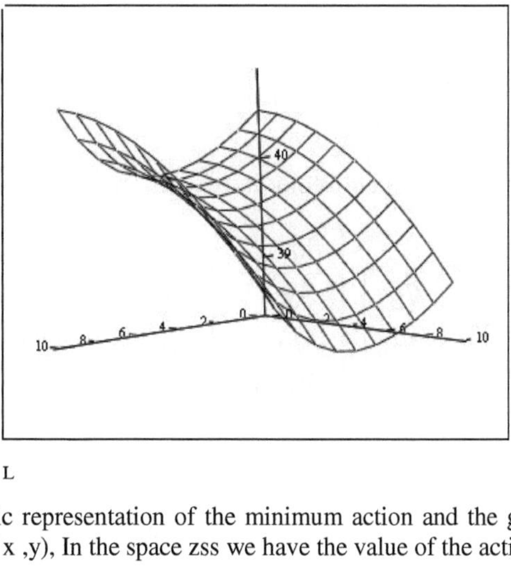

L

Fig. 5.3 Geometric representation of the minimum action and the geodesic in the two dimensional space (x ,y), In the space zss we have the value of the action

For the stationary condition for A is possible to found the Euler Lagrangian ordinary differential equation

$$\frac{d}{dt}\frac{\partial\sqrt{g_{h,k}\frac{dx^h}{dt}\frac{dx^k}{dt}}}{\partial\frac{dx^j}{dt}}-\frac{\partial\sqrt{g_{h,k}\frac{dx^h}{dt}\frac{dx^k}{dt}}}{\partial x^j}=0 \tag{5.17}$$

With sensors S(x,t) active we have the ODE

$$\frac{d}{dt}\frac{\partial\sqrt{g_{h,k}\frac{dx^h}{dt}\frac{dx^k}{dt}}}{\partial\frac{dx^j}{dt}}-\frac{\partial\sqrt{g_{h,k}\frac{dx^h}{dt}\frac{dx^k}{dt}}}{\partial x^j}=S(x,t) \tag{5.18}$$

With the geodesic we can easily found by the Euler Lagrange method [8] the Ordinary differential equations ODE to compute the behaviour of the independent variable x in time.

Example. Given the manifold

$f_1 = q_1^2 + q_2^2 - 1 = 0$, we have

$$\left[\frac{\partial f_1}{\partial q}\right]=\begin{bmatrix}\frac{\partial f_1}{\partial q_1}\\ \frac{\partial f_1}{\partial q_2}\end{bmatrix}=\begin{bmatrix}2q_1\\ 2q_2\end{bmatrix}\simeq\begin{bmatrix}q_1\\ q_2\end{bmatrix},\left[\frac{\partial q}{\partial x}\right]^T=\begin{bmatrix}\frac{\partial q_1}{\partial x}\\ \frac{\partial q_2}{\partial x}\end{bmatrix}^T=(I-\begin{bmatrix}q_1\\ q_2\end{bmatrix}(\begin{bmatrix}q_1\\ q_2\end{bmatrix}^T\begin{bmatrix}q_1\\ q_2\end{bmatrix})^{-1}\begin{bmatrix}q_1\\ q_2\end{bmatrix}^T$$

$$=\frac{1}{q_1^2+q_2^2}\begin{bmatrix}q_2^2 & -q_1q_2\\ -q_1q_2 & q_1^2\end{bmatrix}=\frac{1}{q_1^2+q_2^2}\begin{bmatrix}q_2\\ q_1\end{bmatrix}\begin{bmatrix}q_2 & -q_1\\ -q_2 & q_1\end{bmatrix}\simeq\begin{bmatrix}q_2 & -q_1\\ -q_2 & q_1\end{bmatrix}$$

Now we have two different tangent vector in two different directions. In fact

$$\begin{bmatrix}\frac{\partial q_1}{\partial x}\\ \frac{\partial q_2}{\partial x}\end{bmatrix}=\begin{bmatrix}q_2\\ -q_1\end{bmatrix},\begin{bmatrix}\frac{\partial q_1}{\partial x}\\ \frac{\partial q_2}{\partial x}\end{bmatrix}=\begin{bmatrix}-q_2\\ q_1\end{bmatrix}=-\begin{bmatrix}q_2\\ -q_1\end{bmatrix}$$

In conclusion we have

$$\left[\frac{\partial f_1}{\partial q}\right]^T \left[\frac{\partial q}{\partial x}\right] = \begin{bmatrix} q_1 \\ q_2 \end{bmatrix}^T \begin{bmatrix} q_2 \\ -q_1 \end{bmatrix} = 0$$

and for $x(t) = \omega t$

$$\begin{cases} \dfrac{dq_1}{dt} = \dfrac{\partial q_1}{\partial x}\dfrac{dx}{dt} = \omega q_2 \\ \dfrac{dq_2}{dt} = \dfrac{\partial q_2}{\partial x}\dfrac{dx}{dt} = -\omega q_1 \end{cases}$$

$$\text{So } g_{i,j} = \left[\frac{\partial q}{\partial x}\right]^T \left[\frac{\partial q}{\partial x}\right] = \begin{bmatrix} q_2 \\ -q_1 \end{bmatrix}^T \begin{bmatrix} q_2 \\ -q_1 \end{bmatrix} = q_1^2 + q_2^2 = 1, ds^2 = dx^2$$

Because $x(t) = \omega t$ is the angle in radiant and the radius is equal to one , the infinitesimal transformation ds is a movement on the circle that is the minimum distance in the manifold of the circle. Sensors in (5.9) control the solution of the ODE for the variables x.

5.8 Geodesic in Non Conservative Systems

Ion channel noise asserting that conformational changes in ion channels are actually exposed to two different kinds of fluctuations namely the intrinsic noise and the topological noise. The intrinsic noise is associated with the stochastic nature of the movement of gating particles between the inner and the outer faces of the membrane. The topological noise, on the other hand, is associated with the fluctuations emerging from the uncertainty in accessing the permissible topological states of open gates. In a toy membrane just having three potassium channels (twelve gates), for example, nine open gates can be configured into a variety of topological states with the possible results that none of the channels is open, one is open, or two are open. Now given the join probability for the ion channels given by

$$P = P(q_1, q_2, \ldots\ldots q_n)$$

The join probability to open or close of the channel is computed in the space of the ionic charges

$$q = (q_1, q_2, \ldots\ldots q_n)$$

That move inside the channels. So given a vector q in the charge of the channels we can compute the probability for the given configuration q of charges in the channels. At any configuration we associate a probability. So we have that exist configuration with very low probability and configuration with high probability.

Now given the join probability P, we compute the variation of the probability respect to the charge q_j . So we have

$$D_j P = \frac{\partial P}{\partial q^j}$$

Now given the current

$$i_j = \frac{dq_j(t)}{dt}$$

We can compute the flux of charges for the current as a random variable

$$\Phi_j = P\frac{dq_j(t)}{dt}$$

Now we assume that we have this invariant form

$$\Phi_j + \lambda D_j P = 0$$

When the probability P is a constant distribution value, the flux of the charge is equal to zero. But when the variation of P for the change of charge is positive the flux is negative, when the variation of P is negative the flux is positive. So the flux has a value reverse with the probability. When the probability is a constant value the flux is zero. So the flux is controlled by the probability in a reverse way.

$$i_j = \frac{dq_j}{dt} = -\lambda \frac{1}{P} D_j P = -\lambda \frac{\partial \log P}{\partial q_j} \tag{5.19}$$

Now the current across the channels is the superposition of two current one is the ordinary current I and the other is the random current i. In conclusion the total current is

$$J_i = I_i + i_i = I_i - \lambda_{j,i}(q) \frac{\partial (\log P)}{\partial q^j} \tag{5.20}$$

Tha can write in this way

$$J_i = I_i - \lambda_{i,j} A^j \tag{5.21}$$

Now the noise can be formally presented as a field which value is

$$A^j = \frac{\partial (\log P)}{\partial q^j}$$

For the definition of the current we have

$$J^j = \frac{dQ^j}{dt} = (I_j - \lambda A_j)$$

In conclusion we can change the previous definition of the current and also the geodesic expression in this way

$$W = Z_{k,h} \frac{dQ^k}{dt} \frac{dQ^h}{dt} = ds^2 \tag{5.22}$$

The new geodesic can write also in this way

$$\begin{aligned} dE^2 &= \left(\frac{ds}{dt}\right)^2 = Z_{k,h}(I^h - \lambda^h A^h)(I^k - \lambda^k A^k) \\ &= Z_{h,k}(q) I^j I^k - 2\lambda Z_{h,k}(q) I^h A^k + \lambda^2 Z_{h,k}(q) A^h A^k \\ &= W_1 + W_{1,2} + W_2 \end{aligned} \tag{5.23}$$

So we have three different powers. W_1 is the ordinary power for the ionic current without noise. The second $W_{1,2}$ is the flux of power from current to the noise current. The last W_2 is the power in the noise currents.

5.9 Amari [26] Information Space and Neurodynamic

Now we compute the average of the power as the cost function which value must be minimum. So we have

$$\begin{aligned} E &= \int P[Z_{i,j}(I_i - \lambda A_i)(I_j - \lambda A_j)] dt d^n x \\ &= \int P Z_{i,j}[(I_i I_j + \lambda^2 A_i A_j) - 2\lambda I_i A_j] dt d^n x \\ &= \int P Z_{i,j}[(I_i I_j + \lambda^2 \frac{\partial \log P}{\partial q_i} \frac{\partial \log P}{\partial q_j}) + W_{1,2}] dt d^n x \\ &= \int P[Z_{i,j} I_i I_j + W_{1,2}) dt d^n x + Z_{i,j} \lambda^2 \int P \frac{\partial \log P}{\partial q_i} \frac{\partial \log P}{\partial q_j})] dt d^n x \end{aligned} \tag{5.24}$$

Let us consider a parameterized family of probability distributions

$$S = \{P(x, t, q_1, q_2, \ldots, q_n)\}$$

where x and t are random variables and

$$q = (q_1, q_2, \ldots, q_n)$$

is a real vector parameter to specify a distribution. The family is regarded as an -dimensional manifold having q as a coordinate system is a Riemannian manifold and

$$G(q) = g_{i,j}(q) = \int P \frac{\partial \log P}{\partial q_i} \frac{\partial \log P}{\partial q_j})]dtd^n x \tag{5.25}$$

Play the role of a Fisher information matrix

The idea of information geometry [26] is that a family of probability distributions that are labeled by some continuous parameters as charge q or *state* can be thought as a space, each distribution being a "point", while the parameters q play the role of coordinates. A question that immediately arises is whether there is a natural way to measure the extent to which neighboring "points" can be distinguished from each other. The answer is that such a measure exists, that it is unique, and most remarkably, that it has all the properties one would wish to impose upon a measure of distance. Thus, *distinguishability is distance*, which raises the possibility that perhaps the familiar notion of *physical spatial distance might itself be explained in this way*

$$\begin{aligned} & ds^2 = \sum_{i,j} g_{h,k} dq^h dq^k \\ & where \\ & g_{i,j}(q) = \frac{1}{\lambda^2} E(A^i A^j) \\ & = E(\frac{\partial \log P(x,t,q)}{\partial q_i} \frac{\partial \log P(x,t,q)}{\partial q_j}) = \int P \frac{\partial \log P(x,t,q)}{\partial q_i} \frac{\partial \log P(x,t,q)}{\partial q_j} dxdt \end{aligned} \tag{5.26}$$

We connect the previous distance or metric with the Kullback–Leibler divergence [26] in this way

$$\begin{aligned} & ds^2 = \sum_{i,j} g_{h,k} dq^h dq^k = 2KL(P(x,t,q) : P(x,t,q+dq)) \\ & where \\ & KL(p:q) = \int p \log(\frac{p}{q}) dxdt \end{aligned}$$

It is known as Kullback–Leibler divergence Amari [1].

When the noise is equal to zero the Fisher information assumes the maximum value and the geodesic equals to the classical geodesic. On the other hand, in the case with noise, we assume that the field

$$A^j = \frac{\partial(\log P)}{\partial q^j}$$

interacts with the neural network and destroys the coherence. The information approaches zero and the cost function will be reduced. For the random potential A we have this example of random geodesics

5.10 Electrical Circuit, Percolation and Geodesic [19]

Resistor networks, from which resistors are removed at random, provide the natural generalization of the lattice models for which percolation thresholds and percolation probabilities can be considered. Given the membrane resistances given by the graphs

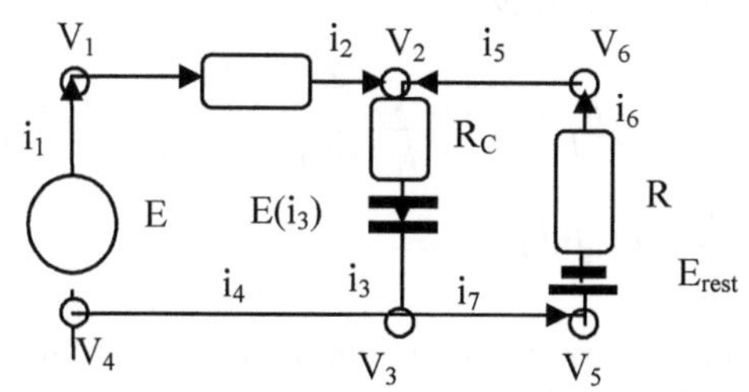

Fig. 5.4 Membrane electrical circuit model

Any resistance can be separate in two parallel resistances as we make in figure 5.4 for the resistance R. The resistance R is separate in two parallel resistances R_1 and R_2. The resistance R_1 is resistance that can be removed or changed at random in the percolation phenomena, R_2 is the crisp resistance. The total resitence R now is given by the traditional expression

$$R = \frac{R_1 R_2}{R_1 + R_2}$$

The total current is I + i where I is the current in the crisp resistance R_2 and i in the random value resistance R_1.

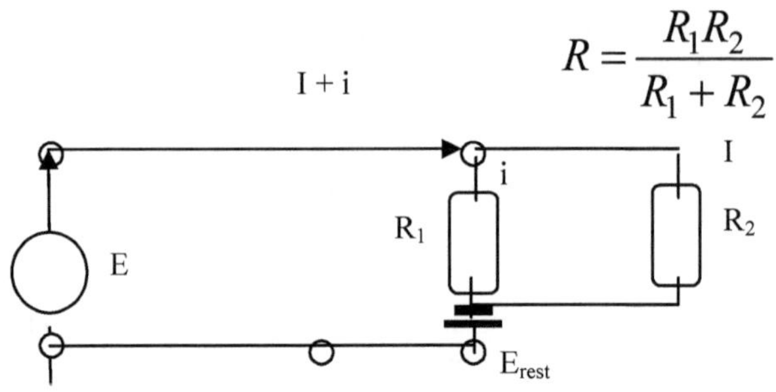

Fig. 5.5 Random R_1 and crisp resistance R_2 generate the equivalent resistance R in a parallel configuration

For the addictively of the power we have

$$W = R_1 i^2 + R_2 I^2 = R(I+i)^2 = \left(\frac{R_1 R_2}{R_1 + R_2}\right)(I+i)^2$$

The geodesic can write in this way

$$\left(\frac{ds}{dt}\right)^2 = \sum_{i,j} Z_{i,j}(I_i + i_i)(I_j + i_j) = \sum_{i,j} \frac{Z_{1,i,j} Z_{2,i,j}}{Z_{1,i,j} + Z_{2,i,j}}(I_i + i_i)(I_j + i_j)$$
$$= \sum_{i,j} Z_{2,i,j} I_i I_j + \sum_{i,j} Z_{1,i,j} i_i i_j \tag{5.27}$$

The geodesic is composed by two parts one is the synchronic and crisp geodesic the other is the noise change of the crisp and synchronic geodesic. The state of the space increase for the contribution of the random currents I. So we have the graph image

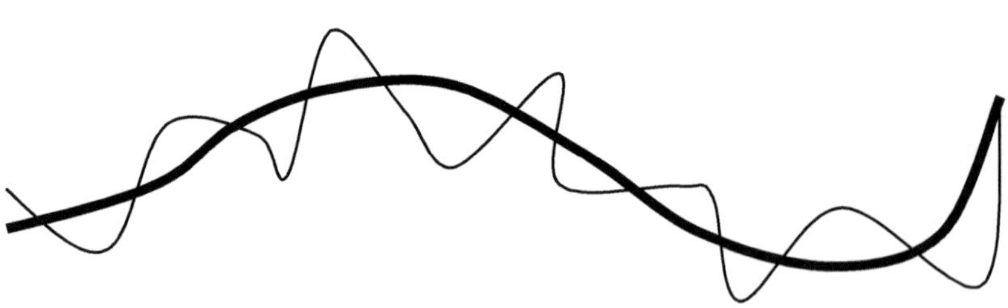

Fig. 5.6 Random geodesic with noise influence on the crisp geodesic

In figure 5.6 in evidence we have the geodesic that are solution of the ODE to obtain wanted transformation. For the noise the geodesic is transformed in a more complex geodesic that is relate to the Fisher information. The total effect is the percolation random geodesic.

5.11 Neural Network Geodesic in the Space of the Electrical Currents [24][25]

In this chapter we show that any electrical neural activity can be represented in the non Euclidean multidimensional space where the minimum path (geodesic) between two points in the multi-space of the currents is function of the neural parameters as conductors and resistor with or without memory. Now we begin with the simplest case given by electrical activity of one little part of the membrane of axons, dendrites or soma. In this case we ignore the presence of the voltage-gated channels in the membrane.

5.12 Toy Example of Geodesic and Electrical Circuit

Given the trivial electrical circuit

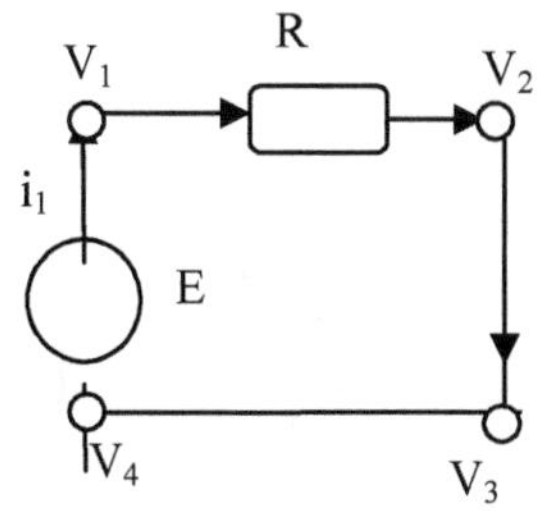

Fig. 5.7 Toy electrical circuit with one generator E and one resistor R

Now we compute the power W that is dissipate at the resistance R in this way

$$W = R(\frac{dq}{dt})^2 = Ri^2$$

Where i is the current in the circuit and q are the charges that move inside the circuit. Now we define the infinitesimal distance ds in this way:

$$(\frac{ds}{dt})^2 = W = R(\frac{dq}{dt})^2 = Ri^2$$

And

$$ds = \sqrt{W}\,dt = \sqrt{R(\frac{dq}{dt})^2}\,dt \qquad (5.28)$$

Now we know that in the electrical circuit the currents flow in the circuit in a way to dissipate the minimum power, The geodesic line in the one dimension current space i is the trajectory in time

For the minimum dissipation of the power or cost C, we have

$$\delta C = \delta \int ds = \delta \int \sqrt{W}\, dt = \delta \int \sqrt{R(\frac{dq}{dt})^2}\, dt = 0 \tag{5.29}$$

Now we can compute the behaviour of the charges for which we have the geodesic condition of the minimum cost. We know that this problem can be solved by the Euler differential equations or ODE

$$\frac{d}{dt}\frac{\partial(\sqrt{R(\frac{dq}{dt})^2})}{\partial \frac{dq}{dt}} - \frac{\partial(\sqrt{R(\frac{dq}{dt})^2})}{\partial q} = 0$$

$$or$$

$$\frac{d}{dt}\frac{\partial R(\frac{dq}{dt})^2}{\partial \frac{dq}{dt}} - \frac{\partial R(\frac{dq}{dt})^2}{\partial q} = 0 \tag{5.30}$$

When R is independent from the charges so R has no memory , we have that the previous equation can write in this way

$$\frac{d(R\frac{dq}{dt})}{dt} = 0$$

When R is independent on the time we have

$$\frac{d(\frac{dq}{dt})}{dt} = \frac{d^2q}{dt^2} = 0$$

$$q(t) = at + b$$

So the current is constant and is equal to

$$i = \frac{dq}{dt} = a = \frac{E}{R}$$

The geodesic is a straight line in the space of the charges and time.

5.12.1 Membrane Electrical Activity and Geodesic

In analogy with the works of Krone, Amari, M. Arbib, A.Pellionisz [17] [18]. For this chapter we refer to the papers [27] [28]. When we ignores the presence of voltage-gated channels in the membrane we can model the membrane electrical activity by this circuit

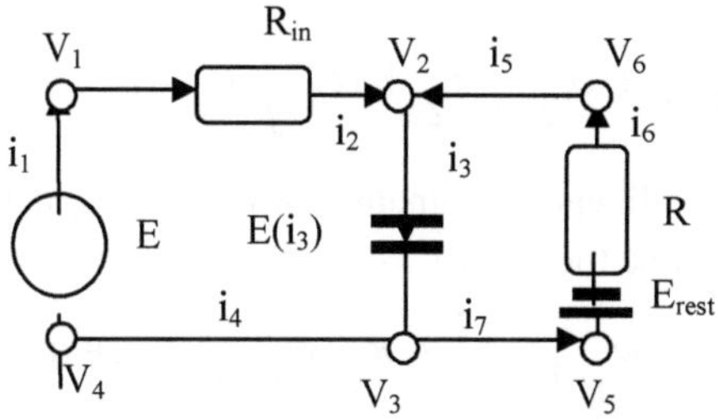

Fig. 5.8 Input voltage source E, Internal to the membrane resistance R_{ins}, Capacitor voltage source $E(i_3)$, R is the membrane resistance, E_{rest} is the rest voltage source generator

We remark that in the figure 5.2 we have three voltages sources. The first generator source is the impulse at the begin of the axon, the second $E(i_3)$ is

$$(V_3 - V_2) = E(i_3) = \frac{q_3(t)}{C} = \frac{1}{C}\int i_3(t)dt$$

The third E_{rest} is the voltages actively generate by ATP when the axon is at the rest state. Any circuit define the relation between the voltages and the currents in this way

$$v_j = f_j(i_1, i_2, \ldots., i_n)$$

Because the expression of the dissipate power W is

$$W = i^1 v_1 + i^2 v_2 + i^3 v_3 + \ldots.. + i^n v_n \qquad (5.31)$$

We have

$$W = i^1 f_1 + i^2 f_2 + i^3 f_3 + \ldots.. + i^n f_n$$

When the edges are separate one from the others we have n free variables so the power can be write in this way

$$W = R_1 i_1^2 + R_2 i_2^2 + R_3 i_3^2 + ... + R_n i_n^2$$

$$for$$

$$v_j = R_j i_j^2$$

Now when we connect the edges in a circuit we reduce the independent currents from n to p < n.

$$i_j = \alpha_{1,j} i_1 + \alpha_{2,j} i_2 + \alpha_{3,j} i_3 + ... + \alpha_{p,j} i_p$$

So all the n currents are linear combination of the independent currents p. In this situation the power assume the form

$$W = R_1 (\alpha_{1,1} i_1 + ... + \alpha_{p,1} i_p)^2 + R_2 (\alpha_{1,1} i_1 + ... + \alpha_{p,1} i_p)^2 \\ + R_3 (\alpha_{1,2} i_1 + ... + \alpha_{p,2} i_p)^2 + ... + R_n (\alpha_{1,n} i_1 + ... + \alpha_{p,n} i_p)^2 \tag{5.32}$$

So the power can write in this form

$$W = \sum_{h,k} g_{h,k} i^h i^k = g_{h,k} i^h i^k$$

Where the metric tensor g is

$$g = \alpha^T R \alpha \tag{5.33}$$

Where

$$\alpha = \begin{bmatrix} \alpha_{1,1} & \alpha_{1,2} & ... & \alpha_{1,p} \\ \alpha_{2,1} & \alpha_{2,2} & ... & \alpha_{2,p} \\ ... & ... & ... & ... \\ \alpha_{n,1} & \alpha_{n,2} & ... & \alpha_{n,p} \end{bmatrix}, \quad \mathrm{R} = \begin{bmatrix} R_1 & 0 & ... & 0 \\ 0 & R_2 & ... & 0 \\ ... & ... & ... & ... \\ 0 & 0 & ... & R_n \end{bmatrix} \tag{5.34}$$

Because

$$i^j = \frac{dq^j}{dt} \quad j = 1, 2, 3, ..., p$$

$$and$$

$$W = \frac{dE}{dt}$$

Where E is the energy and W the power, the geodesic equation can write in this way

$$dE = \sum_{h,k} g_{h,k} dq^h dq^k = \sum_{h,k} (\sum_{j,i} \alpha_{k,j} R_{j,i} \alpha_{i,k}) dq^h dq^k \quad (5.35)$$

We know that W in an electrical circuit assumes the minimum value. So the power is comparable with the Lagrangian in mechanics (Hamilton principle) or the Fermat principle in optics (minimum time). Now for the geodesic in space of the charges is

$$\delta \int \sqrt{W}\, dt = \delta \int \sqrt{g_{h,k} i^h i^k}\, dt = \delta \int \sqrt{g_{h,k} \frac{dq^h}{dt}\frac{dq^k}{dt}}\, dt = 0$$

Now for the Euler differential equations we have that the minimum condition can be found when

$$\frac{d}{dt}\frac{\partial \sqrt{g_{h,k} i^h i^k}}{\partial i^j} - \frac{\partial \sqrt{g_{h,k} i^h i^k}}{\partial q^j} = 0 \quad (5.36)$$

We show in a graphic the geodesic as a minimum cost in this way

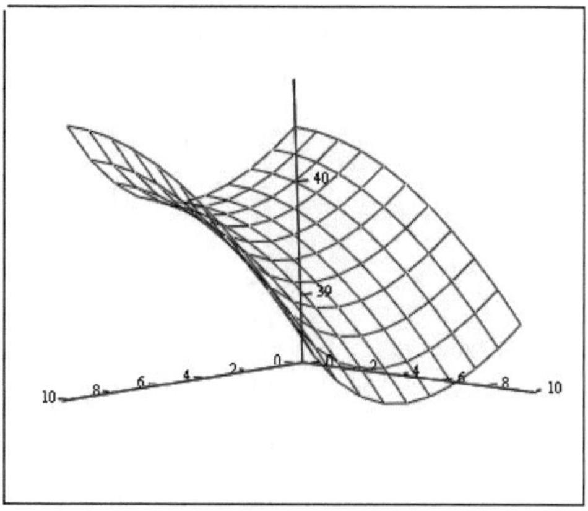

L

Fig. 5.9 A neural network changes the reference and the neurodynamics in a trajectory with minimum dissipation of power or geodesic that we can see in figure where the minimum value is the solution of the ODE in (5.1)

In the context of Freeman's neurodynamics, we hypothesize that the minimum condition in any neural network gives us the meaning of "Intentionality".

Example

In the circuit in figure 5.8 we have 7 currents that for the electrical circuit become only 2 independent currents. Now we compute the parameters $\alpha_{i,j}$ for the circuit in figure 5.1 by the valued graph of the circuit in figure 5.2

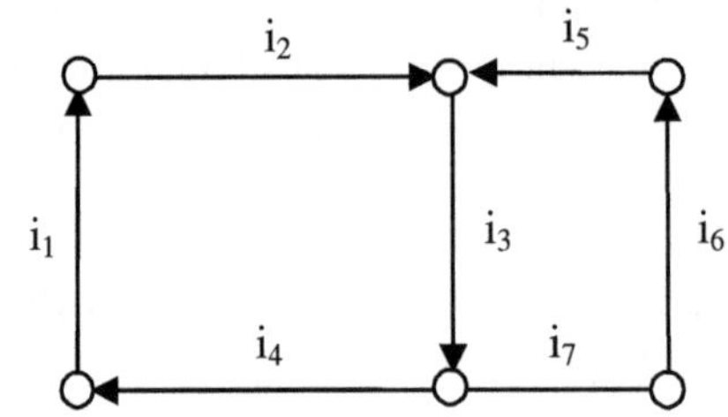

Fig. 5.10 Valued graph of the currents in the circuit in figure 5.8

We can see from the graph in figure 5.8 that all the seven currents can write in this way

$$i_k = \alpha_{k,1} i_1 + \alpha_{k,2} i_5$$

$$k = 1, 2, ..., 7$$

Here, the values of α are 1 or zero. The matrices of the connections α in figure 5.8 is

$$\alpha = \alpha_{i,j} = \begin{bmatrix} \alpha & i_1 & i_5 \\ i_1 & 1 & 0 \\ i_2 & 1 & 0 \\ i_3 & 1 & 1 \\ i_4 & 1 & 0 \\ i_5 & 0 & 1 \\ i_6 & 0 & 1 \\ i_7 & 0 & 1 \end{bmatrix}$$

The resistance matrix is

$$R = \begin{bmatrix} 1 & 0 & 0 & 0 & 0 & 0 & 0 \\ 0 & R_{ins} & 0 & 0 & 0 & 0 & 0 \\ 0 & 0 & 1 & 0 & 0 & 0 & 0 \\ 0 & 0 & 0 & 1 & 0 & 0 & 0 \\ 0 & 0 & 0 & 0 & 1 & 0 & 0 \\ 0 & 0 & 0 & 0 & 0 & R & 0 \\ 0 & 0 & 0 & 0 & 0 & 0 & 1 \end{bmatrix}$$

The voltages in figure 5.8 are

$$v = \begin{bmatrix} v_1 \\ v_2 \\ v_3 \\ v_4 \\ v_5 \\ v_6 \\ v_1 \end{bmatrix} = \begin{bmatrix} V_1 - V_4 \\ V_2 - V_1 \\ V_3 - V_2 \\ V_4 - V_3 \\ V_2 - V_6 \\ V_6 - V_5 \\ V_5 - V_3 \end{bmatrix}$$

The voltage sources or external sources are

$$E = \alpha^T v = \begin{bmatrix} 1 & 0 \\ 1 & 0 \\ 1 & 1 \\ 1 & 0 \\ 0 & 1 \\ 0 & 1 \\ 0 & 1 \end{bmatrix}^T \begin{bmatrix} v_1 \\ v_2 \\ v_3 \\ v_4 \\ v_5 \\ v_6 \\ v_1 \end{bmatrix} = \begin{bmatrix} v_1 + v_2 + v_3 + v_4 \\ v_3 + v_5 + v_6 + v_1 \end{bmatrix} = \begin{bmatrix} E \\ E(i_3) + E_{rest} \end{bmatrix}$$

The metric tensor g is

$$g = \alpha^T R \alpha = \begin{bmatrix} 1 & 0 \\ 1 & 0 \\ 1 & 1 \\ 1 & 0 \\ 0 & 1 \\ 0 & 1 \\ 0 & 1 \end{bmatrix}^T \begin{bmatrix} 1 & 0 & 0 & 0 & 0 & 0 & 0 \\ 0 & R_{ins} & 0 & 0 & 0 & 0 & 0 \\ 0 & 0 & 1 & 0 & 0 & 0 & 0 \\ 0 & 0 & 0 & 1 & 0 & 0 & 0 \\ 0 & 0 & 0 & 0 & 1 & 0 & 0 \\ 0 & 0 & 0 & 0 & 0 & R & 0 \\ 0 & 0 & 0 & 0 & 0 & 0 & 1 \end{bmatrix} \begin{bmatrix} 1 & 0 \\ 1 & 0 \\ 1 & 1 \\ 1 & 0 \\ 0 & 1 \\ 0 & 1 \\ 0 & 1 \end{bmatrix} = \begin{bmatrix} R_{ins}+3 & 1 \\ 1 & R+3 \end{bmatrix}$$

The metric is

$$W = (\frac{ds}{dt})^2 = (R_{ins} + 3)i_1^2 + 2i_1 i_5 + (R + 3)i_5^2$$

$$ds^2 = W dt^2 = (R_{ins} + 3)dq_1^2 + 2dq_1 dq_5 + (R + 3)dq_5^2$$

Following equation:

$$i_5 = i_3 - i_1$$

we have the metric or distance in the space of the current or charges

$$W = (\frac{ds}{dt})^2 = (R_{ins} + 3)i_1^2 + 2i_1(i_3 - i_1) + (R + 3)(i_3 - i_1)^2$$

$$= (R + R_{ins} + 4)i_1^2 - 2(2 + R)i_1 i_3 + (R + 3)i_3^2$$

When the two currents i_1, i_5 are independent one from the others the power is given by the quadratic form

$$W = (\frac{ds}{dt})^2 = i_1^2 + i_3^2$$

and

$$s = \int \sqrt{i_1^2 + i_3^2}\, dt = \int \sqrt{(\frac{dq_1}{dt})^2 + (\frac{dq_3}{dt})^2}\, dt$$

We show in figure 5.4 the quadratic form of the distance that is a circle which radius is the power of the electrical circuit. So in the space of the currents we have

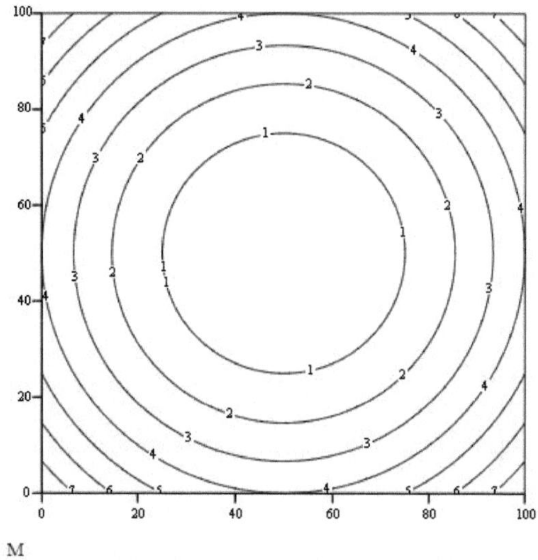

Fig. 5.11 The circles are all points with the same quadratic distance or power W

To obtain the minimum condition

$$\delta s = \delta \int \sqrt{i_1^2 + i_3^2}\, dt = \delta \int \sqrt{\left(\frac{dq_1}{dt}\right)^2 + \left(\frac{dq_3}{dt}\right)^2}\, dt = 0$$

We use the Euler differential equation or ODE which form is

$$\begin{cases} \dfrac{d}{dt} \dfrac{\partial (i_1^2 + i_3^2)}{\partial i_1} = 0 \\[2ex] \dfrac{d}{dt} \dfrac{\partial (i_1^2 + i_3^2)}{\partial i_2} = 0 \end{cases}$$

The solution of the previous equations are the equation of the geodesic that we show in figure 5.5

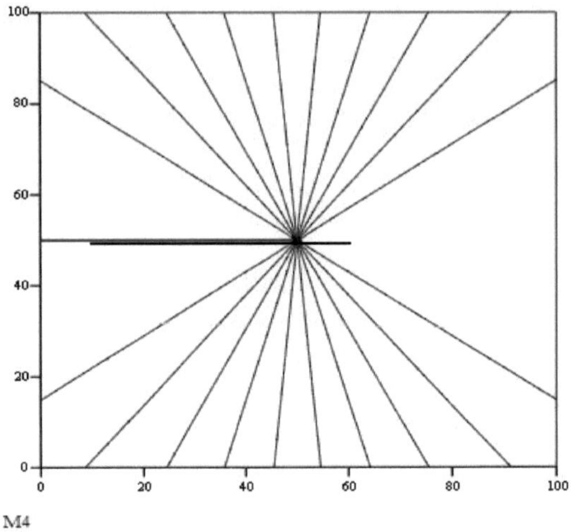

Fig. 5.12 Geodesic as straight lines when the currents are one separate from the others

The electrical circuit in figure 5.8 with the value of the resistances

$R = 3\ \Omega$, $R_{ins} = 5\ \Omega$

Deform the space and the geodesic in figure 5.9 in this geodesic

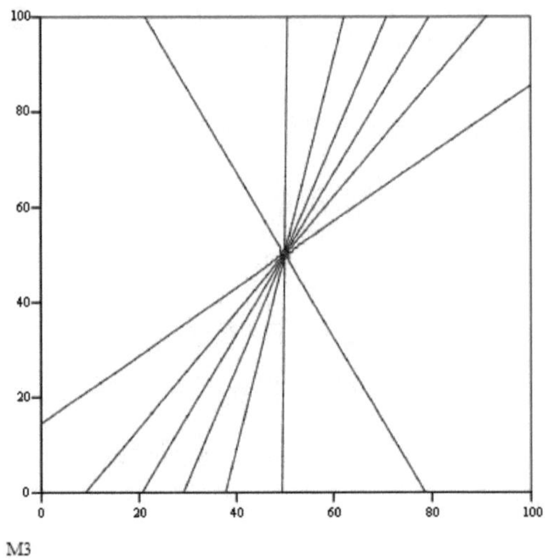

Fig. 5.13 New geodesics obtained by the circuit in figure 5.2 as deformation of the space and geodesic in figure 5.9

In conclusion any neural network or the equivalent electrical circuit generate a deformation of the currents space and geodesic trajectories.

5.13 Relation between Voltage Sources and Currents

The relation between voltages sources and currents for the circuit in figure 5.2 is

$$\begin{bmatrix} i_1 \\ i_5 \end{bmatrix} = \begin{bmatrix} i_1 \\ i_3 - i_1 \end{bmatrix} I = (\alpha^T Z \alpha)^{-1} E = \begin{bmatrix} R_{ins}+3 & 1 \\ 1 & R+3 \end{bmatrix}^{-1} \begin{bmatrix} E \\ E(i_3)+E_{rest} \end{bmatrix}$$

$$= \frac{1}{3R+3R_{ins}+8} \begin{bmatrix} R+3 & -1 \\ -1 & R_{ins}+3 \end{bmatrix} \begin{bmatrix} E \\ E(i_3)+E_{rest} \end{bmatrix}$$

$$= \frac{1}{3R+3R_{ins}+8} \begin{bmatrix} (R+3)E-(E(i_3)+E_{rest}) \\ (R_{ins}+3)(E(i_3)+E_{rest})-E \end{bmatrix}$$

Now when we know E and E_{rest} we have a recursive learning to compute the vector currents

$$\begin{bmatrix} i_1 \\ i_3 \end{bmatrix}$$

The analog electrical circuit compute the current vectors without any algorithm but by the electrical circuit. Now for all parts of the neural network is possible to give the analog electrical circuit in this way.

5.14 Geodesic in Presence of Voltage-Gated Channels in the Membrane

When we have the voltage gate channels in the membranes, we can introduce resistor with memory or memristors that represented by a variable resistor in figure 5.14.

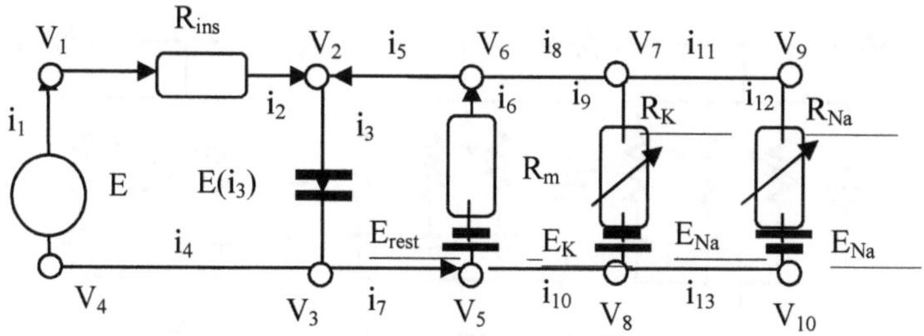

Fig. 5.14 Electrical membrane circuit with voltage-gated channels in the membrane

The graph (topology) of the electrical circuit is

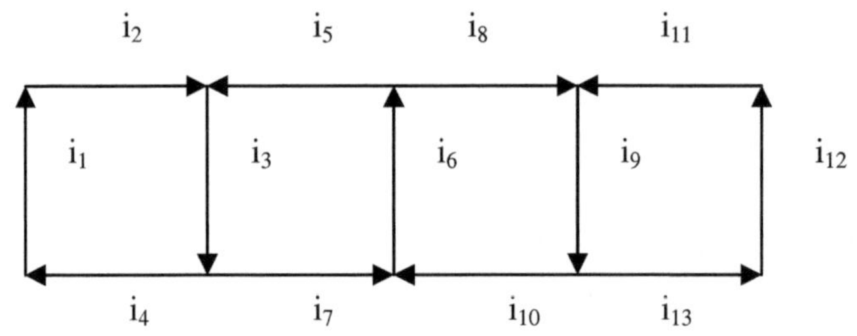

The independent currents are

$$(i_2, i_5, i_8, i_{11})$$

The metric tensor is

$$g = \begin{bmatrix} R_{ins} + 3 & 1 & 0 & 0 \\ 1 & R_m + 3 & R_m & 0 \\ 0 & R_m & R_K + R_m + 2 & R_K \\ 0 & 0 & R_K & R_K + R_{Na} + 2 \end{bmatrix}$$

The geodesic trajectory is

$$W = (R_{ins} + 3)i_2^2 + (R_m + 3)i_5^2 + (R_K + R_m + 2)i_8^2 + (R_K + R_{Na} + 2)i_{11}^2 + \\ + 2i_2 i_5 + 2R_m i_5 i_8 + 2R_K i_8 i_{11}$$

The geodesic assume the minimum value for the currents inside the circuit. For the axon we repeat the same simple unity many times as follows:

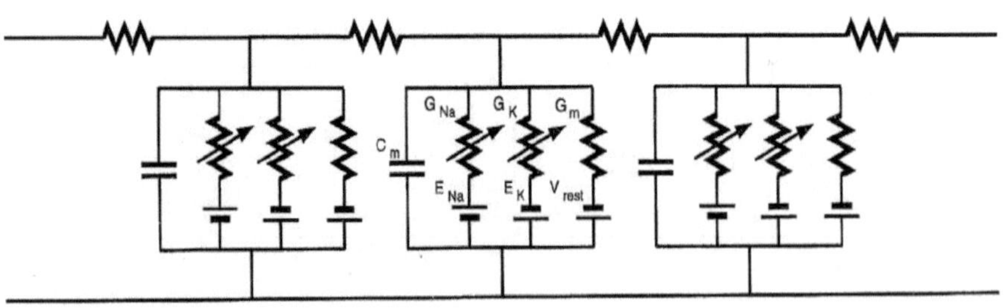

Fig. 5.15 Axon and electrical circuit

The topology of the axon with three elements is

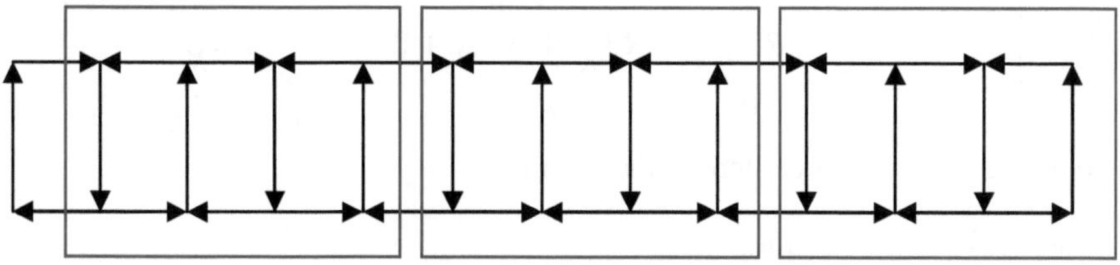

Fig. 5.16 Topology of the three steps of the axons element. At any loop we have one current so we have 1+3+1+3+1+3 = 4*3 =12 loops and 12 free currents.

In figure 5.16 we have three unities of the axon. The space of the currents has dimension equal to

$$1+3+1+3+1+3 = 12$$

If G_1 is the space of the currents in the first four loops, the total space for the three steps is given by the direct sum of three equal spaces

$$G = G_1 \oplus G_2 \oplus G_3$$

The metric tensor is in the 12 space of the currents and is given by the matrix

$$g = \begin{bmatrix} g_1 & 0 & 0 \\ 0 & g_2 & 0 \\ 0 & 0 & g_3 \end{bmatrix}$$

where

$$g_1 = \begin{bmatrix} R1_{ins} + 3 & 1 & 0 & 0 \\ 1 & R1_m + 3 & R1_m & 0 \\ 0 & R1_m & R1_K + R1_m + 2 & R1_K \\ 0 & 0 & R1_K & R1_K + R1_{Na} + 2 \end{bmatrix}$$

$$g_2 = \begin{bmatrix} R2_{ins} + 3 & 1 & 0 & 0 \\ 1 & R2_m + 3 & R2_m & 0 \\ 0 & R2_m & R2_K + R2_m + 2 & R2_K \\ 0 & 0 & R2_K & R2_K + R2_{Na} + 2 \end{bmatrix} \quad (5.37)$$

$$g_3 = \begin{bmatrix} R3_{ins} + 3 & 1 & 0 & 0 \\ 1 & R3_m + 3 & R3_m & 0 \\ 0 & R3_m & R3_K + R3_m + 2 & R3_K \\ 0 & 0 & R3_K & R3_K + R3_{Na} + 2 \end{bmatrix}$$

So the geodesic for three steps in the axon is

$$W = \left(\frac{ds}{dt}\right)^2 = W_1 + W_2 + W_3$$

where

$$W_1 = (R1_{ins} + 3)i_2^2 + (R1_m + 3)i_5^2 + (R1_K + R1_m + 2)i_8^2 + (R1_K + R1_{Na} + 2)i_{11}^2 + \\ + 2i_2 i_5 + 2R1_m i_5 i_8 + 2R1_K i_8 i_{11}$$

$$W_2 = (R2_{ins} + 3)i_2^2 + (R2_m + 3)i_5^2 + (R2_K + R2_m + 2)i_8^2 + (R2_K + R2_{Na} + 2)i_{11}^2 + \\ + 2i_2 i_5 + 2R2_m i_5 i_8 + 2R2_K i_8 i_{11}$$

$$W_3 = (R3_{ins} + 3)i_2^2 + (R3_m + 3)i_5^2 + (R3_K + R3_m + 2)i_8^2 + (R3_K + R3_{Na} + 2)i_{11}^2 + \\ + 2i_2 i_5 + 2R3_m i_5 i_8 + 2R3_K i_8 i_{11}$$

For axon with myelin we have a similar structure but only at the node.

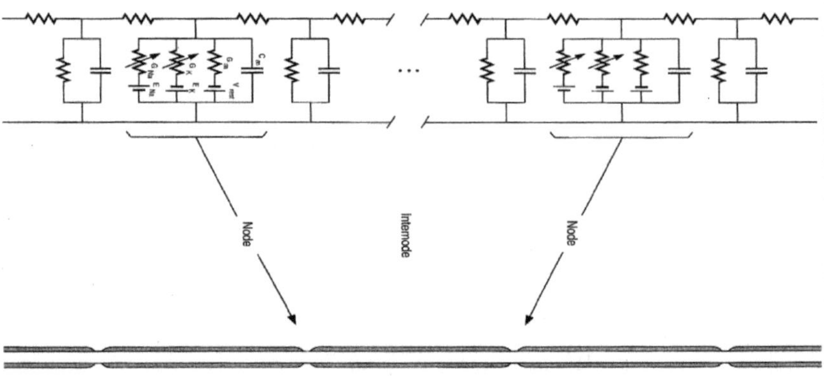

Fig. 5.17 Axons with myelin and electrical circuit

5.15 Geodesic Image of the Synapses and Dendrites

We remember that any synapse is a little part of the three of dendrite that we show in figure 5.15

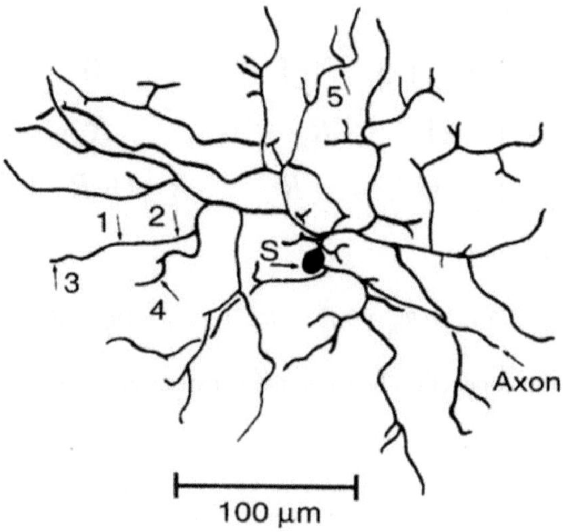

Fig. 5.18 Dendritic tree with synapses, soma and axon

The dendrite tree is useful to simulate a complex tensor metric in this general form

$$g = \begin{bmatrix} g_{1,1} & g_{1,2} & \cdots & g_{1,n} \\ g_{2,1} & g_{2,2} & \cdots & g_{2,n} \\ \cdots & \cdots & \cdots & \cdots \\ g_{n,1} & g_{n,2} & \cdots & g_{n,n} \end{bmatrix}$$

The electrical activity of the synapse is given by the circuit

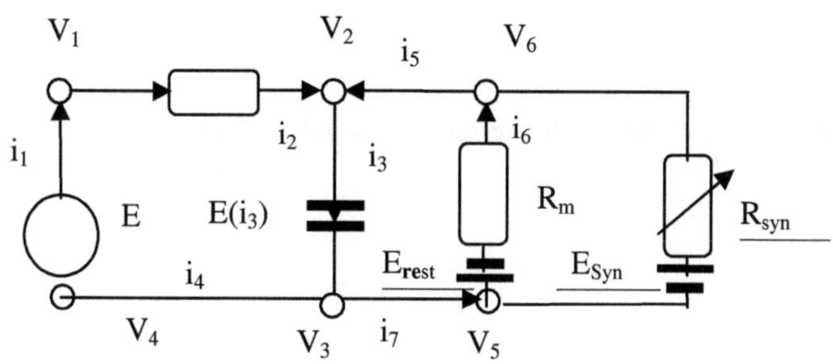

Fig. 5.19 Electrical circuit for synapses

The metric tensor is

$$g = \begin{bmatrix} R_{ins} + 3 & 1 & 0 \\ 1 & R_m + 3 & R_m \\ 0 & R_m & R_{syn} + R_m + 2 \end{bmatrix}$$

The geodesic trajectory is

$$W = \left(\frac{ds}{dt}\right)^2 = (R_{ins} + 3)i_2^2 + (R_m + 3)i_5^2 + (R_m + R_{sys} + 2)i_8^2 + 2i_2 i_5 + 2R_m i_5 i_8 \quad (5.38)$$

Now we show the connection between the synapses and the dendrite by this electrical devise

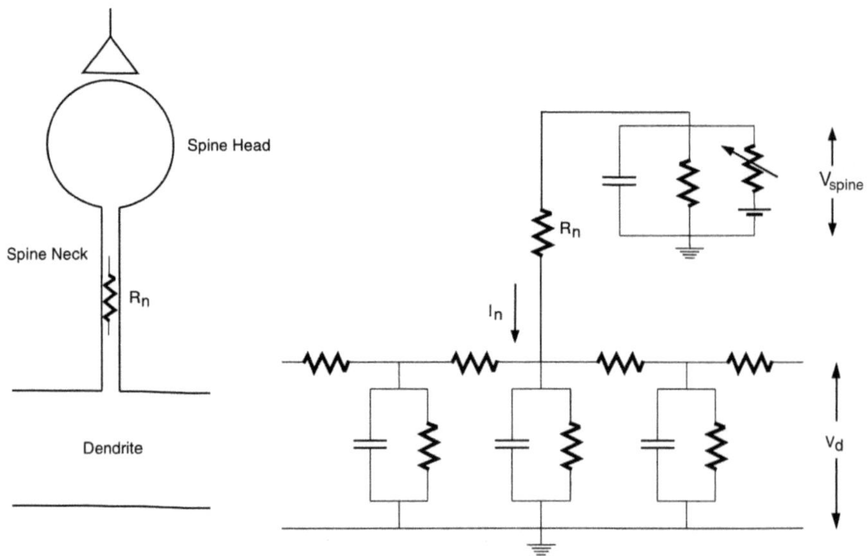

Fig. 5.20 Electrical connection between synapses and dendrite

5.16 Geodesic Image of Shunting Inhibition

For the shunting inhibition we have

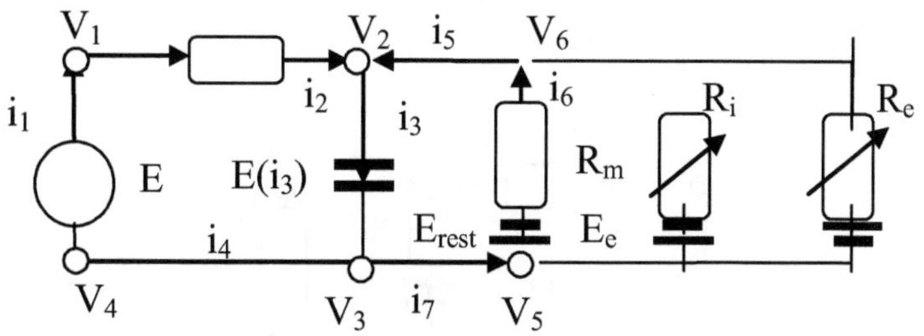

Fig. 5.21 Electrical scheme for shunting inhibition. R_e is the excitatory resistance, R_i is the inhibitory resistance, R_m is the resistance of the membrane.

The geodesic is

$$W = (R_{ins} + 3)i_2^2 + (R_m + 3)i_5^2 + (R_i + R_m + 2)i_8^2 + (R_i + R_e + 2)i_{11}^2 + \quad (5.39)$$
$$+ 2i_2 i_5 + 2R_m i_5 i_8 + 2R_i i_8 i_{11}$$

5.17 Example of Implementation of Wanted Function in the CNS System

Given the simple system

$$\begin{cases} q_1 = x_1 x_2 \\ q_2 = x_1 + x_2 \end{cases}$$

The metric tensor is

$$\frac{\partial q}{\partial x} = \begin{bmatrix} \frac{\partial q_1}{\partial x_1} & \frac{\partial q_1}{\partial x_2} \\ \frac{\partial q_2}{\partial x_1} & \frac{\partial q_2}{\partial x_2} \end{bmatrix} = \begin{bmatrix} x_2 & x_1 \\ 1 & 1 \end{bmatrix} \quad (5.40)$$

and

$$g = \begin{bmatrix} \frac{\partial q_1}{\partial x_1} & \frac{\partial q_1}{\partial x_2} \\ \frac{\partial q_2}{\partial x_1} & \frac{\partial q_2}{\partial x_2} \end{bmatrix}^T \begin{bmatrix} \frac{\partial q_1}{\partial x_1} & \frac{\partial q_1}{\partial x_2} \\ \frac{\partial q_2}{\partial x_1} & \frac{\partial q_2}{\partial x_2} \end{bmatrix} = \begin{bmatrix} x_2 & x_1 \\ 1 & 1 \end{bmatrix}^T \begin{bmatrix} x_2 & x_1 \\ 1 & 1 \end{bmatrix} = \begin{bmatrix} 1+x_2^2 & 1+x_1x_2 \\ 1+x_1x_2 & 1+x_1^2 \end{bmatrix} \quad (5.41)$$

The geodesic is

$$\left(\frac{ds}{dt}\right)^2 = (1 + x_2^2)i_1^2 + 2(1 + x_1 x_2)i_1 i_2 + (1 + x_1^2)i_2^2 \tag{5.42}$$

With the circuit of the membrane

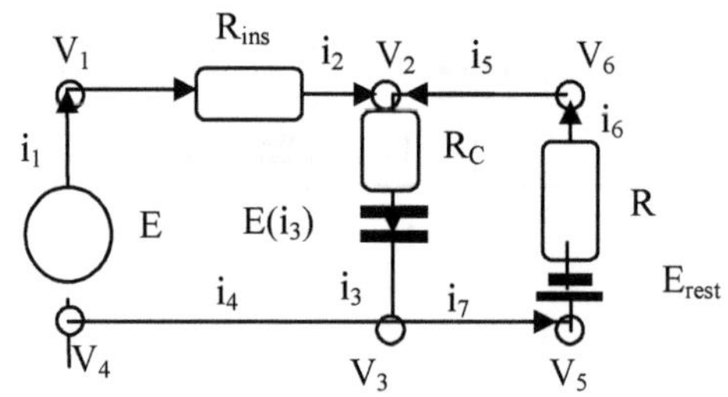

Fig. 5.22 The electrical circuit of the neural unit for the non inertial change of reference

We have the geodesic

$$\left(\frac{ds}{dt}\right)^2 = (R_C + R_{ins} + 2)i_1^2 + 2R_C i_1 i_2 + (R + R_C + 2)i_2^2 \tag{5.43}$$

Now when the resistance are function to the charges (adaptable conductance) in this way

$$\begin{cases} R_C + R_{ins} + 2 = 1 + x_2^2 \\ R_C = (1 + x_1 x_2) \\ R + R_C + 2 = 1 + x_1^2 \end{cases} \tag{5.44}$$

We can compute the resistances that simulate change of the currents inside the space q_1 and q_2,

5.18 Conclusion

We want conclude with the Freeman definition and explanation of the intentionality

"The biological basis of mental activity can be explored with two assumptions: that animals think in ways less complex than humans do, and that the neural mechanisms of mental processes are basically the same across species. The widespread concern for the property of consciousness does not offer a good biological target, because it remains a matter of conjecture whether it exists and what forms

it might take in animals. An alternative concept is intentionality, which can be recognized and studied in its manifestations of goal-directed behavior. Three accepted meanings of the term are the 'aboutness' of beliefs and ideas according to analytic philosophers, intent as

conceived by cognitive psychologists, and wholeness in the process of healing from injury in

medicine and surgery. Intention is interpreted as, respectively, an attribute of mental representations, the expression of motivations and biological drives, or the proclivity of biological tissue to grow toward wholeness. An experimental search for representations in CNSs has failed to find evidence for them. Analysis of CNS Intentionality 10 Walter J Freeman dynamics reveals neural mechanisms for the construction of large-scale space-time patterns of CNS activity, which emerge by interactions of neurons in the limbic system, and which coordinate the operations of sensory and motor cortices. The conclusion is offered that the three characterizations of the concept of intentionality, namely unity, intent, and wholeness, can be incorporated into a nonlinear dynamical model of CNS function centered in the construction of meaning rather than representations. In this view, consciousness is the dynamical process in which meanings are continually under construction in a chaotic trajectory through CNS state space, and awareness is

the subjective experience of the momentary focus of the activity that constitutes a meaning.

In this paper we argue that the multidimensional space of the charges and currents reflect the geometric structure of the nonlinear neurodynamics corresponding to the optimal behaviour of the CNS. The memory in the neural network is not a passive element where we can store the information. Rather, the memory is in the deformation of the original space of the currents and charges as manifested in the neural network. Memory is embodied in the neural parameters as synaptic conductances that give the geodesic trajectories in the deformed, non-Euclidean space. The optimal nonlinear dynamics is a geodesic inside the deformed space that guides the neural computation. Intentionality reflects this deformation, which gives us the correct reference where the dynamics is in agreement with the goals of the animal. Noise is an important component of neural computation, which opens the possibility of exploring new reference systems and new points of views to achieve novel aims and goals. The trajectories obtained through optimal movements with noise can explore new optimal situations that cannot be obtained without the noise. Intention with noise as percolation process opens the possibly to a change of the intent itself.

References

[1] Freeman, W.J.: Mass Action In The Nervous System. In: Examination of the Neurophysiological Basis of Adaptive Behavior through the EEG. Academic Press, New York (1975)

[2] Kozma, R., Freeman, W.J.: Intermittent spatio-temporal desynchronization and sequenced synchrony in ECoG signals. Interdispl. J. Chaos 18, 037131 (2008); Bollobas, B., Kozma, R., Miklos, D. (eds.): Handbook of Large-Scale Random Networks, Bolyai Society Mathematical Studies. Springer, New York (2009) ISBN: 978-3-540-69394-9; Balister, P., Bollobas, B., Kozma, R.: Mean field models of probabilistic cellular automata. Random Structures and Algorithms 29, 399–415 (2006); Kozma, R., Puljic, M., Bollobas, B., Balister, P., Freeman, W.J.: Phase Transitions in the Neuropercolation Model of Neural Populations with Mixed Local and Non-Local Interactions. Biol. Cybernetics 92(6), 367–379 (2005); Kozma, R., Freeman, W.J.: Basic Principles of the KIV Model and its application to the Navigation Problem. Journal of Integrative Neuroscience 2(1), 125–146 (2003)

[3] Freeman, W.J.: A pseudo-equilibrium thermodynamic model of information processing in nonlinear CNS dynamics. Neural Networks (2008), doi:10.1016/j.neunet.2007.12.011

[4] Silverberg, J., Widom, A.: Classical Analytical Mechanics and En-tropy Production. ArXiv:Physics/0701096v1 (January 8, 2007)

[5] Kohno, T., Aihara, K.: A Design Method for Analog and Digital Silicon Neurons-Mathematical-Model-Based Method. In: AIP Conference Proceedings, vol. 1028, pp. 113–128 (2008)

[6] Hindmarsh, J., Rose, R.: A model of neuronal bursting using three coupled first order differential equations. Proceedings of the Royal Society of London B221, 87–102 (1984)

[7] Rinzel, J., Ermentrout, B.: Analysis of Neural Excitability and Oscillations. In: Koch, C., Segev, I. (eds.) Methods in Neural Modeling, pp. 251–291. MIT Press (1998)

[8] David Brown, J., York Jr., J.W.: Jacobi's Action and the Density of States ArXiv: gr –qe /9301018v2 (January 19, 1993)

[9] Resconi, G.: Modelling Fuzzy Cognitive Map By Electrical and Chemical Equivalent Circuits. In: Joint Conference on Information Science, Salt lake City Center USA, July 8-24 (2007)

[10] Resconi, G., Srini, V.P.: Electrical Circuit As A Morphogenetic System. GEST International Transactions on Computer Science and Engineering 53(1), 47–92 (2009)

[11] Todorov, E.: Optimality principles in sensorimotor control (review). Nature Neuroscience 7(9), 907–915 (2004)

[12] Todorov, E., Jordan, M.: Optimal feedback control as a theory of motor coordination. Nature Neuroscience 5(11), 1226–1235 (2002)

[13] Mandzel, A.A., Flach, T.: Geometric Methods in the study of human motor control. Cognitive Studies 6(3), 309–321 (1999)

[14] Haridas, M., Patil, S., Manjunath, T.C.: Recent ATC's in the design of Memristors. International Journal of Computer and Electrical Engineering 2(3) (2010)

[15] Hindmarsh, J.L., Rose, R.M.: A model of the nerve impulse using two first-order differential equations. Nature 269, 162–164 (1982)

[16] Wilke, S.D., Eurich, C.W.: Neural spike statistics modify the impact of background noise. Neurocomputing 38(40), 445–450 (2001)

[17] Arbib, M.A., Amari, S.I.: J. Theoretical Biology 112, 123–155 (1985)

[18] Ostriker, G., Pellionisz, A., Llinas, R.: Tensor Computer model of gaze – I. oculomotor activity is expressed in non-orthogonal natural coordinates. Neuroscience 14(2), 483–500 (1985)
[19] Deward, L.: Classical mechanics, Lecture Physics 3001 Deaprtment of Theoretical Physics research School of Physical Sciences & Engineering. The Australia national University Camberra ACT 02000
[20] Kirkpatrick, S.: Percolation and Conduction. Rev. Mod. Phys. 45, 574–588 (1973); IBM Thomas J. Watson Research Center, Yorktown Heights, New York 10598
[21] London, M.: The effect of the synaptic input on computational processes in neurons Thesis, University of Jerusalem (2001)
[22] Poggio, T., Koch, C.: An analog model of computation for the ill-posed problems of early vision, MIT Artificial Intelligence laboratory (May 1984)
[23] Poggio, T., Koch, C.: III Posed Problems in early vision from Computational Theory of Analogue Network. Proceeding of the Royal Society of London, Series B. Biology Science 226(1244), 303–323 (1985)
[24] Freeman, W.J.: Journal of Mind and Behaviour 18(2-3), 291–304 (1997)
[25] Krone, G.: Tensorial Analysis of network. General Electrical Publication (1939)
[26] Maurice, O.: Compatibilité electromagnetique des systemes complexes. Heimes Sciences (2007)
[27] Amari, S.-I.: Infromation Geometry on Hierarchy of probability. IEEE Transaction on Information Theory 47(5), 1701 (2001)
[28] Treyer, W.E.: Caltech laboratory, CNS 221 Spring 2006 Lecture 1 (2006), `http://www.klab.caltech.edu/~cns221`
[29] Young, E.: 5-3164 System Biology II Neural Network (580-422) Lecture 9 nonlinear cable theory, eyoung@jhu.edu
[30] Chua, L., Kang, S.M.: Memresistive devise and systems. Proceeding of IEEE 64(2), 209–223 (1976)

Appendix A

Under the geometry and geodesic model we can study physical and geometric properties. In fact

We know that in physics of the possible configuration we have free variables (independent) and conditioned variables (dependent). Now between the free variables and conditioned variables we have the relation or constrains. We know that the kinetic energy for all the variables, dependent or independent is the ordinary expression

$$T = \frac{1}{2} m \frac{dx^i}{dt} \frac{dx^j}{dt}$$

Now the relation between independent q variables with the dependent variables x is

$$\begin{cases} x_1 = x_1(q_1, q_2, \ldots, q_n) \\ x_2 = x_2(q_1, q_2, \ldots, q_n) \\ \ldots \\ x_m = x_m(q_1, q_2, \ldots, q_n) \end{cases}$$

So

$$a_{i,j} = \sum_{h,k} \frac{\partial x_i}{\partial q_h} \frac{\partial x_j}{\partial q_k}$$

For [8] we can write the kinetic energy for independent variables q as a geodesic

$$T = \frac{1}{2} m a_{i,j}(q) \frac{dq^i}{dt} \frac{dq^j}{dt}$$

Where the tensor a is function of q

$$a_{i,j}(q)$$

Include the action of field on the given n dimensional configuration space. The cost or geodesic function for the physical phenomena is

$$C = \int dt \sqrt{\frac{1}{2} m a_{i,j}(q) \frac{dq^i}{dt} \frac{dq^j}{dt}}$$

The cost function can substitute in a geometric way the classical Hamiltonians and Lagrangians, as well as the Hamilton principle. We can say that also in physics we have "Intentionality," but a physical system does not have the possibility to control its intentionality (geodesic). In the CNS, we have the intentionality that we can control.

Examples
For the spherical surface we have the constrains

$$y_1^2 + y_2^2 + y_3^2 = r^2$$

We show the geometric image of the constrain

Fig. 5.23 Illustration of the geodesic principle. In green we have the geodesic has line with minimum cost.

The radium of the sphere is constant the free variables (Independent variables) are α and β all the variables are x_1 , x_2 , x_3 . The relation between independent variables and the others is

$$\begin{cases} x_1 = r\sin(\alpha)\cos(\beta) \\ x_2 = r\sin(\alpha)\sin(\beta) \\ x_3 = r\cos(\alpha) \end{cases}$$

This relation is a constraint that we can represent as the surface in figure 5.23. Now we compute the geodesic in the space (x_1 , x_2 , x_3) and the reduce space of the independent variables α and β. So we have

$$\begin{aligned} T &= \frac{1}{2}m[(\frac{dx_1}{dt})^2 + (\frac{dx_2}{dt})^2 + (\frac{dx_3}{dt})^2] \\ &= \frac{1}{2}m[(\frac{dx_1}{d\alpha}\frac{d\alpha}{dt} + \frac{dx_1}{d\alpha}\frac{d\beta}{dt})^2 + (\frac{dx_2}{d\alpha}\frac{d\alpha}{dt} + \frac{dx_2}{d\alpha}\frac{d\beta}{dt})^2 + (\frac{dx_3}{d\alpha}\frac{d\alpha}{dt} + \frac{dx_3}{d\alpha}\frac{d\beta}{dt})^2] \\ &= \frac{1}{2}m[r^2(\frac{d\alpha}{dt})^2 + r^2\sin^2(\alpha)(\frac{d\beta}{dt})^2] \end{aligned}$$

And the function cost that must be at the minimum value is

$$C = \int dt\sqrt{\frac{1}{2}m[(\frac{dx_1}{dt})^2 + (\frac{dx_2}{dt})^2 + (\frac{dx_3}{dt})^2]} = \int dt\sqrt{\frac{1}{2}m[r^2(\frac{d\alpha}{dt})^2 + r^2\sin^2(\alpha)(\frac{d\beta}{dt})^2]}$$

Now for neural network or electrical circuit and the physics we have the similarity relation

Physics	Neural network
Position q	Charges q
Momentum p	Voltage E
velocity	Current I

In fact between velocity and momentum we have the well known relation [8].

$$p_j = a_{i,j} \frac{dq^i}{dt} = a_{i,j} v^i$$

That is the same relation between the voltages sources E and the current

$$E_j = Z_{i,j} \frac{dq^i}{dt} = Z_{i,j} I^i$$

Chapter 6
Electrical Circuit as Constrain in the Multidimensional Space of the Voltages or Currents

6.1 Geometry of Voltage, Current, and Electrical Power

In this chapter we will show that electrical circuit can be represented as a projection of a vector in the voltage space into a subspace where defined constrain is satisfy. Let us consider the electrical circuit shown in figure 6.1 where e_k are the edges of the circuit, V are the electrical potentials in the nodes, Z_k are the impedances and i_k are the currents.

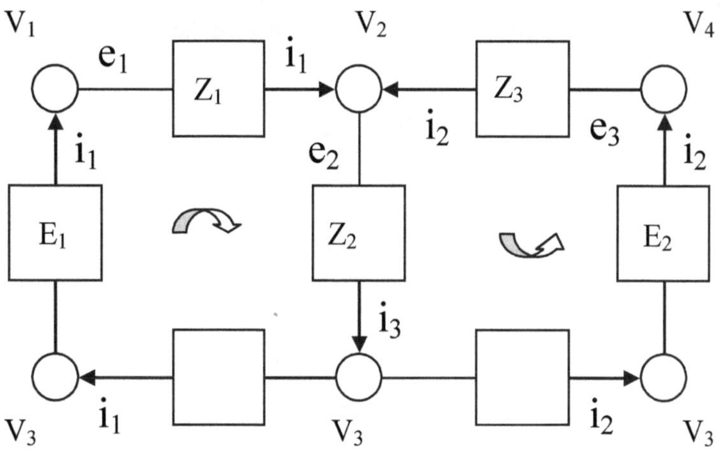

Fig. 6.1 An electrical circuit where V_j are the potential, Z_k the impedances, e_h the edges of the circuit.

For the EC by the Ohm's law we have the relation between the current and the voltages in this way

G. Resconi: Geometry of Knowledge for Intelligent Systems, SCI 407, pp. 159–206.
springerlink.com

$$V = \begin{bmatrix} V_1 - V_2 \\ V_3 - V_2 \\ V_4 - V_2 \end{bmatrix} = \begin{bmatrix} v_1 \\ v_2 \\ v_3 \end{bmatrix} = \begin{bmatrix} Z_1 i_1 \\ Z_2(i_1 + i_2) \\ Z_3 i_2 \end{bmatrix}$$
$$= \begin{bmatrix} Z_1 & 0 \\ Z_2 & Z_2 \\ 0 & Z_3 \end{bmatrix} \begin{bmatrix} i_1 \\ i_2 \end{bmatrix} = \begin{bmatrix} Z_1 & 0 & 0 \\ 0 & Z_2 & 0 \\ 0 & 0 & Z_3 \end{bmatrix} \begin{bmatrix} 1 & 0 \\ 1 & 1 \\ 0 & 1 \end{bmatrix} \begin{bmatrix} i_1 \\ i_2 \end{bmatrix} = ZHi \tag{6.1}$$

So we have

$$H^T V = H^T ZHi$$

Where

$$H^T ZH$$

Is a quadratic matrix so we have

$$i = (H^T ZH)^{-1} H^T V \tag{6.2}$$

Where

$$E = H^T V$$

Are the voltages sources. Given the voltages sources we can compute the currents sources and from the currents sources we can compute the voltages sources in this way

$$(H^T ZH)\, i = H^T V = E$$

Now given

$$V = ZHi$$

We have

$$V = \begin{bmatrix} Z_1 & 0 \\ Z_2 & Z_2 \\ 0 & Z_3 \end{bmatrix} \begin{bmatrix} i_1 \\ i_2 \end{bmatrix} = \begin{bmatrix} \begin{bmatrix} Z_1 & 0 \\ Z_2 & Z_2 \end{bmatrix} \\ [0 \quad Z_3] \end{bmatrix} \begin{bmatrix} i_1 \\ i_2 \end{bmatrix}$$
$$= \begin{bmatrix} Z \\ Z_C \end{bmatrix} i = \begin{bmatrix} Zi \\ Z_C i \end{bmatrix} = \begin{bmatrix} Zi \\ RZi \end{bmatrix} = \begin{bmatrix} v \\ Rv \end{bmatrix} = \begin{bmatrix} v \\ v_C \end{bmatrix} = \begin{bmatrix} v_1 \\ v_2 \\ v_3 \end{bmatrix}$$

For the EC in figure 6.1, the three voltages are not independent but there is relation between the voltages given by the equation

$$Rv = v_C = v_3 \tag{6.3}$$

where

$$R = Z_C Z^{-1}$$

For the EC in figure 6.1 we have

$$R = \begin{bmatrix} 0 & Z_3 \end{bmatrix} \begin{bmatrix} Z_1 & 0 \\ Z_2 & Z_2 \end{bmatrix}^{-1} = \begin{bmatrix} -\frac{Z_3}{Z_1} & \frac{Z_3}{Z_2} \end{bmatrix} \tag{6.4}$$

The relation between the voltages is

$$\begin{bmatrix} -\frac{Z_3}{Z_1} & \frac{Z_3}{Z_2} \end{bmatrix} \begin{bmatrix} v_1 \\ v_2 \end{bmatrix} = v_3$$

That can write in this way

$$-\frac{Z_3}{Z_1} v_1 + \frac{Z_3}{Z_2} v_2 = v_3$$

and

$$\frac{v_2}{Z_2} = \frac{v_3}{Z_3} + \frac{v_1}{Z_1}$$

That for the Ohm rule is the relation between the currents. Now given a generic voltage vector with the relation

$$V = ZHi$$

Given the current source and for (6.1) and (6.2), we have

$$V = ZHi = ZH(H^T ZH)^{-1} H^T v \tag{6.5}$$

Now when (6.3) is true then V = v, but when (6.3) is false then V is different from v but the distance between V and v is the minimum distance. In fact we have

$$\begin{bmatrix} V_1 \\ V_2 \end{bmatrix} = \begin{bmatrix} Z \\ Z_C \end{bmatrix} i$$

where

$$i = (\begin{bmatrix} H_1 \\ H_2 \end{bmatrix}^T \begin{bmatrix} Z \\ Z_C \end{bmatrix})^{-1} \begin{bmatrix} H_1 \\ H_2 \end{bmatrix}^T \begin{bmatrix} v_1 \\ v_2 \end{bmatrix} = (H_1^T Z + H_2^T Z_C)^{-1}(H_1^T v_1 + H_2^T v_2)$$

$$= (H_1^T Z + H_2^T RZ)^{-1}(H_1^T v_1 + H_2^T v_2) = Z^{-1}(H_1^T + H_2^T R)^{-1}(H_1^T v_1 + H_2^T v_2)$$

when

$$v_2 = R v_1$$

$$i = Z^{-1} v_1$$

and

$$\begin{bmatrix} V_1 \\ V_2 \end{bmatrix} = \begin{bmatrix} Z \\ Z_C \end{bmatrix} Z^{-1} v_1 = \begin{bmatrix} v_1 \\ Z_C Z^{-1} v_1 \end{bmatrix} = \begin{bmatrix} v_1 \\ R v_1 \end{bmatrix} = \begin{bmatrix} v_1 \\ v_2 \end{bmatrix}$$

Now for

$$v_C = Rv + \Gamma v$$

$$i = Z^{-1}(H_1^T + H_2^T R)^{-1}(H_1^T v_1 + H_2^T (Rv + \Gamma v))$$

$$= Z^{-1}(I + (H_1^T + H_2^T R)^{-1} H_2^T \Gamma) v = Z^{-1}(I + A) v$$

where

$$A = (H_1^T + H_2^T R)^{-1} H_2^T \Gamma$$

And

$$\begin{bmatrix} V_1 \\ V_2 \end{bmatrix} = \begin{bmatrix} Z \\ Z_C \end{bmatrix} Z^{-1}(I + A) v = \begin{bmatrix} (I + A) v \\ Z_C Z^{-1}(I + A) v \end{bmatrix} = \begin{bmatrix} (I + A) v \\ R(I + A) v \end{bmatrix} \tag{6.6}$$

Now for R in (6.4) we have

$$v_C = Rv + \Gamma v = \begin{bmatrix} -\dfrac{Z_3}{Z_1} & \dfrac{Z_3}{Z_2} \end{bmatrix} \begin{bmatrix} v_1 \\ v_2 \end{bmatrix} + \begin{bmatrix} \Gamma_1 & \Gamma_2 \end{bmatrix} \begin{bmatrix} v_1 \\ v_2 \end{bmatrix}$$

$$= v_1(\Gamma_1 - \frac{Z_3}{Z_1}) + v_2(\Gamma_2 + \frac{Z_3}{Z_2}) = v_3$$

So we have

$$\Gamma_1 = \frac{v_3}{v_1} + \frac{Z_3}{Z_1} - \frac{v_2}{v_1}(\Gamma_2 + \frac{Z_3}{Z_2})$$

When

$$\Gamma_2 = 0$$

$$\Gamma_1 = \frac{v_3}{v_1} + \frac{Z_3}{Z_1} - \frac{v_2}{v_1}\frac{Z_3}{Z_2}$$

We have

$$A = (H_1^T + H_2^T R)^{-1} H_2^T \Gamma = \frac{Z_1 Z_2 v_3 - Z_1 Z_3 v_2 + Z_2 Z_3 v_1}{v_1 (Z_1 Z_2 + Z_1 Z_3 + Z_2 Z_3)} \begin{bmatrix} 1 & 0 \\ 1 & 0 \end{bmatrix}$$

And for the (6.5) we have

$$\begin{aligned} i &= Z^{-1}(I + A)v \\ &= \begin{bmatrix} Z_1 & 0 \\ Z_2 & Z_2 \end{bmatrix}^{-1} \begin{bmatrix} 1 & 0 \\ 0 & 1 \end{bmatrix} v + \frac{Z_1 Z_2 v_3 - Z_1 Z_3 v_2 + Z_2 Z_3 v_1}{(Z_1 Z_2 + Z_1 Z_3 + Z_2 Z_3) v_1} \begin{bmatrix} 1 & 0 \\ 1 & 0 \end{bmatrix} v \\ &= \begin{bmatrix} Z_1 v_1 + Z_3 v_1 - Z_2 v_3 + Z_3 v_2 \\ Z_1 v_2 - Z_2 v_1 + Z_1 v_3 + Z_2 v_3 \end{bmatrix} \frac{1}{(Z_1 Z_2 + Z_1 Z_3 + Z_2 Z_3)} \\ &= (\begin{bmatrix} 1 & 0 \\ 1 & 1 \\ 0 & 1 \end{bmatrix}^T \begin{bmatrix} Z_1 & 0 \\ Z_2 & Z_2 \\ 0 & Z_3 \end{bmatrix})^{-1} \begin{bmatrix} 1 & 0 \\ 1 & 1 \\ 0 & 1 \end{bmatrix}^T \begin{bmatrix} v_1 \\ v_2 \\ v_3 \end{bmatrix} \end{aligned}$$

6.1.1 Geometric Representation of the EC

We present the EC as a subspace of the voltage space. Given the operator

$$Q = B(B^T B)^{-1} B^T \tag{6.7}$$

Where B is a rectangular matrix with m rows and n colons with m > n, for a vector X in the m dimensional space we have

$$QX = B(B^T B)^{-1} B^T X$$

Now we compute the vector difference of X with QX so we obtain

$$QX - X = B(B^T B)^{-1} B^T X - X = (B(B^T B)^{-1} B^T - I)X$$

Now we compute the scalar product of Q X with X – QX, so we obtain

$$(QX)^T (QX - X) = X^T Q^T (QX - X) = X^T Q^T QX - X^T Q^T X$$

but

$$Q^T = B(B^T B)^{-1} B^T = Q$$

and

$$Q^T Q = Q^2 = B(B^T B)^{-1} B^T B(B^T B)^{-1} B^T = B(B^T B)^{-1} B^T = Q$$

So

$$(QX)^T (QX - X) = 0$$

In conclusion QX is the orthogonal projection of X into the subspace B. In graph way we have

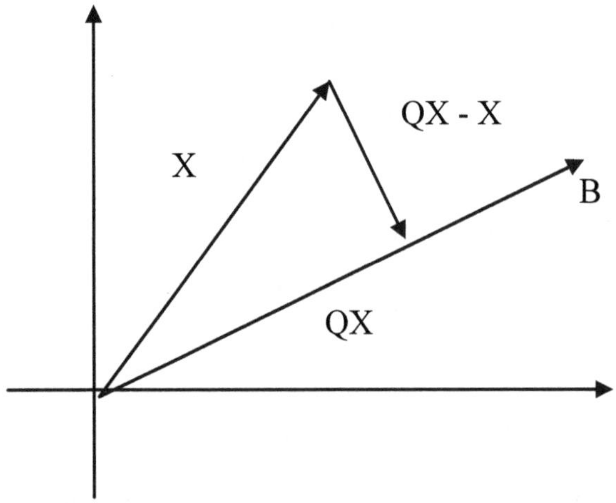

Fig. 6.2 Projection of the vector X into the subspace B

Now given two vectors X and Y we compute the projection of X and Y on the same subspace B with the condition

$$QX = QY$$
$$B(B^T B)^{-1} B^T X = B(B^T B)^{-1} B^T Y$$

Now we assume that Y belong to the subspace A so

$$Y = A\alpha$$

Y is the linear combination with coefficient α of the vectors that form the subspace A.

Now for the previous identity we have

$$B(B^T B)^{-1} B^T X = B(B^T B)^{-1} B^T A\alpha$$

So we have

$$B^T X = B^T A\alpha$$

$$and$$

$$\alpha = (B^T A)^{-1} B^T X$$

And

$$Y = A(B^T A)B^T X = (A(B^T A)B^T)X = PX$$

Y is the oblique projection of X into A. In fact we have

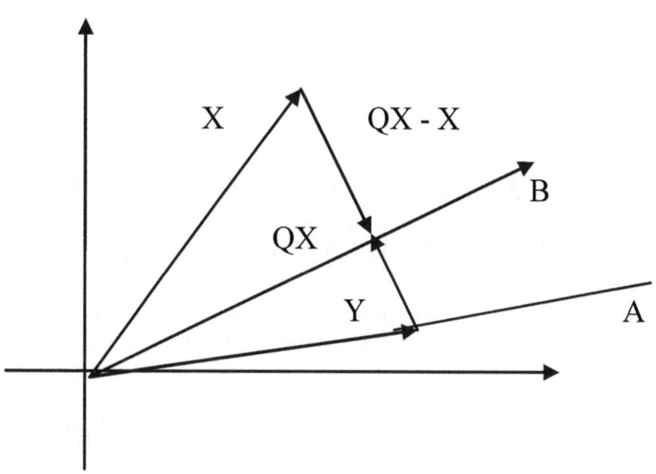

Fig. 6.3 Projection of X and Y into the subspace B

And

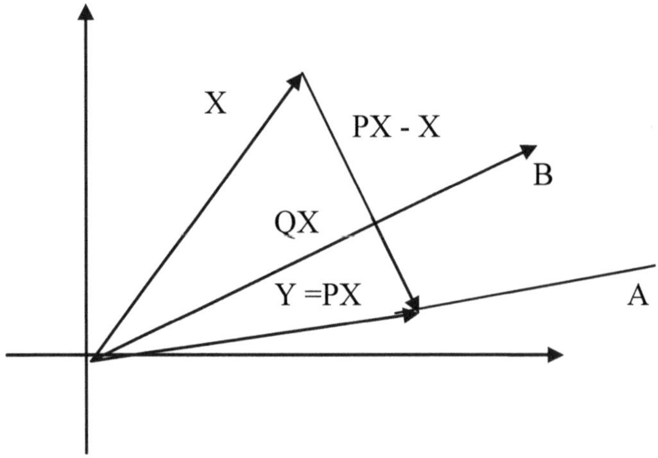

Fig. 6.4 Oblique projection of the vector X into the subspace A

We remark that

$$(PX)^T(X - PX) = X^T P^T (X - PX) = X^T P^T X - X^T P^T PX$$

but

$$P^T = (A(B^T A)^{-1} B^T)^T = B(A^T B)^{-1} A^T$$

and

$$P^T P = B(A^T B)^{-1} A^T A(B^T A)^{-1} B^T$$

So

$$(PX)^T(X - PX) = X^T B(A^T B)^{-1} A^T X - X^T B(A^T B)^{-1} A^T A(B^T A)^{-1} B^T X$$
$$= X^T B(A^T B)^{-1} A^T (I - A(B^T A)^{-1} B^T)X$$

The scalar product of the oblique projection of PX – X and X is not equal to zero. With the scalar product we can compute the angle of the projection. We remark also that

$$P^2 = A(B^T A)^{-1} B^T A(B^T A)^{-1} B^T = A(B^T A)^{-1} B^T = P$$

So P is a projection. When we repeat P two times the result is again P.

Now we came back to the EC if we put

$$A = ZH, B = H$$
$$Y = A(B^T A)B^T X = ZH(H^T ZH)^{-1} H^T v = V$$

The EC relation (6.5) is the oblique projection operator of the voltage vector v into the impedance matrices ZH. When we change the impedance value we change the angles of the oblique projection. In a graphic way we have

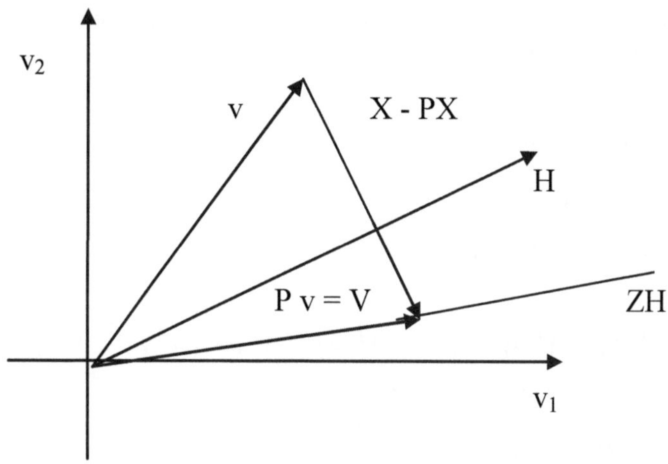

Fig. 6.5 The Electrical Circuit EC as the oblique projection of the voltage vector v into the subspace of the impedance ZH

With EC is possible to define in the m voltages space any type of reflection and refraction as we show in this figure

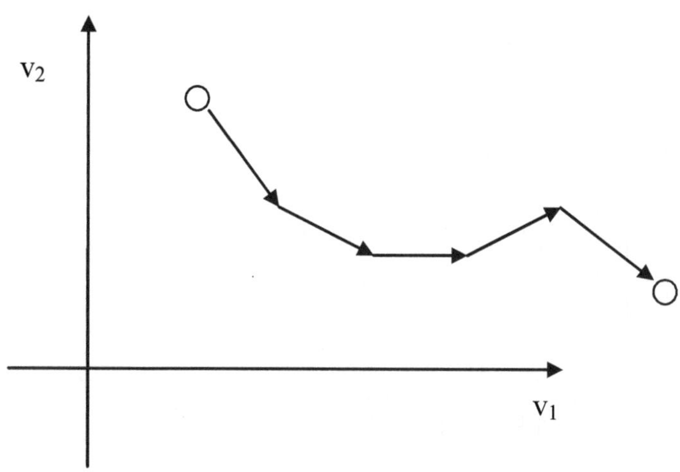

Fig. 6.6 Refraction and reflection movement into the space of the voltages

With EC we can create geodetic movement in the m dimensional space of the voltages. We create a new m dimension optical system similar to the **quantum optic.**

6.1.2 Electrical Power as Logistic Function in the Voltages m Dimensional Space

Now given the flow f that in the electrical circuit is the current I and the effort e that in the electrical circuit is e we have the power P that have a lot of properties that we study in this part.

$$
\begin{aligned}
&for\\
&f = (A^T A)^{-1} A^T e\\
&(A^T A) f = A^T e\\
&P = f^T (A^T e) = f^T (A^T A) f\\
&and\\
&P = ((A^T A)^{-1} A^T e)^T (A^T A)(A^T A)^{-1} A^T e = (A^T e)^T (A^T A)^{-1} (A^T e)
\end{aligned}
$$

Now for (6.4) we have

$$
\begin{aligned}
P &= e^T A(A^T A)^{-1} A^T e = e^T e\\
&= \begin{bmatrix} E_1 \\ RE_1 \end{bmatrix}^T \begin{bmatrix} E_1 \\ RE_1 \end{bmatrix} = E_1^T E_1 + (RE_1)^T (RE_1) = P_S
\end{aligned}
$$

And for

$$
e_L = \begin{bmatrix} E \\ RE+L \end{bmatrix}
$$

We have

$$
\begin{aligned}
&P_1 = e^T A(A^T A)^{-1} A^T e\\
&\begin{bmatrix} E_1 \\ RE_1+L \end{bmatrix}^T \begin{bmatrix} E_1 + A_1 (A^T A)^{-1} A_2{}^T L \\ R(E_1 + A_1 (A^T A)^{-1} A_2{}^T L) \end{bmatrix}\\
&= \begin{bmatrix} E_1 \\ RE_1+L \end{bmatrix}^T \begin{bmatrix} E_1+M \\ R(E_1+M) \end{bmatrix} = E_1^T (E_1 + M) + (RE_1 + L)^T R(E_1 + M)\\
&P_2 = (A(A^T A)^{-1} A^T e)^T A(A^T A)^{-1} A^T e\\
&= e^T A(A^T A)^{-1} A^T A(A^T A)^{-1} A^T e\\
&= e^T A(A^T A)^{-1} A^T e = P_1
\end{aligned}
$$

Given any vector of effort e, the power P is the same when we change e into

$$e' = A(A^T A)^{-1} A^T e$$

So we have

$$P = (A^T e')^T (A^T A)^{-1} (A^T e') = (e')^T (e') = (A^T e)^T (A^T A)^{-1} (A^T e)$$

Now we prove that the optimal solution can write in this way

$$\begin{cases} \min\limits_{\beta} P = f^T (A^T A) f = f^T G f \\ \\ E = A^T e \end{cases}$$

The previous expression means that we search among all possible value of the flow *f* that minimise the power P given $E = A^T e$ as a constrain. It is obvious that without fixed value for E the minimum condition for P is zero and this has no meaning. Now to solve the minimum problem with constrains we use the Lagrange multipliers. So we have

$$D = f^T Gf + \lambda(E - A^T e) = \beta^T G\beta + \lambda(E - A^T Af) = f^T Gf + \lambda(E - Gf)$$

Now we compute the derivative in this way

$$\frac{\partial D}{\partial f_j} = \frac{\partial f^T}{\partial f_j} Gf + f^T G \frac{\partial f}{\partial f_j} - \lambda G \frac{\partial f}{\partial f_j} = 0$$

That can write in this way

$$v_j^T Gf + f^T Gv_j = 2f^T Gv_j = \lambda Gv_j$$

For which we have $\lambda = 2f^T$ and

$$D = f^T Gf + 2f^T (E - Gf) = f^T Gf + 2f^T E - 2f^T Gf = 2f^T E - f^T Gf$$

Now the minimum condition is

$$\frac{\partial D}{\partial f_j} = 2\frac{\partial f^T}{\partial f_j}E - (\frac{\partial f^T}{\partial f_j}Gf + f^T G\frac{\partial f}{\partial f_j}) = 2v_j^T E - 2v_j^T Gf = 0$$

The solution is $E = Gf$ for which we have

$$f = (A^T A)^{-1} E = (A^T A)^{-1} A^T e$$

That is solution of the equation

$$(A^T A)f = A^T e = E$$

Proposition

The solution of the equation $A^T Af = A^T e = E$ that is

$$f = (A^T A)^{-1} A^T e$$

Among all possible values of f minimise the power

$$P = f^T A^T Af$$

For given values of the constrain E.

6.1.3 Classical Parallel and Series Method to Compute Currents and Geometric Method

Now we compare the quantum approach to electrical circuit that uses the projection operator and the traditional calculus by the method of the equivalent circuit. Let us consider an electrical circuit with a voltage source, shown in figure 6.6.

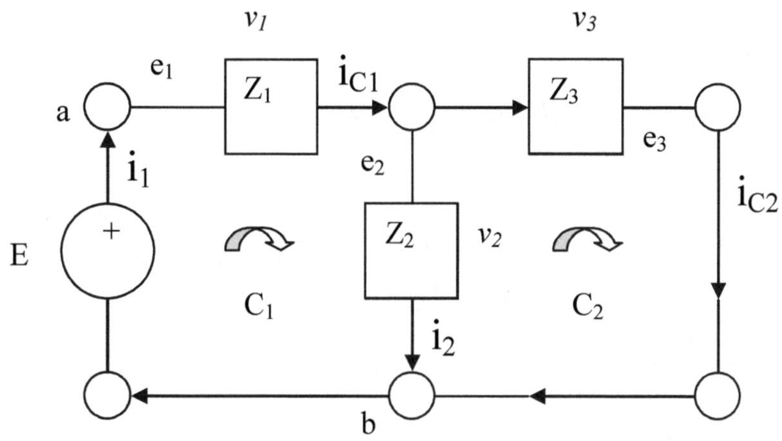

Fig. 6.7 Electrical Circuit with voltage source E

For the three edge e_1, e_2 , e_3 and the two cycles C_1 , C_2 the cycle matrix is

$$H = \begin{bmatrix} 1 & 0 \\ 1 & -1 \\ 0 & 1 \end{bmatrix}$$

for the three edges the impedance matrix is

$$Z = \begin{bmatrix} Z_1 & 0 & 0 \\ 0 & Z_2 & 0 \\ 0 & 0 & Z_3 \end{bmatrix}$$

and for the previous expressions we have

$$\begin{bmatrix} v_{C_1} \\ v_{C_2} \end{bmatrix} = \begin{bmatrix} E \\ 0 \end{bmatrix} = \begin{bmatrix} 1 & 0 \\ 1 & -1 \\ 0 & 1 \end{bmatrix} \begin{bmatrix} v_1 \\ v_2 \\ v_3 \end{bmatrix} = \begin{bmatrix} v_1 + v_2 \\ -v_2 + v_3 \end{bmatrix}$$

and

$$i_C = Z_C{}^{-1} v_C = \begin{bmatrix} i_{C_1} \\ i_{C_2} \end{bmatrix} = (H^T Z H)^{-1} \begin{bmatrix} v_{C_1} \\ v_{C_2} \end{bmatrix} =$$

$$\left(\begin{bmatrix} 1 & 0 \\ 1 & -1 \\ 0 & 1 \end{bmatrix}^T \begin{bmatrix} Z_1 & 0 & 0 \\ 0 & Z_2 & 0 \\ 0 & 0 & Z_3 \end{bmatrix} \begin{bmatrix} 1 & 0 \\ 1 & -1 \\ 0 & 1 \end{bmatrix} \right)^{-1} \begin{bmatrix} v_{C_1} \\ v_{C_2} \end{bmatrix} = \begin{bmatrix} Z_1 + Z_2 & -Z_2 \\ -Z_2 & Z_2 + Z_3 \end{bmatrix}^{-1} \begin{bmatrix} E \\ 0 \end{bmatrix} =$$

$$\begin{bmatrix} E \dfrac{Z_2 + Z_3}{Z_1 Z_2 + Z_1 Z_3 + Z_2 Z_3} \\ E \dfrac{Z_2}{Z_1 Z_2 + Z_1 Z_3 + Z_2 Z_3} \end{bmatrix}$$

The two currents can be obtained without the matrix form by the traditional equivalent circuits. So we have

$$Z_{a,b} = Z_1 + \frac{1}{\dfrac{1}{Z_2} + \dfrac{1}{Z_3}} = \frac{Z_1 Z_2 + Z_1 Z_3 + Z_2 Z_3}{Z_2 + Z_3}$$

So we have

$$i_{C_1} = \frac{E}{Z_{a,b}} = E \frac{Z_2 + Z_3}{Z_1 Z_2 + Z_1 Z_3 + Z_2 Z_3}$$

that is the same result of the matrix method. Now with the classical method for i_{C2} we have

$$i_{C_2} = i_{C_1} - \frac{E - Z_1 i_{C_1}}{Z_2} = E \frac{Z_2}{Z_1 Z_2 + Z_1 Z_3 + Z_2 Z_3}$$

with the same result of the matrix method. For the computation of the power we can use the sources of the voltage. So we obtain

$$P = \begin{bmatrix} v_{C_1} \\ v_{C_2} \end{bmatrix} \begin{bmatrix} Z_1 + Z_2 & -Z_2 \\ -Z_2 & Z_2 + Z_3 \end{bmatrix}^{-1} \begin{bmatrix} v_{C_1} \\ v_{C_2} \end{bmatrix} = \begin{bmatrix} E \\ 0 \end{bmatrix} \begin{bmatrix} Z_1 + Z_2 & -Z_2 \\ -Z_2 & Z_2 + Z_3 \end{bmatrix}^{-1} \begin{bmatrix} E \\ 0 \end{bmatrix} =$$

$$E_1^2 \frac{Z_2 + Z_3}{Z_1 Z_2 + Z_1 Z_3 + Z_2 Z_3}$$

With the classical method we obtain

$$P = \frac{E_1^{\,2}}{Z_{a,b}} = E_1^{\,2} \frac{Z_2 + Z_3}{Z_1 Z_2 + Z_1 Z_3 + Z_2 Z_3}$$

Now we define the projection operator Q in the following way

$$v = ZH(Z_C^{\,-1})H^T v = Qv \tag{6.8}$$

Remark

The projection operator give us the interaction between the free vector v and the context or electrical circuit. In fact given three independent voltages

$$v = \begin{bmatrix} v_1 \\ v_2 \\ v_3 \end{bmatrix}$$

Now we project this vector inside the space of the electrical circuit in figure 6.6 (interaction) the circuit 6 is similar to an instrument in the quantum mechanics. So we obtain

$$v = \begin{bmatrix} Z_1 & 0 & 0 \\ 0 & Z_2 & 0 \\ 0 & 0 & Z_3 \end{bmatrix} \begin{bmatrix} 1 & 0 \\ 1 & -1 \\ 0 & 1 \end{bmatrix} (\begin{bmatrix} Z_1 + Z_2 & -Z_2 \\ -Z_2 & Z_2 + Z_3 \end{bmatrix}^{-1} \begin{bmatrix} 1 & 0 \\ 1 & -1 \\ 0 & 1 \end{bmatrix}^T \begin{bmatrix} v_1 \\ v_2 \\ v_3 \end{bmatrix} =$$

$$\frac{1}{Z_1 Z_2 + Z_1 Z_3 + Z_2 Z_3} \begin{bmatrix} v_1(Z_2 + Z_3) + v_2 Z_3 + v_3 Z_2 \\ v_2(Z_1 + Z_3) + v_1 Z_3 - v_3 Z_1 \\ -v_3(Z_1 + Z_2) + v_2 Z_1 - v_1 Z_2 \end{bmatrix} = Qv$$

The vector v bounded with the electrical circuit has the following properties

$$\begin{bmatrix} v_1 \\ v_2 \\ v_3 \end{bmatrix} = Qv = \frac{1}{Z_1 Z_2 + Z_1 Z_3 + Z_2 Z_3} \begin{bmatrix} v_1(Z_2 + Z_3) + v_2 Z_3 + v_3 Z_2 \\ v_2(Z_1 + Z_3) + v_1 Z_3 - v_3 Z_1 \\ -v_3(Z_1 + Z_2) + v_2 Z_1 - v_1 Z_2 \end{bmatrix}$$

and the solution is

$$\begin{bmatrix} v_1 \\ v_2 \\ v_3 \end{bmatrix} = \begin{bmatrix} Z_1(\frac{v_2}{Z_2} + \frac{v_3}{Z_3}) \\ Z_2(\frac{v_1}{Z_2} - \frac{v_3}{Z_3}) \\ Z_3(\frac{v_1}{Z_1} - \frac{v_2}{Z_2}) \end{bmatrix} \tag{6.9}$$

we remark that Q is a projection operator. In fact

$$Q^2 v = ZH(H^T ZH)^{-1} H^T ZH(H^T ZH)^{-1} H^T v = ZH(H^T ZH)^{-1} H^T v = Qv$$

So we have that

$$\frac{1}{Z_1Z_2 + Z_1Z_3 + Z_2Z_3} \begin{bmatrix} v_1(Z_2 + Z_3) + v_2Z_3 + v_3Z_2 \\ v_2(Z_1 + Z_3) + v_1Z_3 - v_3Z_1 \\ -v_3(Z_1 + Z_2) + v_2Z_1 - v_1Z_2 \end{bmatrix} =$$

$$Q \frac{1}{Z_1Z_2 + Z_1Z_3 + Z_2Z_3} \begin{bmatrix} v_1(Z_2 + Z_3) + v_2Z_3 + v_3Z_2 \\ v_2(Z_1 + Z_3) + v_1Z_3 - v_3Z_1 \\ -v_3(Z_1 + Z_2) + v_2Z_1 - v_1Z_2 \end{bmatrix}$$

So equation (6.6) is true for the projection Qv. In fact we can see that for

$$\frac{1}{Z_1Z_2 + Z_1Z_3 + Z_2Z_3} \begin{bmatrix} v_1(Z_2 + Z_3) + v_2Z_3 + v_3Z_2 \\ v_2(Z_1 + Z_3) + v_1Z_3 - v_3Z_1 \\ -v_3(Z_1 + Z_2) + v_2Z_1 - v_1Z_2 \end{bmatrix} = \begin{bmatrix} v_1 \\ v_2 \\ v_3 \end{bmatrix}$$

In conclusion, given a free vector v this vector become part of the electrical circuit with its constraints set by the projection operator. The difference $v - Qv$ is the vector orthogonal with the electrical circuit. The difference is a vector which *values are incompatible or incoherent with the electrical circuit as a context.*

For our computation we write

$$v = ZH(H^T ZH)^{-1} H^T v = Qv = Q_{Cycle} v$$

6.2 A New Method to Compute the Inverse Matrix

Given the matrix

$$G = \begin{bmatrix} G_{1,1} & G_{1,2} & \dots & G_{1,n} \\ G_{2,1} & G_{2,2} & \dots & G_{2,n} \\ \dots & \dots & \dots & \dots \\ G_{n,1} & G_{n,2} & \dots & G_{n,n} \end{bmatrix}$$

For the matrix calculus we have

$$G^{-1} = \begin{bmatrix} (G^{-1})_{1,1} & (G^{-1})_{1,2} & \dots & (G^{-1})_{1,n} \\ (G^{-1})_{2,1} & (G^{-1})_{2,2} & \dots & (G^{-1})_{2,n} \\ \dots & \dots & \dots & \dots \\ (G^{-1})_{n,1} & (G^{-1})_{n,2} & \dots & (G^{-1})_{n,n} \end{bmatrix} =$$

$$\begin{bmatrix} \frac{(-1)^{1+1}\det(M_{1,1})}{\det(G)} & \frac{(-1)^{1+2}\det(M_{1,2})}{\det(G)} & \dots & \frac{(-1)^{1+n}\det(M_{1,n})}{\det(G)} \\ \frac{(-1)^{2+1}\det(M_{2,1})}{\det(G)} & \frac{(-1)^{2+2}\det(M_{2,2})}{\det(G)} & \dots & \frac{(-1)^{2+n}\det(M_{2,n})}{\det(G)} \\ \dots & \dots & \dots & \dots \\ \frac{(-1)^{n+1}\det(M_{n,1})}{\det(G)} & \frac{(-1)^{n+2}\det(M_{n,2})}{\det(G)} & \dots & \frac{(-1)^{n+n}\det(M_{n,n})}{\det(G)} \end{bmatrix}$$

Where

$$M_{k,j} = \begin{bmatrix} G_{1,1} & \dots & G_{1,j-1} & G_{1,j+1} & \dots & G_{1,n} \\ \dots & \dots & \dots & \dots & \dots & \dots \\ G_{k-1,1} & \dots & G_{k-1,j-1} & G_{k-1,j+1} & \dots & G_{k-1,n} \\ G_{k+1,1} & \dots & G_{k+1,j-1} & G_{k+1,j+1} & \dots & G_{k+1,n} \\ \dots & \dots & \dots & \dots & \dots & \dots \\ G_{k,n} & \dots & G_{k,j-1} & G_{k,j+1} & \dots & G_{k,n} \end{bmatrix}$$

Theorem 6.1

The inverse of the matrix G can write in this way

$$(G^{-1})_{j,k} = \frac{(-1)^{k+j}\det(M_{k,j})}{\det(G)} = \frac{(-1)^{k+j}\det(M_{k,j})}{\det(M_{k,j})((-1)^{k+j}C_{k,j}+(-1)^{k+j-1}W^T(M_{k,j})^{-1}V)} =$$

$$\frac{1}{(C_{k,j}-W^T(M_{k,j})^{-1}V)}$$

Where

$$W = \begin{bmatrix} G_{k,1} \\ \cdots \\ G_{k,j-1} \\ G_{k,j+1} \\ \cdots \\ G_{k,n} \end{bmatrix}, V = \begin{bmatrix} G_{1,j} \\ \cdots \\ G_{k-1,j} \\ G_{k+1,j} \\ \cdots \\ G_{n,j} \end{bmatrix}$$

Proof

Because we use the inverse matrix is useful to compute the inverse matrix with a particular algorithm that we prove in this chapter.

For the *Laplace theorem* we have

$$det(G) = G_{k,1}(-1)^{k+1}\det(M_{k,1}) + G_{k,2}(-1)^{k+2}\det(M_{k,2}) + ... + G_{k,n}(-1)^{k+n}\det(M_{k,n})$$

For example given the matrix

$$G = \begin{bmatrix} G_{1,1} & G_{2,1} & G_{3,1} \\ G_{1,2} & G_{2,2} & G_{3,2} \\ G_{1,3} & G_{2,3} & G_{3,3} \end{bmatrix}$$

Where

$$G = \begin{bmatrix} \begin{bmatrix} G_{1,1} \\ G_{2,1} \end{bmatrix} & \begin{bmatrix} G_{2.1} & G_{3,1} \\ G_{2,2} & G_{3,2} \end{bmatrix} \\ G_{3,1} & \begin{bmatrix} G_{2,3} & G_{3,3} \end{bmatrix} \end{bmatrix} = \begin{bmatrix} \begin{bmatrix} G_{1,1} \\ G_{2,1} \end{bmatrix} & M_{3,1} \\ G_{3,1} & \begin{bmatrix} G_{2,3} & G_{3,3} \end{bmatrix} \end{bmatrix}$$

We have the determinant

$$det(G) = G_{3,1}(-1)^{3+1}\det(M_{3,1}) + G_{3,2}(-1)^{3+2}\det(M_{3,2}) + G_{3,3}(-1)^{3+3}\det(M_{3,3}) \tag{6.10}$$

and

$$det(G) = (G_{3,1})\det(\begin{bmatrix} G_{1,2} & G_{1,3} \\ G_{2,2} & G_{2,3} \end{bmatrix}) + G_{3,2}(-1)\det(\begin{bmatrix} G_{1,1} & G_{1,3} \\ G_{2,1} & G_{3,3} \end{bmatrix}) + G_{3,3}\det(\begin{bmatrix} G_{1,1} & G_{1,2} \\ G_{2,1} & G_{2,2} \end{bmatrix}) \tag{6.11}$$

The determinant can write in this way

$$det(G) = det(M_{k,j})[(-1)^{k+j} G_{k,j} +$$

$$\frac{G_{k,1}(-1)^{k+1}\det(M_{k,1}) + \ldots G_{k,j-1}(-1)^{k+j-1}\det(M_{k,j-1}) + G_{k,j+1}(-1)^{k+j+1}\det(M_{k,j+1}) + \ldots G_{k,n}(-1)^{k+n}\det(M_{k,n})}{det(M_{k,j})}]$$

For example the (10.22) can write in this way

$$det(G) = \det(M_{3,2})\,[G_{3,2}(-1)^{3+2} + \frac{G_{3,1}(-1)^{3+1}\det(M_{3,1}) + G_{3,3}(-1)^{3+3}\det(M_{3,3})}{\det(M_{3,2})}]$$

The term T

$$T = G_{k,1}(-1)^{k+1}\det(M_{k,1}) + \ldots G_{k,j-1}(-1)^{k+j-1}\det(M_{k,j-1}) + G_{k,j+1}(-1)^{k+j+1}\det(M_{k,j+1}) + \ldots G_{k,n}(-1)^{k+n}\det(M_{k,n})$$

That in our example is

$$T = G_{3,1}(-1)^{3+1}\det(M_{3,1}) + G_{3,3}(-1)^{3+3}\det(M_{3,3})$$

Can write in this way

$$T = \begin{bmatrix} G_{k,1} & \dots & G_{k,j-1} & G_{k,j+1} & \dots & G_{k,n} \end{bmatrix} \begin{bmatrix} (-1)^{k+1}\det(M_{k,1}) \\ \dots \\ (-1)^{k+j-1}\det(M_{k,j-1}) \\ (-1)^{k+j+1}\det(M_{k,j+1}) \\ \dots \\ (-1)^{k+n}\det(M_{k,n}) \end{bmatrix}$$

In our example we have

$$T = G_{3,1}(-1)^{3+1}\det(M_{3,1}) + G_{3,3}(-1)^{3+3}\det(M_{3,3}) = \begin{bmatrix} G_{3,1} & G_{3,3} \end{bmatrix} \begin{bmatrix} (-1)^{3+1}\det(M_{3,1}) \\ (-1)^{3+3}\det(M_{3,3}) \end{bmatrix}$$

Bur now we know that

$$M_{k,1} = \begin{bmatrix} G_{1,2} & G_{1,3} & \dots & G_{1,j} & \dots & G_{1,n} \\ \dots & \dots & \dots & \dots & \dots & \dots \\ G_{k-1,2} & G_{k-1,3} & \dots & G_{k-1,j} & \dots & G_{k-1,n} \\ G_{k+1,2} & G_{k+1,3} & \dots & G_{k+1,j} & \dots & G_{k+1,n} \\ \dots & \dots & \dots & \dots & \dots & \dots \\ G_{n,2} & G_{n,3} & \dots & G_{n,j} & \dots & G_{n,n} \end{bmatrix},$$

$$M_{k,j-1} = \begin{bmatrix} G_{1,1} & \dots & G_{1,j-2} & G_{1,j} & \dots & G_{1,n} \\ \dots & \dots & \dots & \dots & \dots & \dots \\ G_{k-1,1} & \dots & G_{k-1,j-2} & G_{k-1,j} & \dots & G_{k-1,n} \\ G_{k+1,1} & \dots & G_{k+1,j-2} & G_{k+1,j} & \dots & G_{k+1,n} \\ \dots & \dots & \dots & \dots & \dots & \dots \\ G_{n,1} & \dots & G_{n,j-2} & G_{n,j} & \dots & G_{n,n} \end{bmatrix}$$

$$M_{k,j+1} = \begin{bmatrix} G_{1,1} & \dots & G_{1,j} & G_{1,j+2} & \dots & G_{1,n} \\ \dots & \dots & \dots & \dots & \dots & \dots \\ G_{k-1,1} & \dots & G_{k-1,j} & G_{k-1,j+2} & \dots & G_{k-1,n} \\ G_{k+1,1} & \dots & G_{k+1,j} & G_{k+1,j+2} & \dots & G_{k+1,n} \\ \dots & \dots & \dots & \dots & \dots & \dots \\ G_{n,1} & \dots & G_{n,j} & G_{n,j+2} & \dots & G_{n,n} \end{bmatrix}$$

$$M_{k,n} = \begin{bmatrix} G_{1,1} & \dots & G_{1,j} & G_{1,j+1} & \dots & G_{1,n-1} \\ \dots & \dots & \dots & \dots & \dots & \dots \\ G_{k-1,1} & \dots & G_{k-1,j} & G_{k-1,j+1} & \dots & G_{k-1,n-1} \\ G_{k+1,1} & \dots & G_{k+1,j} & G_{k+1,j+1} & \dots & G_{k+1,n-1} \\ \dots & \dots & \dots & \dots & \dots & \dots \\ G_{n,1} & \dots & G_{n,j} & G_{n,j+1} & \dots & G_{n,n-1} \end{bmatrix}$$

Now again for the *Laplace theorem* we have

$$\det(M_{k,1}) = G_{1,j}(-1)^{j-1+1}\det(A_{1,j}) + \dots + G_{k-1,j}(-1)^{j-1+k-1}\det(A_{k-1,j}) +$$
$$G_{k+1,j}(-1)^{j-1+k}\det(A_{k+1,j}) + \dots + G_{n,j}(-1)^{j-1+n-1}\det(A_{n,j})$$
$$\det(M_{k,j-1}) = G_{1,j}(-1)^{j-1+1}\det(A_{1,j}) + \dots + G_{k-1,j}(-1)^{j-1+k-1}\det(A_{k-1,j}) +$$
$$G_{k+1,j}(-1)^{j-1+k}\det(A_{k+1,j}) + \dots + G_{n,j}(-1)^{j-1+n-1}\det(A_{n,j})$$
$$\det(M_{k,j+1}) = G_{1,j}(-1)^{j+1}\det(A_{1,j}) + \dots + G_{k-1,j}(-1)^{j+k-1}\det(A_{k-1,j}) +$$
$$G_{k+1,j}(-1)^{j+k}\det(A_{k+1,j}) + \dots + G_{n,j}(-1)^{j+n-1}\det(A_{n,j})$$
$$\det(M_{k,n}) = G_{1,j}(-1)^{j+1}\det(A_{1,j}) + \dots + G_{k-1,j}(-1)^{j+k-1}\det(A_{k-1,j}) +$$
$$G_{k+1,j}(-1)^{j+k}\det(A_{k+1,j}) + \dots + G_{n,j}(-1)^{j+n-1}\det(A_{n,j})$$

Because

$$M_{3,1} = \begin{bmatrix} G_{1,2} & G_{1,3} \\ G_{2,2} & G_{2,3} \end{bmatrix}, M_{3,3} = \begin{bmatrix} G_{1,1} & G_{1,2} \\ G_{2,1} & G_{2,2} \end{bmatrix}$$

We have

$$\det(M_{3,1}) = G_{1,2}(-1)^{2+1-1}G_{2,3} + G_{2,2}(-1)^{2+2-1}G_{1,3},$$
$$\det(M_{3,3}) = G_{1,2}(-1)^{1+2}G_{2,1} + G_{2,2}(-1)^{2+2}G_{1,1}$$

Now the previous determinant can write in this matrix form

$$\det(M_{k,1}) = \begin{bmatrix} (-1)^{j-1+1}A_{1,1} & \dots & (-1)^{j-1+k-1}A_{k-1,1} & (-1)^{j-1+k}A_{k,1} & \dots & (-1)^{j-1+n-1}A_{n-1,1} \end{bmatrix} \begin{bmatrix} G_{1,j} \\ \dots \\ G_{k-1,j} \\ G_{k+1,j} \\ \dots \\ G_{n,j} \end{bmatrix}$$

$$\det(M_{k,j-1}) = \begin{bmatrix} (-1)^{j-1+1}A_{1,j-1} & \dots & (-1)^{j-1+k-1}A_{k-1,j-1} & (-1)^{j-1+k}A_{k+1,j-1} & \dots & (-1)^{j-1+n-1}A_{n,j-1} \end{bmatrix} \begin{bmatrix} G_{1,j} \\ \dots \\ G_{k-1,j} \\ G_{k+1,j} \\ \dots \\ G_{n,j} \end{bmatrix}$$

$$\det(M_{k,j+1}) = \begin{bmatrix} (-1)^{j+1}A_{1,j+1} & \dots & (-1)^{j+k-1}A_{k-1,j+1} & (-1)^{j+k}A_{k+1,j+1} & \dots & (-1)^{j+n-1}A_{n,j+1} \end{bmatrix} \begin{bmatrix} G_{1,j} \\ \dots \\ G_{k-1,j} \\ G_{k+1,j} \\ \dots \\ G_{n,j} \end{bmatrix}$$

$$\det(M_{k,n}) = \begin{bmatrix} (-1)^{j+1}A_{1,n} & \dots & (-1)^{j+k-1}A_{k-1,n} & (-1)^{j+k}A_{k+1,n} & \dots & (-1)^{j+n-1}A_{n,n} \end{bmatrix} \begin{bmatrix} G_{1,j} \\ \dots \\ G_{k-1,j} \\ G_{k+1,j} \\ \dots \\ G_{n,j} \end{bmatrix}$$

For example we have

$$\det(M_{3,1}) = \begin{bmatrix} (-1)^{2+1-1}G_{1,2} & (-1)^{2+2-1}G_{2,2} \end{bmatrix} \begin{bmatrix} G_{2,3} \\ G_{1,3} \end{bmatrix}$$

$$\det(M_{3,3}) = \begin{bmatrix} (-1)^{1+2}G_{1,2} & (-1)^{2+2}G_{2,2} \end{bmatrix} \begin{bmatrix} G_{2,1} \\ G_{1,1} \end{bmatrix}$$

Or

$$\det(M_{3,1}) = \begin{bmatrix} G_{1,2} & -G_{2,2} \end{bmatrix} \begin{bmatrix} G_{2,3} \\ G_{1,3} \end{bmatrix}$$

$$\det(M_{3,3}) = \begin{bmatrix} -G_{1,2} & G_{2,2} \end{bmatrix} \begin{bmatrix} G_{2,1} \\ G_{1,1} \end{bmatrix}$$

and

$$(-1)^{k+1}\det(M_{k,1}) = (-1)^{k+j-1}\begin{bmatrix} (-1)^{1+1}A_{1,1} & \cdots & (-1)^{k-1+1}A_{k-1,1} & (-1)^{k+1}A_{k,1} & \cdots & (-1)^{n-1+1}A_{n-1,1} \end{bmatrix}\begin{bmatrix} G_{1,j} \\ \cdots \\ G_{k-1,j} \\ G_{k+1,j} \\ \cdots \\ G_{n,j} \end{bmatrix}$$

$(-1)^{k+j-1}\det(M_{k,j-1}) =$

$$(-1)^{k+j-1}\begin{bmatrix} (-1)^{j-1+1}A_{1,j-1} & \cdots & (-1)^{j-1+k-1}A_{k-1,j-1} & (-1)^{j-1+k}A_{k,j-1} & \cdots & (-1)^{j-1+n-1}A_{n-1,j-1} \end{bmatrix}\begin{bmatrix} G_{1,j} \\ \cdots \\ G_{k-1,j} \\ G_{k+1,j} \\ \cdots \\ G_{n,j} \end{bmatrix}$$

Because

$(-1)^{k+j+1} = (-1)^{k+j+1-1+1} = (-1)^{k+j-1+2} = ((-1)^{k+j-1})(-1)^2 = (-1)^{k+j-1}$), *we* have

$(-1)^{k+j+1}\det(M_{k,j+1}) =$

$$(-1)^{k+j-1}\begin{bmatrix} (-1)^{j+1}A_{1,j} & \cdots & (-1)^{j+k-1}A_{k-1,j} & (-1)^{j+k}A_{k,j} & \cdots & (-1)^{j+n-1}A_{n-1,j} \end{bmatrix}\begin{bmatrix} G_{1,j} \\ \cdots \\ G_{k-1,j} \\ G_{k+1,j} \\ \cdots \\ G_{n,j} \end{bmatrix}$$

Becasue

$k + n = (k + j - 1) + n - j + 1$,$(-1)^{k+n} = (-1)^{k+n}(-1)^{-2} = (-1)^{k-1+n-1}$,$(-1)^{n+1+k} = (-1)^{n-1+2+k} = (-1)^{n-1+k}$,

$(-1)^n = (-1)^{n-1+n-1+2} = (-1)^{n-1+n-1}$

we have

$(-1)^{k+n}\det(M_{k,n}) =$

$$(-1)^{k+j-1}\begin{bmatrix} (-1)^{n}A_{1,n-1} & \cdots & (-1)^{k-1+n-1}A_{k-1,n-1} & (-1)^{k+n-1}A_{k,n-1} & \cdots & (-1)^{n-1+n-1}A_{n-1,n-1} \end{bmatrix}\begin{bmatrix} G_{1,j} \\ \cdots \\ G_{k-1,j} \\ G_{k+1,j} \\ \cdots \\ G_{n,j} \end{bmatrix}$$

Because the minor of $M_{h,k}$ are $A_{i,j}$ we have

$$T = (-1)^{(k+j-1)} \det(M_{k,j}) \begin{bmatrix} G_{k,1} & \cdots & G_{k,j-1} & G_{k,j+1} & \cdots & G_{k,n} \end{bmatrix} M_{k,j}^{-1} \begin{bmatrix} G_{1,j} \\ \cdots \\ G_{k-1,j} \\ G_{k+1,j} \\ \cdots \\ G_{n,j} \end{bmatrix}$$

And the final form is

$$det(G) = det(M_{k,j})((-1)^{k+j} G_{k,j} +$$

$$(-1)^{k+j-1} \begin{bmatrix} G_{k,1} & \cdots & G_{k,j-1} & G_{k,j+1} & \cdots & G_{k.n} \end{bmatrix} M_{k,j}^{-1} \begin{bmatrix} G_{1,j} \\ \cdots \\ G_{k-1,j} \\ G_{k+1,j} \\ \cdots \\ G_{n,j} \end{bmatrix})$$

So we have

$$(G^{-1})_{j,k} = \frac{(-1)^{k+j} \det(M_{k,j})}{\det(G)} = \frac{(-1)^{k+j} \det(M_{k,j})}{\det(M_{k,j})((-1)^{k+j} C_{k,j} + (-1)^{k+j-1} W^T (M_{k,j})^{-1} V)} =$$

$$\frac{1}{(C_{k,j} - W^T (M_{k,j})^{-1} V)}$$

Where

$$W = \begin{bmatrix} G_{k,1} \\ \cdots \\ G_{k,j-1} \\ G_{k,j+1} \\ \cdots \\ G_{k,n} \end{bmatrix}, V = \begin{bmatrix} G_{1,j} \\ \cdots \\ G_{k-1,j} \\ G_{k+1,j} \\ \cdots \\ G_{n,j} \end{bmatrix}$$

Example of the use of the new method in the Theorem 6.1. Given the matrix

$$G=\begin{bmatrix} G_{1,1} & G_{1,2} \\ G_{2,1} & G_{2,2} \end{bmatrix}$$

For the classical method the inverse matrix is

$$G^{-1}=\begin{bmatrix} \dfrac{G_{2,2}}{G_{2,2}G_{1,1}-G_{1,2}G_{2,1}} & \dfrac{-G_{1,2}}{G_{2,2}G_{1,1}-G_{1,2}G_{2,1}} \\ \dfrac{-G_{2,1}}{G_{2,2}G_{1,1}-G_{1,2}G_{2,1}} & \dfrac{G_{1,1}}{G_{2,2}G_{1,1}-G_{1,2}G_{2,1}} \end{bmatrix}$$

For the new theorem we obtain

$$G^{-1}=\begin{bmatrix} \dfrac{1}{G_{1,1}-G_{1,2}\dfrac{1}{G_{2,2}}G_{2,1}} & \dfrac{1}{G_{1,2}-G_{1,1}\dfrac{1}{G_{2,1}}G_{2,2}} \\ \dfrac{1}{G_{2,1}-G_{1,1}\dfrac{1}{G_{1,2}}G_{2,2}} & \dfrac{1}{G_{2,2}-G_{2,1}\dfrac{1}{G_{1,1}}G_{1,2}} \end{bmatrix}^{T}=$$

$$\begin{bmatrix} (G_{1,1}-G_{1,2}\dfrac{1}{G_{2,2}}G_{2,1})^{-1} & (G_{1,2}-G_{1,1}\dfrac{1}{G_{2,1}}G_{2,2})^{-1} \\ (G_{2,1}-G_{1,1}\dfrac{1}{G_{1,2}}G_{2,2})^{-1} & (G_{2,2}-G_{2,1}\dfrac{1}{G_{1,1}}G_{1,2})^{-1} \end{bmatrix}^{T}=$$

$$\begin{bmatrix} (G_{1,1}-e_{1,1})^{-1} & (G_{1,2}-e_{1,2})^{-1} \\ (G_{2,1}-e_{2,1})^{-1} & (G_{2,2}-e_{2,2})^{-1} \end{bmatrix}^{T}$$

Where the matrix

$$e = \begin{bmatrix} e_{1,1} & e_{1,2} \\ e_{2,1} & e_{2,2} \end{bmatrix} = \begin{bmatrix} G_{1,2}\dfrac{1}{G_{2,2}}G_{2,1} & G_{1,1}\dfrac{1}{G_{2,1}}G_{2,2} \\ G_{1,1}\dfrac{1}{G_{1,2}}G_{2,2} & G_{2,1}\dfrac{1}{G_{1,1}}G_{1,2} \end{bmatrix}$$

is the matrix that give us the difference of G^{-1} with the local inverse values

$$LG^{-1} = \begin{bmatrix} \dfrac{1}{G_{1,1}} & \dfrac{1}{G_{1,2}} \\ \dfrac{1}{G_{2,1}} & \dfrac{1}{G_{2,2}} \end{bmatrix} = \begin{bmatrix} G_{1,1}^{-1} & G_{1,2}^{-1} \\ G_{2,1}^{-1} & G_{2,2}^{-1} \end{bmatrix}$$

For the matrix G,

$$G = \begin{bmatrix} G_{1,1} & G_{2,1} & G_{3,1} \\ G_{1,2} & G_{2,2} & G_{3,2} \\ G_{1,3} & G_{2,3} & G_{3,3} \end{bmatrix}$$

using the theorem we have

$$G^{-1} = \begin{bmatrix} (G_{1,1} - W_{1,1}^T M_{1,1}^{-1} V_{1,1})^{-1} & (G_{1,2} - W_{1,2}^T M_{1,2}^{-1} V_{1,2})^{-1} & (G_{1,3} - W_{1,3}^T M_{1,3}^{-1} V_{1,3})^{-1} \\ (G_{2,1} - W_{2,1}^T M_{2,1}^{-1} V_{2,1})^{-1} & (G_{2,2} - W_{2,2}^T M_{2,2}^{-1} V_{2,2})^{-1} & (G_{2,3} - W_{2,3}^T M_{2,3}^{-1} V_{2,3})^{-1} \\ (G_{3,1} - W_{3,1}^T M_{3,1}^{-1} V_{3,1})^{-1} & (G_{3,2} - W_{3,2}^T M_{3,2}^{-1} V_{3,})^{-1} & (G_{3,3} - W_{3,3}^T M_{3,3}^{-1} V_{3,3})^{-1} \end{bmatrix}^T$$

where

$$e = \begin{bmatrix} W_{1,1}^T M_{1,1}^{-1} V_{1,1} & W_{1,2}^T M_{1,2}^{-1} V_{1,2} & W_{1,3}^T M_{1,3}^{-1} V_{1,3} \\ W_{2,1}^T M_{2,1}^{-1} V_{2,1} & W_{2,2}^T M_{2,2}^{-1} V_{2,2} & W_{2,3}^T M_{2,3}^{-1} V_{2,3} \\ W_{3,1}^T M_{3,1}^{-1} V_{3,1} & W_{3,2}^T M_{3,2}^{-1} V_{3,} & W_{3,3}^T M_{3,3}^{-1} V_{3,3} \end{bmatrix} =$$

$$\begin{bmatrix} \begin{bmatrix} G_{1,2} \\ G_{1,3} \end{bmatrix}^T \begin{bmatrix} G_{2,2} & G_{2,3} \\ G_{3,2} & G_{3,3} \end{bmatrix}^{-1} \begin{bmatrix} G_{2,1} \\ G_{3,1} \end{bmatrix} & \begin{bmatrix} G_{1,1} \\ G_{1,3} \end{bmatrix}^T \begin{bmatrix} G_{2,1} & G_{2,3} \\ G_{3,1} & G_{3,3} \end{bmatrix}^{-1} \begin{bmatrix} G_{2,2} \\ G_{3,2} \end{bmatrix} & \begin{bmatrix} G_{1,1} \\ G_{1,2} \end{bmatrix}^T \begin{bmatrix} G_{2,1} & G_{2,2} \\ G_{3,1} & G_{3,2} \end{bmatrix}^{-1} \begin{bmatrix} G_{2,3} \\ G_{3,3} \end{bmatrix} \\ \begin{bmatrix} G_{2,2} \\ G_{2,3} \end{bmatrix}^T \begin{bmatrix} G_{1,2} & G_{1,3} \\ G_{3,2} & G_{3,3} \end{bmatrix}^{-1} \begin{bmatrix} G_{1,1} \\ G_{3,1} \end{bmatrix} & \begin{bmatrix} G_{1,2} \\ G_{3,2} \end{bmatrix}^T \begin{bmatrix} G_{1,1} & G_{1,3} \\ G_{3,1} & G_{3,3} \end{bmatrix}^{-1} \begin{bmatrix} G_{2,1} \\ G_{2,3} \end{bmatrix} & \begin{bmatrix} G_{2,1} \\ G_{2,2} \end{bmatrix}^T \begin{bmatrix} G_{1,1} & G_{1,2} \\ G_{3,1} & G_{3,2} \end{bmatrix}^{-1} \begin{bmatrix} G_{1,3} \\ G_{3,3} \end{bmatrix} \\ \begin{bmatrix} G_{3,2} \\ G_{3,3} \end{bmatrix}^T \begin{bmatrix} G_{1,2} & G_{1,3} \\ G_{2,2} & G_{2,3} \end{bmatrix}^{-1} \begin{bmatrix} G_{1,1} \\ G_{2,1} \end{bmatrix} & \begin{bmatrix} G_{3,1} \\ G_{3,3} \end{bmatrix}^T \begin{bmatrix} G_{1,1} & G_{1,3} \\ G_{2,1} & G_{2,3} \end{bmatrix}^{-1} \begin{bmatrix} G_{1,2} \\ G_{2,2} \end{bmatrix} & \begin{bmatrix} G_{3,1} \\ G_{3,2} \end{bmatrix}^T \begin{bmatrix} G_{1,1} & G_{1,2} \\ G_{2,1} & G_{2,2} \end{bmatrix}^{-1} \begin{bmatrix} G_{1,3} \\ G_{2,3} \end{bmatrix} \end{bmatrix}$$

6.3 Electrical Circuit and the New Method for Inverse of the Matrix

For the electrical circuit in figure 6.6 we have that

$$H_{Cycle} = \begin{bmatrix} 1 & 0 & 0 \\ 0 & 1 & 0 \\ 0 & 0 & 1 \\ -1 & 0 & 1 \\ 1 & -1 & 0 \\ 0 & -1 & 1 \end{bmatrix}$$

$$Z_{Cycle} = \begin{bmatrix} 1 & 0 & 0 \\ 0 & 1 & 0 \\ 0 & 0 & 1 \\ -1 & 0 & 1 \\ 1 & -1 & 0 \\ 0 & -1 & 1 \end{bmatrix}^T \begin{bmatrix} Z_1 & 0 & 0 & 0 & 0 & 0 \\ 0 & Z_2 & 0 & 0 & 0 & 0 \\ 0 & 0 & Z_3 & 0 & 0 & 0 \\ 0 & 0 & 0 & Z_4 & 0 & 0 \\ 0 & 0 & 0 & 0 & Z_5 & 0 \\ 0 & 0 & 0 & 0 & 0 & Z_6 \end{bmatrix} \begin{bmatrix} 1 & 0 & 0 \\ 0 & 1 & 0 \\ 0 & 0 & 1 \\ -1 & 0 & 1 \\ 1 & -1 & 0 \\ 0 & -1 & 1 \end{bmatrix} =$$

$$\begin{bmatrix} Z_1 + Z_4 + Z_5 & -Z_5 & -Z_4 \\ -Z_5 & Z_2 + Z_5 + Z_6 & -Z_6 \\ -Z_4 & -Z_6 & Z_3 + Z_4 + Z_6 \end{bmatrix}$$

Where the diagonal elements collect all the impedance in the mesh (union of the impedances). In the other parts we have the intersection of the impedances between two meshes. The sign is negative because the directions of the meshes are opposite one to the other. Given the points E and D in figure 6.10 we have

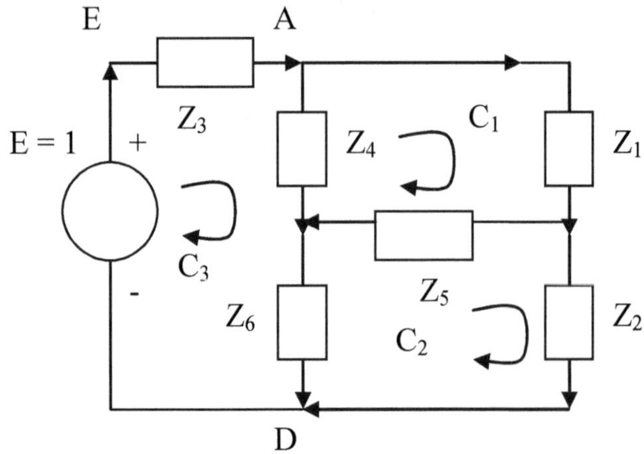

Fig. 6.8 Local voltage source and external influence

For the power

$$W = v_C^{\,T} Z_C^{-1} v_C = \begin{bmatrix} 0 \\ 0 \\ E \end{bmatrix}^T Z_C^{-1} \begin{bmatrix} 0 \\ 0 \\ E \end{bmatrix} = \begin{bmatrix} 0 \\ 0 \\ 1 \end{bmatrix}^T Z_C^{-1} \begin{bmatrix} 0 \\ 0 \\ 1 \end{bmatrix}$$

Or

$$W = \begin{bmatrix} 0 \\ 0 \\ 1 \end{bmatrix}^T \begin{bmatrix} Z_1 + Z_4 + Z_5 & -Z_5 & -Z_4 \\ -Z_5 & Z_2 + Z_5 + Z_6 & -Z_6 \\ -Z_4 & -Z_6 & Z_3 + Z_4 + Z_6 \end{bmatrix}^{-1} \begin{bmatrix} 0 \\ 0 \\ 1 \end{bmatrix}$$

Now because we have only one voltage source in the cycle C_3, and the value of the sources is equal to one, for the classical electrical circuit we have

$$W = \frac{E^2}{Z_{ED}} = \frac{1}{Z_{ED}}$$

So we can conclude that

$$\frac{1}{Z_{E,D}} = \begin{bmatrix} 0 \\ 0 \\ 1 \end{bmatrix}^T \begin{bmatrix} Z_1+Z_4+Z_5 & -Z_5 & -Z_4 \\ -Z_5 & Z_2+Z_5+Z_6 & -Z_6 \\ -Z_4 & -Z_6 & Z_3+Z_4+Z_6 \end{bmatrix}^{-1} \begin{bmatrix} 0 \\ 0 \\ 1 \end{bmatrix} =$$

$$\frac{(-1)^{3+3} M_{3,3}}{\det(G)} = \frac{1}{(G_{3,3} - \begin{bmatrix} G_{3,1} \\ G_{3,2} \end{bmatrix}^T \begin{bmatrix} G_{1,1} & G_{1,2} \\ G_{2,1} & G_{2,2} \end{bmatrix}^{-1} \begin{bmatrix} G_{1,3} \\ G_{2,3} \end{bmatrix}} =$$

$$\frac{1}{(Z_3+Z_4+Z_6 - \begin{bmatrix} -Z_4 \\ -Z_6 \end{bmatrix}^T \begin{bmatrix} Z_1+Z_4+Z_5 & -Z_5 \\ -Z_5 & Z_2+Z_5+Z_6 \end{bmatrix}^{-1} \begin{bmatrix} -Z_4 \\ -Z_6 \end{bmatrix}}$$

So we have

$$Z_{E,D} = Z_3 + Z_4 + Z_6 - \begin{bmatrix} -Z_4 \\ -Z_6 \end{bmatrix}^T \begin{bmatrix} Z_1+Z_4+Z_5 & -Z_5 \\ -Z_5 & Z_2+Z_5+Z_6 \end{bmatrix}^{-1} \begin{bmatrix} -Z_4 \\ -Z_6 \end{bmatrix}$$

That give us the impedance among the points E and D in a matrix way. We remark that

$$Z_3 + Z_4 + Z_6$$

is the impedance in the cycle C_3 and that

$$e_{3,3} = \begin{bmatrix} -Z_4 \\ -Z_6 \end{bmatrix}^T \begin{bmatrix} Z_1+Z_4+Z_5 & -Z_5 \\ -Z_5 & Z_2+Z_5+Z_6 \end{bmatrix}^{-1} \begin{bmatrix} -Z_4 \\ -Z_6 \end{bmatrix}$$

is the correction for all the other impedances connected with the cycle C_3 Now we can show that the previous result can be obtained in a more complex way by the classical electrical circuit analysis. Now the circuit in figure 6.10 can write in the following way

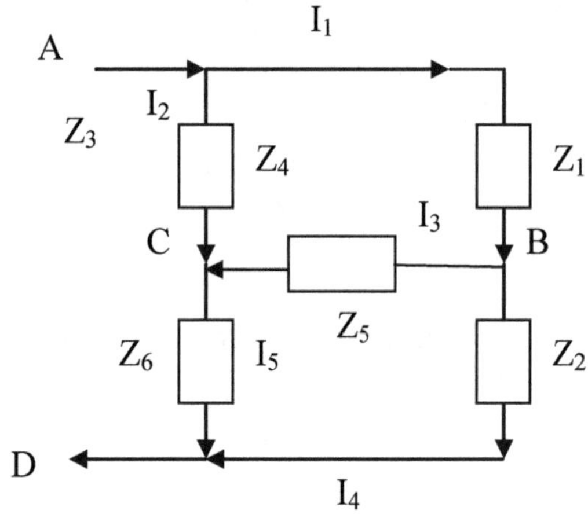

Fig. 6.9 Electrical circuit without the sources

For the circuit in figure 6.7 using the classical way we have

$$V_{B,A} = Z_1 I_1 , Z_4 I_2 = V_{C,A}$$
$$V_{D,B} = Z_2 I_4 , V_{D,C} = Z_6 I_5 , V_{B,C} = Z_5 I_3$$
$$I_4 + I_5 = I_1 + I_2 = I , I_1 - I_3 - I_4 = 0 , I_2 + I_3 - I_5 = 0$$
$$V_{A,B} + V_{B,C} - V_{C,A} = 0, V_{B,D} - V_{D,C} - V_{C,B} = 0$$
$$V_{A,B} + V_{B,D} - V_{D,C} - V_{C,A} = 0$$

So we have

$$Z_1 I_1 + Z_5 I_3 - Z_4 I_2 = 0$$
$$Z_1 I_1 + Z_2 (I_1 - I_3) - Z_6 (I_2 + I_3) - Z_4 I_2 = 0$$
$$I_1 + I_2 = I$$

Which solutions are

$$I_1 = I\frac{Z_1Z_6 + Z_4Z_2 + Z_4Z_6 + Z_4Z_5}{(Z_4Z_6 + Z_4Z_5 + Z_5Z_6 + Z_4Z_2 + Z_6Z_1 + Z_1Z_5 + Z_2Z_5 + Z_2Z_1)}$$

$$I_2 = I\frac{Z_2Z_5 + Z_2Z_1 + Z_6Z_1 + Z_1Z_5}{(Z_4Z_6 + Z_4Z_5 + Z_5Z_6 + Z_4Z_2 + Z_6Z_1 + Z_1Z_5 + Z_2Z_5 + Z_2Z_1)}$$

$$I_3 = I\frac{Z_4Z_2 - Z_6Z_1}{(Z_4Z_6 + Z_4Z_5 + Z_5Z_6 + Z_4Z_2 + Z_6Z_1 + Z_1Z_5 + Z_2Z_5 + Z_2Z_1)}$$

And

$$V_{AD} = V_{AB} + V_{BD} =$$

$$Z_1I_1 + Z_2(I_1 - I_3) = (Z_4 + Z_6 - \begin{bmatrix} -Z_4 \\ -Z_6 \end{bmatrix}^T \begin{bmatrix} Z_1 + Z_4 + Z_5 & -Z_5 \\ -Z_5 & Z_2 + Z_5 + Z_6 \end{bmatrix}^{-1} \begin{bmatrix} -Z_4 \\ -Z_6 \end{bmatrix})I$$

and

$$Z_{AD} = \frac{V_{AD}}{I} = (Z_4 + Z_6 - \begin{bmatrix} -Z_4 \\ -Z_6 \end{bmatrix}^T \begin{bmatrix} Z_1 + Z_4 + Z_5 & -Z_5 \\ -Z_5 & Z_2 + Z_5 + Z_6 \end{bmatrix}^{-1} \begin{bmatrix} -Z_4 \\ -Z_6 \end{bmatrix})$$

and

$$Z_{ED} = (Z_3 + Z_4 + Z_6 - \begin{bmatrix} -Z_4 \\ -Z_6 \end{bmatrix}^T \begin{bmatrix} Z_1 + Z_4 + Z_5 & -Z_5 \\ -Z_5 & Z_2 + Z_5 + Z_6 \end{bmatrix}^{-1} \begin{bmatrix} -Z_4 \\ -Z_6 \end{bmatrix})$$

So the matrix method and the traditional method give the same result.

When we fixed the value of the current as $I = -\frac{1}{2}$ we have

$$V_{A,D} = Z_{A,D}I = -\frac{1}{2}Z_{A,S} =$$

$$-\frac{1}{2}(Z_3 + Z_4 + Z_6 - \begin{bmatrix} -Z_4 \\ -Z_6 \end{bmatrix}^T \begin{bmatrix} Z_1 + Z_4 + Z_5 & -Z_5 \\ -Z_5 & Z_2 + Z_5 + Z_6 \end{bmatrix}^{-1} \begin{bmatrix} -Z_4 \\ -Z_6 \end{bmatrix}$$

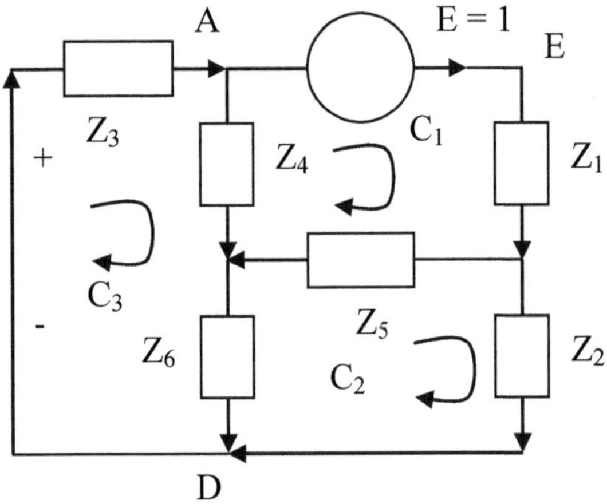

Fig. 6.10 Change of the voltage source position inside the electrical circuit

For which we have

$$W = \begin{bmatrix} 1 \\ 0 \\ 0 \end{bmatrix}^T \begin{bmatrix} Z_1 + Z_4 + Z_5 & -Z_5 & -Z_4 \\ -Z_5 & Z_2 + Z_5 + Z_6 & -Z_6 \\ -Z_4 & -Z_6 & Z_3 + Z_4 + Z_6 \end{bmatrix}^{-1} \begin{bmatrix} 1 \\ 0 \\ 0 \end{bmatrix}$$

And

$$Z_{E,A} = (Z_1 + Z_4 + Z_5 - \begin{bmatrix} -Z_5 \\ -Z_4 \end{bmatrix}^T \begin{bmatrix} Z_2 + Z_5 + Z_6 & -Z_6 \\ -Z_6 & Z_3 + Z_4 + Z_6 \end{bmatrix}^{-1} \begin{bmatrix} -Z_5 \\ -Z_4 \end{bmatrix}$$

And for

$$W = \begin{bmatrix} 0 \\ 1 \\ 0 \end{bmatrix}^T \begin{bmatrix} Z_1 + Z_4 + Z_5 & -Z_5 & -Z_4 \\ -Z_5 & Z_2 + Z_5 + Z_6 & -Z_6 \\ -Z_4 & -Z_6 & Z_3 + Z_4 + Z_6 \end{bmatrix}^{-1} \begin{bmatrix} 0 \\ 1 \\ 0 \end{bmatrix}$$

And

$$Z_{E,D} = (Z_2 + Z_5 + Z_6 - \begin{bmatrix} -Z_5 \\ -Z_6 \end{bmatrix}^T \begin{bmatrix} Z_1 + Z_4 + Z_5 & -Z_4 \\ -Z_4 & Z_3 + Z_4 + Z_6 \end{bmatrix}^{-1} \begin{bmatrix} -Z_5 \\ -Z_6 \end{bmatrix}$$

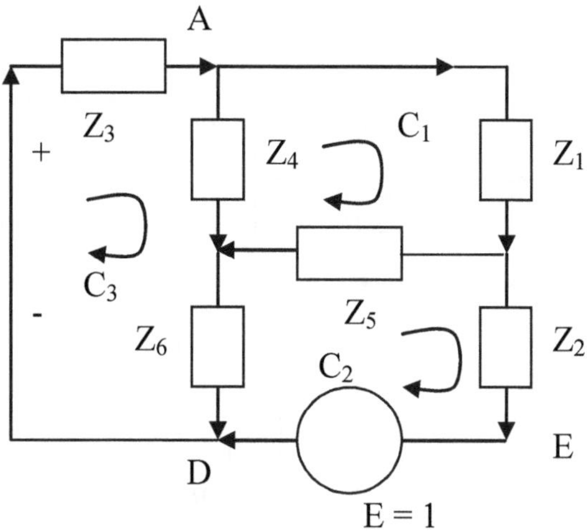

Fig. 6.11 Another position of the voltage source inside the same circuit

The symmetry principle consist in the possibility from the source of voltage to compute the currents and by the currents came back to the original sources of the voltage. In a formal way we have

The dual is

$$v = H_N (H^T{}_N C H_N)^{-1} H^T{}_N i = C^{-1} Q_N i \tag{6.12}$$

Where

$$i = H_C I \ \ and \ \ v = H_N E \tag{6.13}$$

and E are the sources of the voltages and I are the sources of the currents.

The relation between the sources of the current and the sources of the voltage is

$$(H^T{}_C Z H_C) I = E \tag{6.14}$$

6.4 Transistor and Amplifier by Morphogenetic System

In this chapter we present the morphognetic system for electronic devise

Now for the transistor in figure 6.13

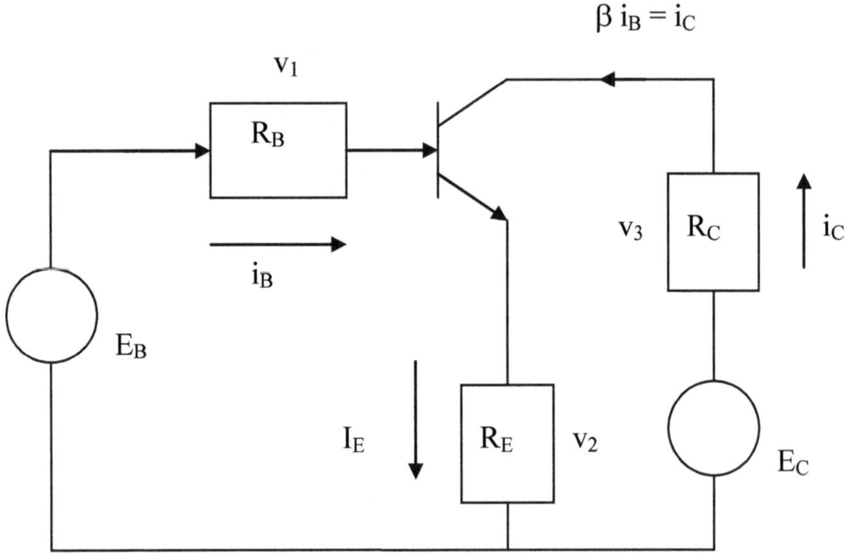

Fig. 6.12 Transistor with the sources of potential

The graph that represent the transistor is

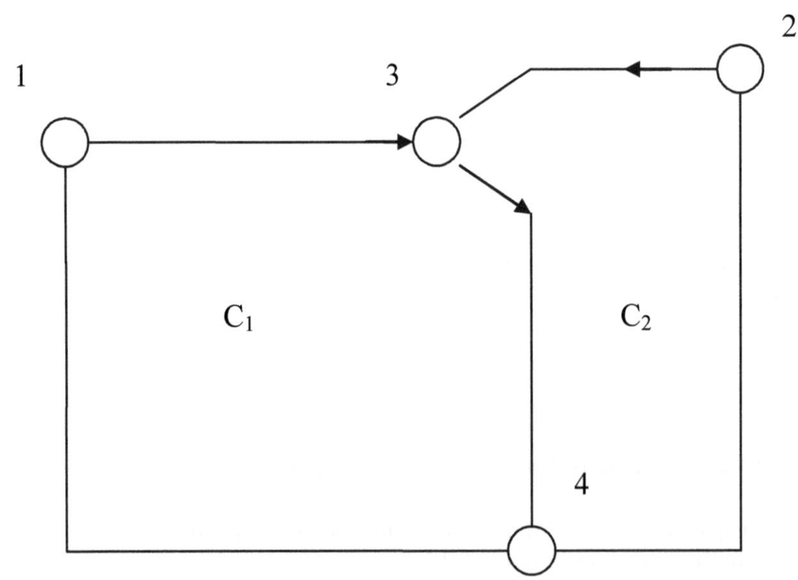

Fig. 6.13 Graph of the circuit in figure 6.13

The Matrix of the Cycles is

$$H_{Cycle} = H_C = \begin{bmatrix} & C_1 & C_2 \\ e_{13} & 1 & 0 \\ e_{34} & 1 & 1 \\ e_{23} & 0 & 1 \end{bmatrix}$$

Now for the (6.4) we have

$$I = (H^T{}_C Z H_C)^{-1} H_C^T v = (\begin{bmatrix} 1 & 0 \\ 1 & 1 \\ 0 & 1 \end{bmatrix}^T \begin{bmatrix} R_B & 0 & 0 \\ 0 & R_E & 0 \\ 0 & 0 & R_C \end{bmatrix} \begin{bmatrix} 1 & 0 \\ 1 & 1 \\ 0 & 1 \end{bmatrix})^{-1} \begin{bmatrix} 1 & 0 \\ 1 & 1 \\ 0 & 1 \end{bmatrix}^T \begin{bmatrix} v_1 \\ v_2 \\ v_3 \end{bmatrix} =$$

$$(\begin{bmatrix} 1 & 0 \\ 1 & 1 \\ 0 & 1 \end{bmatrix}^T \begin{bmatrix} R_B & 0 & 0 \\ 0 & R_E & 0 \\ 0 & 0 & R_C \end{bmatrix} \begin{bmatrix} 1 & 0 \\ 1 & 1 \\ 0 & 1 \end{bmatrix})^{-1} \begin{bmatrix} v_1 + v_2 \\ v_2 + v_3 \end{bmatrix} = (\begin{bmatrix} 1 & 0 \\ 1 & 1 \\ 0 & 1 \end{bmatrix}^T \begin{bmatrix} R_B & 0 & 0 \\ 0 & R_E & 0 \\ 0 & 0 & R_C \end{bmatrix} \begin{bmatrix} 1 & 0 \\ 1 & 1 \\ 0 & 1 \end{bmatrix})^{-1} \begin{bmatrix} E_B \\ E_C \end{bmatrix} =$$

$$(\begin{bmatrix} 1 & 0 \\ 1 & 1 \\ 0 & 1 \end{bmatrix}^T \begin{bmatrix} R_B & 0 & 0 \\ 0 & R_E & 0 \\ 0 & 0 & R_C \end{bmatrix} \begin{bmatrix} 1 & 0 \\ 1 & 1 \\ 0 & 1 \end{bmatrix})^{-1} \begin{bmatrix} E_B \\ \eta E_B \end{bmatrix} = \begin{bmatrix} I_B \\ \beta I_B \end{bmatrix}$$

So we have

$$J = (\begin{bmatrix} 1 & 0 \\ 1 & 1 \\ 0 & 1 \end{bmatrix}^T \begin{bmatrix} R_B & 0 & 0 \\ 0 & R_E & 0 \\ 0 & 0 & R_C \end{bmatrix} \begin{bmatrix} 1 & 0 \\ 1 & 1 \\ 0 & 1 \end{bmatrix})^{-1} \begin{bmatrix} E_B \\ E_C \end{bmatrix} = \begin{bmatrix} I_B \\ \beta I_B \end{bmatrix}$$

The voltages and the currents are

$$i = \begin{bmatrix} i_B \\ i_E \\ i_C \end{bmatrix} = \begin{bmatrix} 1 & 0 \\ 1 & 1 \\ 0 & 1 \end{bmatrix} I = \begin{bmatrix} 1 & 0 \\ 1 & 1 \\ 0 & 1 \end{bmatrix} \begin{bmatrix} I_B \\ I_C \end{bmatrix} = \begin{bmatrix} I_B \\ I_B + I_C \\ I_C \end{bmatrix}$$

$$v = \begin{bmatrix} v_1 \\ v_2 \\ v_3 \end{bmatrix} = \begin{bmatrix} R_B & 0 & 0 \\ 0 & R_E & 0 \\ 0 & 0 & R_C \end{bmatrix} \begin{bmatrix} I_B \\ I_B + I_C \\ I_C \end{bmatrix} = \begin{bmatrix} R_B I_B \\ R_E (I_B + I_C) \\ R_C I_C \end{bmatrix}$$

So we have

$$\begin{bmatrix} E_B \\ \eta E_B \end{bmatrix} = (H^T{}_C Z H_C) I = (\begin{bmatrix} 1 & 0 \\ 1 & 1 \\ 0 & 1 \end{bmatrix}^T \begin{bmatrix} R_B & 0 & 0 \\ 0 & R_E & 0 \\ 0 & 0 & R_C \end{bmatrix} \begin{bmatrix} 1 & 0 \\ 1 & 1 \\ 0 & 1 \end{bmatrix}) \begin{bmatrix} I_B \\ \beta I_B \end{bmatrix} =$$

$$\begin{bmatrix} R_E(I_B + \beta I_B) + R_B I_B \\ R_E(I_B + \beta I_B) + R_C \beta I_B \end{bmatrix} = \begin{bmatrix} E_B \\ \eta E_B \end{bmatrix}$$

For the first equation we have

$$R_E(I_B + \beta I_B) + R_B I_B = E_B$$

and

$$I_B = \frac{E_B}{R_E(1+\beta) + R_B}$$

Because the power is

$$P = I^T G I = \begin{bmatrix} I_B \\ \beta I_B \end{bmatrix}^T \begin{bmatrix} 1 & 0 \\ 1 & 1 \\ 0 & 1 \end{bmatrix}^T \begin{bmatrix} R_B & 0 & 0 \\ 0 & R_E & 0 \\ 0 & 0 & R_C \end{bmatrix} \begin{bmatrix} 1 & 0 \\ 1 & 1 \\ 0 & 1 \end{bmatrix} \begin{bmatrix} I_B \\ \beta I_B \end{bmatrix} = (R_E + R_B) I_B^2 + (R_E + R_C) I_B^2 \beta^2 + 2 R_E I_B \beta$$

The power can be obtained also in this way (invariant by covariant and contravariant component of the vector of potential.

$$P = \begin{bmatrix} I_B \\ I_C \end{bmatrix}^T \begin{bmatrix} E_B \\ E_C \end{bmatrix} = E_B I_B + E_C I_C$$

we have that P is invariant for the transformation

$$UE_1 \text{ and } UE_2 \quad \text{with } U^T U = I$$

In fact we have

$$P' = (\begin{bmatrix} u_{1,1} & u_{1,2} \\ u_{2,1} & u_{2,2} \end{bmatrix} \begin{bmatrix} I_B \\ I_C \end{bmatrix})^T \begin{bmatrix} u_{1,1} & u_{1,2} \\ u_{2,1} & u_{2,2} \end{bmatrix} \begin{bmatrix} E_B \\ E_C \end{bmatrix} =$$

$$\begin{bmatrix} I_B \\ I_C \end{bmatrix}^T \begin{bmatrix} u_{1,1} & u_{1,2} \\ u_{2,1} & u_{2,2} \end{bmatrix}^T \begin{bmatrix} u_{1,1} & u_{1,2} \\ u_{2,1} & u_{2,2} \end{bmatrix} \begin{bmatrix} E_B \\ E_C \end{bmatrix} =$$

$$\begin{bmatrix} I_B \\ I_C \end{bmatrix}^T \begin{bmatrix} 1 & 0 \\ 0 & 1 \end{bmatrix} \begin{bmatrix} E_B \\ E_C \end{bmatrix} = P$$

So to have the power invariant we must have the transformations

$$\begin{bmatrix} I'_B \\ I'_C \end{bmatrix} = \begin{bmatrix} u_{1,1} & u_{1,2} \\ u_{2,1} & u_{2,2} \end{bmatrix} \begin{bmatrix} I_B \\ I_C \end{bmatrix} = \begin{bmatrix} u_{1,1} I_B + u_{1,2} I_C \\ u_{2,1} I_B + u_{2,2} I_C \end{bmatrix}$$

$$\begin{bmatrix} E'_B \\ E_C' \end{bmatrix} = \begin{bmatrix} u_{1,1} & u_{1,2} \\ u_{2,1} & u_{2,2} \end{bmatrix} \begin{bmatrix} E_B \\ E_C \end{bmatrix} = \begin{bmatrix} u_{1,1} E_B + u_{1,2} E_C \\ u_{2,1} E_B + u_{2,2} E_C \end{bmatrix}$$

Now because

$$I_C = \beta I_B \text{ and } I'_C = \beta' I_B'$$

With

$$I_B = \frac{E_B}{R_E(1+\beta)+R_B} \text{ and } I_B' = \frac{E_B'}{R_E'(1+\beta')+R_B'}$$

We have

$$I'_C = u_{2,1} I_B + u_{2,2} I_C = u_{2,1} I_B + u_{2,2} \beta I_B = \beta' I'_B =$$

$$\beta'(u_{1,1} I_B + u_{1,2} \beta I_B) = \beta' \frac{u_{1,1} E_B + u_{1,2} E_C}{R_B' + (1+\beta')R_E'}$$

and

$$u_{1,1} I_B + u_{1,2} \beta I_B = \frac{u_{1,1} E_B + u_{1,2} E_C}{R_B' + (1+\beta')R_E'}$$

$$u_{2,1} I_B + u_{2,2} \beta I_B = \beta' \frac{u_{1,1} E_B + u_{1,2} E_C}{R_B' + (1+\beta')R_E'}$$

Which solutions are

$$\begin{bmatrix} \beta' \\ R_B' \end{bmatrix} = \begin{bmatrix} \dfrac{u_{21} + u_{22}\beta}{u_{11} + u_{12}\beta} \\ \Gamma R_B - (\Gamma(1+\beta) + \dfrac{u_{21} + u_{22}\beta}{u_{11} + u_{12}\beta} + 1) R_E \end{bmatrix}$$

$$\Gamma = \frac{E_B'}{E_B(u_{11} + u_{12}\beta)}$$

$$E_B' = u_{11} + u_{12} E_B$$

For the second equation we have

$$R_E(I_B+\beta I_B)+R_C\beta I_B = E_C = \eta E_B$$

For the computation of the currents we have

$$\eta = \frac{R_E(1+\beta)+\beta R_C}{R_E(1+\beta)+R_B}$$

Now for the transistor property we have

$$v_2 + v_3 = \eta E_B$$
$$v_2 = \eta E_B - v_3 = \eta E_B - R_C i_C = V_{CC} - V_{CC} + \eta E_B - R_C i = V_{CC} - R_C i + (-V_{CC} + \eta E_B)$$

Now for

$$E_{EXT} = V_{CC}, E_{INT} = (\eta E_B - V_{CC})$$
$$v_2 = V_{CC} - R_C I_C + (-V_{CC} + \eta E_B) = E_{EXT} - R_C I_C + E_{INT}$$

In conclusion we obtain the circuit

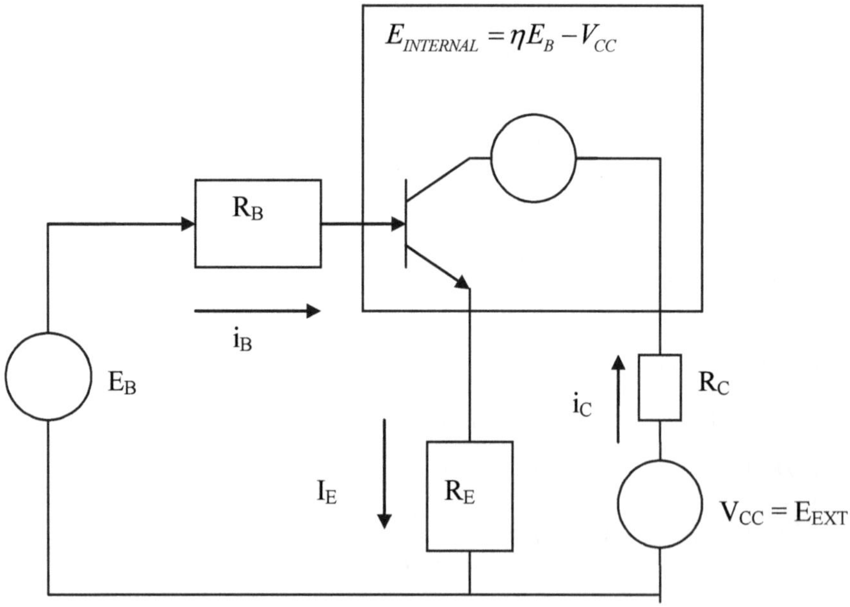

Fig. 6.14 External generators E_{EXT} and internal generator E_{IN}

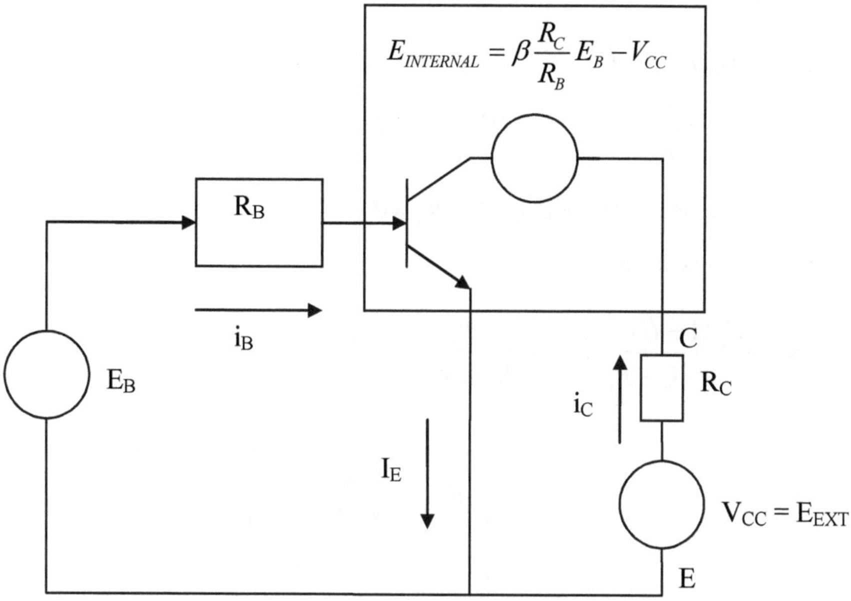

Fig. 6.15 Common emitter and generators at the collector

When $R_E = 0$ or common emitter we have

$$\eta = \beta \frac{R_C}{R_B}, I_B = \frac{E_B}{R_B}$$

Now we remark that

$$E = \begin{bmatrix} 1 & 0 \\ 1 & 1 \\ 0 & 1 \end{bmatrix}^T \begin{bmatrix} v_1 \\ 0 \\ v_3 \end{bmatrix} = \begin{bmatrix} v_1 \\ v_3 \end{bmatrix} = \begin{bmatrix} E_B \\ E_C \end{bmatrix} = \begin{bmatrix} E_B \\ \eta E_B \end{bmatrix}$$

So we obtain

$$E_C = v_3 = R_C I_C$$

In fact we have

$$v_3 = R_C I_C = R_C \beta I_B = R_C \beta \frac{E_B}{R_B}$$

$$E_C = \eta E_B = \beta \frac{R_C}{R_B} E_B$$

Now we have

$$E_{EXT} + E_{INT} = R_C I_C$$

$$E_{EXT} - R_C \beta \frac{E_B}{R_B} = V_{CC} - R_C \beta \frac{E_B}{R_B} = -E_{INT}$$

1) Common collector amplifier $R_C = 0$

In this case we have

$$\eta = \frac{R_E(1+\beta)}{R_E(1+\beta)+R_B}, I_B = \frac{E_B}{R_E(1+\beta)}$$

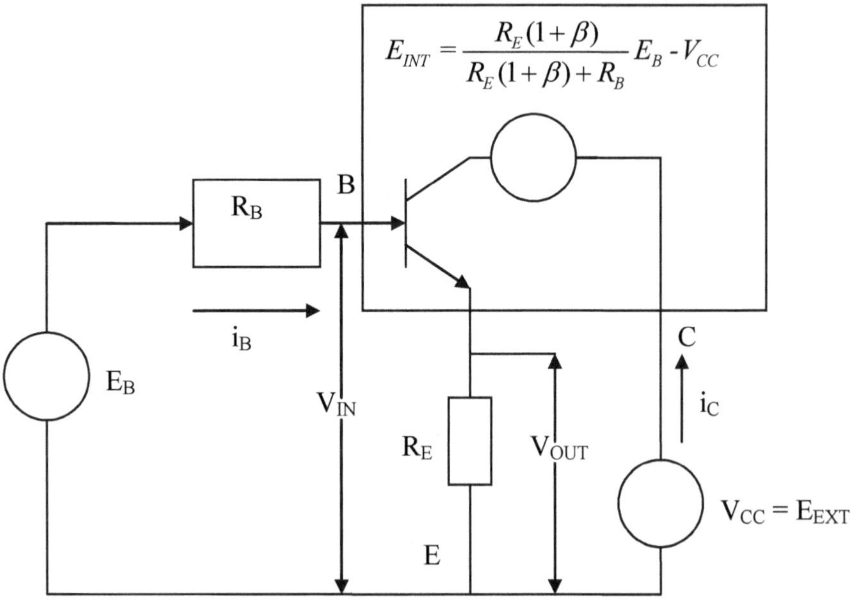

Fig. 6.16 Common collector transistor

and

$$V_{OUT} = R_E(I_B + I_C) = R_E(I_B + \beta I_B) = R_E(1+\beta)I_B =$$
$$R_E(1+\beta)\frac{E_B}{R_E(1+\beta)+R_B} = R_E(1+\beta)\frac{V_{IN}}{R_E(1+\beta)+R_B}$$
$$\frac{V_{OUY}}{V_{IN}} = R_E(1+\beta)\frac{1}{R_E(1+\beta)+R_B}$$

When $R_B = 0$, we have

$$\frac{V_{OUY}}{V_{IN}} = R_E(1+\beta)\frac{1}{R_E(1+\beta)} = 1$$

Now for the differential amplifier

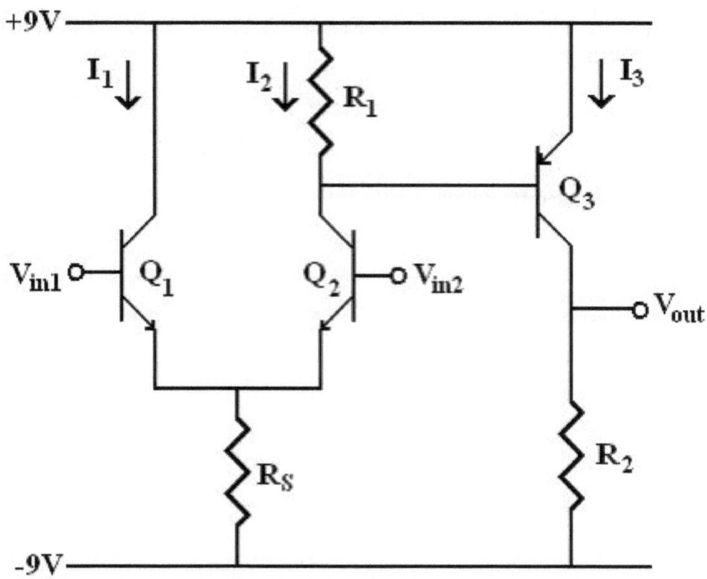

Fig. 6.17 Differential amplifier

We have the circuit

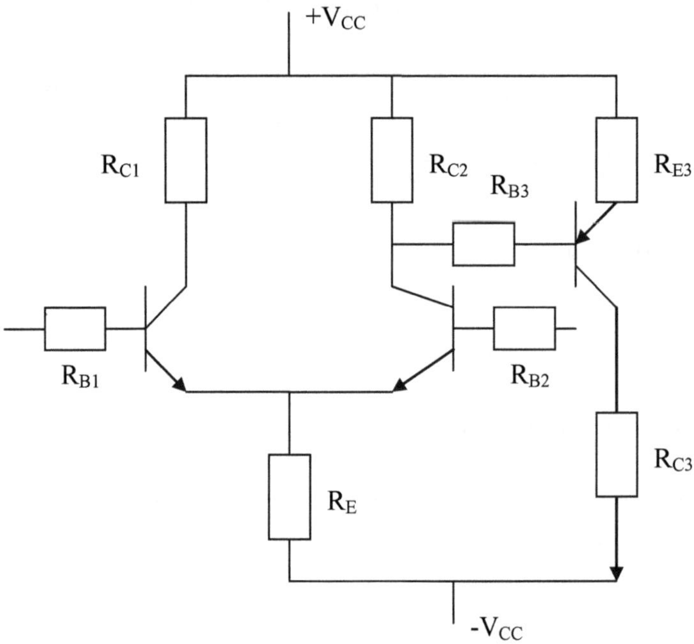

Fig. 6.18 Scheme of the differential amplifier

Now we study the mirror set of transistors

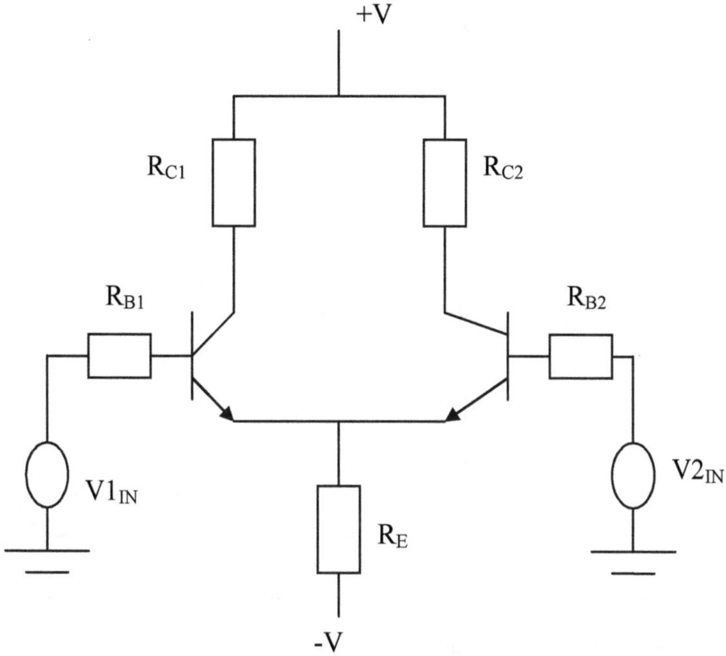

Fig. 6.19 Scheme of the principal part of the differential amplifier

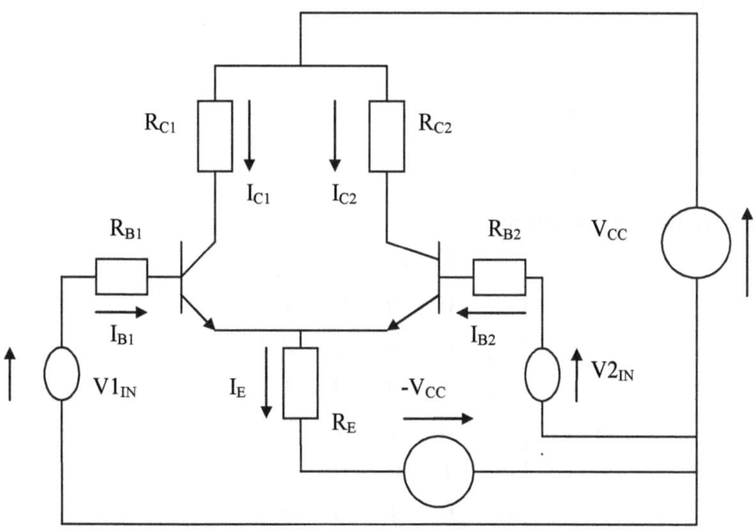

Fig. 6.20 Specification of the differential amplifier

Cycle circuit

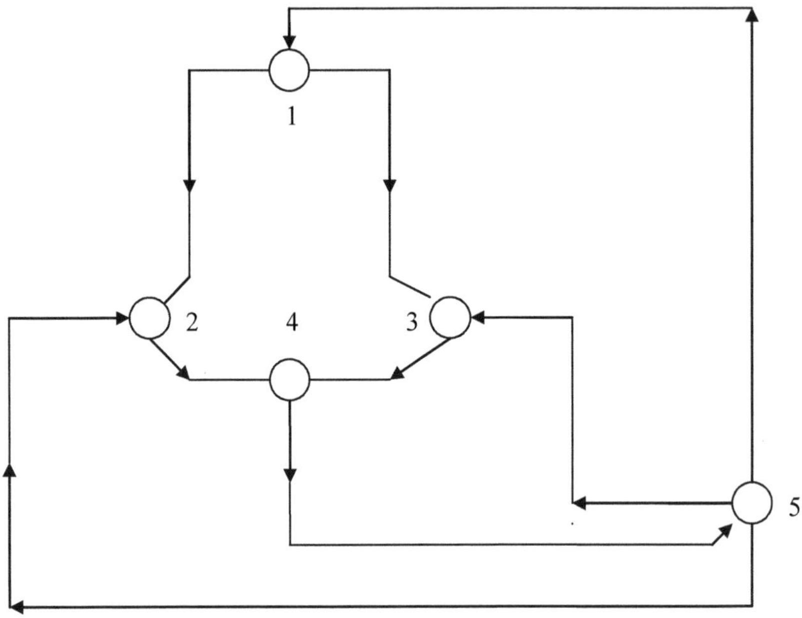

Fig. 6.21 Graph of the differential amplifier

Now we have four Cycles

$C_1 = (5 , 2 ,4 ,5)$

$C_2 = (3 , 4 ,5 ,3)$

$C_3 = (1 , 3 ,4 ,5 , 1)$

$C_4 = (1 , 2 ,4 ,5 , 1)$

So we have the context matrix

$$H_{Cycle} = \begin{bmatrix} & C_1 & C_2 & C_3 & C_4 \\ e_{12} & 0 & 0 & 0 & 1 \\ e_{13} & 0 & 0 & 1 & 0 \\ e_{24} & 1 & 0 & 0 & 1 \\ e_{34} & 0 & 1 & 1 & 0 \\ e_{52} & 1 & 0 & 0 & 0 \\ e_{53} & 0 & 1 & 0 & 0 \\ e_{45} & 1 & 1 & 1 & 1 \\ e_{51} & 0 & 0 & 1 & 1 \end{bmatrix} \tag{6.15}$$

With the H_{Cycle} we can compute the voltage sources

$$E = H^T{}_{Cycle} v = \begin{bmatrix} 0 & 0 & 0 & 1 \\ 0 & 0 & 1 & 0 \\ 1 & 0 & 0 & 1 \\ 0 & 1 & 1 & 0 \\ 1 & 0 & 0 & 0 \\ 0 & 1 & 0 & 0 \\ 1 & 1 & 1 & 1 \\ 0 & 0 & 1 & 1 \end{bmatrix}^T \begin{bmatrix} v_{12} \\ v_{13} \\ v_{24} \\ v_{34} \\ v_{52} \\ v_{53} \\ v_{45} \\ v_{51} \end{bmatrix} = \begin{bmatrix} v_{24} + v_{52} + v_{45} \\ v_{34} + v_{53} + v_{45} \\ v_{13} + v_{34} + v_{45} + v_{51} \\ v_{12} + v_{24} + v_{45} + v_{51} \end{bmatrix} = \begin{bmatrix} E_{B1} \\ E_{B2} \\ E_{C2} \\ E_{C1} \end{bmatrix} = \begin{bmatrix} E_{B1} \\ E_{B2} \\ \eta E_{B2} \\ \eta E_{B1} \end{bmatrix} \tag{6.16}$$

Now for the theory we have

$$I = \begin{bmatrix} I_{B1} \\ I_{B2} \\ \beta_2 I_{B2} \\ \beta_1 I_{B1} \end{bmatrix} = (H^T{}_{Cycle} Z H_{Cycle})^{-1} \begin{bmatrix} E_{B1} \\ E_{B2} \\ \eta_2 E_{B2} \\ \eta_1 E_{B1} \end{bmatrix} \tag{6.17}$$

Where

$$Z = \begin{bmatrix} R_{12} & 0 & 0 & 0 & 0 & 0 & 0 & 0 \\ 0 & R_{13} & 0 & 0 & 0 & 0 & 0 & 0 \\ 0 & 0 & R_{24} & 0 & 0 & 0 & 0 & 0 \\ 0 & 0 & 0 & R_{34} & 0 & 0 & 0 & 0 \\ 0 & 0 & 0 & 0 & R_{52} & 0 & 0 & 0 \\ 0 & 0 & 0 & 0 & 0 & R_{53} & 0 & 0 \\ 0 & 0 & 0 & 0 & 0 & 0 & R_{45} & 0 \\ 0 & 0 & 0 & 0 & 0 & 0 & 0 & R_{51} \end{bmatrix} \tag{6.18}$$

For the (6.31) we can compute the currents

$$\begin{bmatrix} I_{B1} \\ I_{B2} \end{bmatrix} = \frac{1}{D} \begin{bmatrix} (R_{53} + R_{45}(1+\beta_2))E_{B1} - R_{45}(1+\beta_2)E_{B2} \\ (R_{52} + R_{45}(1+\beta_1))E_{B2} - R_{45}(1+\beta_1)E_{B1} \end{bmatrix} \tag{6.19}$$

where

$$D = R_{53}R_{52} + R_{53}R_{45}(1+\beta_1) + R_{52}R_{45}(1+\beta_2) \tag{6.20}$$

So at the collector we have the currents

$$\begin{bmatrix} I_{C1} \\ I_{C2} \end{bmatrix} = \begin{bmatrix} \beta_1 I_{B1} \\ \beta_2 I_{B2} \end{bmatrix} = \frac{1}{D}\begin{bmatrix} \beta_1((R_{53} + R_{45}(1+\beta_2))E_{B1} - R_{45}(1+\beta_2)E_{B2}) \\ \beta_2((R_{52} + R_{45}(1+\beta_1))E_{B2} - R_{45}(1+\beta_1)E_{B1}) \end{bmatrix} \tag{6.21}$$

For the particular case

$R_{52} = R_{53} = R_B$, $\beta_1 = \beta_2$, $R_{45} = R_E >> R_B$

We have

$$\begin{bmatrix} I_{B1} \\ I_{B2} \end{bmatrix} \simeq \frac{R_E(1+\beta)}{2R_E R_B(1+\beta)}\begin{bmatrix} E_{B1} - E_{B2} \\ E_{B2} - E_{B1} \end{bmatrix} = \frac{1}{2R_B}\begin{bmatrix} E_{B1} - E_{B2} \\ E_{B2} - E_{B1} \end{bmatrix}$$

So

$$\begin{bmatrix} I_{C1} \\ I_{C2} \end{bmatrix} = \begin{bmatrix} \beta I_{B1} \\ \beta I_{B2} \end{bmatrix} \simeq \frac{\beta}{2R_B}\begin{bmatrix} E_{B1} - E_{B2} \\ E_{B2} - E_{B1} \end{bmatrix}$$

$$I_{C1} + I_{C2} = 0$$

When we increase I_{C1} the current I_{C2} decrease and vice verse. For the relation between the collector currents, the two currents cannot decrease or increase together, so the system amplify only the difference between the base currents.

The voltage sources at the collectors are

$$\begin{bmatrix} E_{C1} \\ E_{C2} \end{bmatrix} = \begin{bmatrix} \eta_1 E_{B1} \\ \eta_2 E_{B2} \end{bmatrix} = \begin{bmatrix} (R_{45}(1+\beta_1) + \beta_1 R_{12})I_{B1} + (R_{45}(1+\beta_2) + \beta_2 R_{51})I_{B2} \\ R_{45}(1+\beta_1)I_{B1} + (R_{45}(1+\beta_2) + \beta_2(R_{13} + R_{51})I_{B2} \end{bmatrix} \tag{6.22}$$

6.5 Discussion

We have shown an approach to analyzing electrical circuit using some of the quantum mechanics principles - superposition and projection operator. Recently electrical circuits have been used to solve different problems in other domains. We list some of them in this section. Physical particles can be simulated by electrical circuits using the following analogy:

Position x of the particle	represented by potential V
Momentum p	represented by current i
Proper time	represented by resistance
Action S	represented by power
External force and momentum	represented by voltage and current sources

In field theory we have the following analogy:

Gradient	represented by the incidence matrix H in the current projection
Divergence	represented by the transpose of the incidence matrix H^T
Laplacian	represented by $H^T H$
Extension of Laplacian in general coordinates (Beltrami)	represented by $G = H^T C H$

Electrical circuits have been used to simulate potential theory in non homogeneous media and random walks, wave equation, optics and wave field control, tensor calculus in general coordinates, neural network processes, and complex biolochemical systems.

In conclusion the new approach to electrical circuit analysis opens the possibility to create a form of computing based on the projection operator extensively used in quantum mechanics.

References

1. Zupanovic, P., Juretic, D.: The chemical Cycle Kinetics close to the Equilibrium State and Elecrrical Circuit Analogy. Croatia Chemical Acta CCACAA 77(4), 561–571 (2004)
2. Perus, M., Bischof, H., John Caulfield, H., Loo, C.K.: Quantum – implementable selective reconstruction of high-resolution images. Applied Optics 43(33) (November 20, 2004)
3. Chaczko, Z.: Autopoietics of Biomimetic Middleware System. Private Correspondence (November 2007)
4. Bruers, S.: Classification and discussion of macroscopic entropy production principles, arXiv: cond-mat0604482v3 [cond-mat.stat-mech] (May 2, 2007)
5. Bejan, A.: Shape and Structure, from Engineering to Nature. Cambridge University Press, Cambridge (2000)
6. Newman, M.E.J.: The Structure and Function of Complex Networks. Santa Fe Institute Publication (2004)
7. Tarakanov, A.O., Skormin, V.A., Sokolova, S.S.: Immunocomputing: Principles and Applications. Springer, Heidelberg (2003)
8. Resconi, G., Nikravesh, M.: Morphic Computing: Concepts and Foundation. In: Nikravesh, M., Zadeh, L.A., Kacprzyk, J. (eds.) Forging the New Frontieres: Fuzzy Pioneers I. Studies in Fuzziness and Soft Computing. Springer, Heidelberg (2007)

9. Resconi, G., Nikravesh, M.: Morphic Computing: Quantum and Field. In: Nikravesh, M., Zadeh, L.A., Kacprzyk, J. (eds.) Forging the New Frontieres: Fuzzy Pioneers II. Studies in Fuzziness and Soft Computing. Springer, Heidelberg (2007)
10. Resconi, G., Nikravesh, M.: Morphic Computing Applied Soft Computing Journal (July 2007)
11. Resconi, G., Nikravesh, M.: Morphic Computing part 1 Foundation. In: IFSA 2007 World Congress Cancun, Mexico, June 18-21 (2007)
12. Resconi, G.: Modelling Fuzzy Cognitive Map By Electrical and Chemical Equivalent Circuits. In: Joint Conference on Information Science, Salt lake City Center USA, July 8-24 (2007)
13. Resconi, G.: The Morphogenetic Systems in Risk Analysis. In: Proceeding of the 1st International Conference on Risk Analysis and Crisis Response, Shangai China, September 25-26, pp. 161–165 (2007)
14. Jaeger, R.C.: Microelectronic Circuit Deign. The McGraw-Hill Companies Inc., New York (1997)
15. Johnson, D.E., Johnson, J.R., Hilburn, J.L.: Electric Circuit Analyais. Prentice Hall, Englewood Cliffs (1989)
16. Resconi, G., Srini, V.P.: Electrical Circuit as a Morphogenetic System. GESTS International Journal Trans. Computer Science and Engineering 53(1), 47–92

Chapter 7
Superposition and Geometry for Evidence and Quantum Mechanics in the Tensor Calculus

7.1 Introduction

In this chapter we prove that the interference in Coherent Quantum Mechanics is represented by a deformed space of the intensity for different particle beams. In the interference the complex number representation of the quantum mechanics is substituted by general real coordinates where the angles between general coordinates are the difference of the phases. To reformulate the traditional quantum model, we use the evidence theory and its geometric image. The evidence theory defines a non additive probability denoted basic probability assignment. We assume that in quantum mechanics for interference and entanglement phenomena, the probability is not the traditional probability but is the basic probability assignment in the evidence theory.

Operation in quantum mechanics that represent Energy or Density Matrix or momentum are considered as operators that change the geometry of the basic probabilities associates with the states in quantum mechanics. We reformulate the Feynman Path Integral in quantum mechanics and also the Entanglement Definition.

7.2 Evidence Theory and Geometry

Evidence Theory or Dempster Shafer Theory is based on the definition of a non additive probability measure denoted *basic probability assignment,* it is defined as follows.

Given a set X and the power set P(X) in which the elements are all possible subsets of X are contained , we define a non additive measure m

$$m : P(X) \rightarrow [\, 0 \,, 1 \,] \tag{7.1}$$

for which we have

$$\sum_{A \subseteq X} m(A) = 1 \tag{7.2}$$

G. Resconi: Geometry of Knowledge for Intelligent Systems, SCI 407, pp. 207–228.
springerlink.com

The non additive measure m is denoted as a *basic probability assignment* Shafer [4]

For example given the set X = { a , b , c } then the set of all the subsets or the power set is

$$P(X) = \{ \varnothing , \{ a \} , \{ b \} , \{ c \}, \{ a , b \} , \{ a , c \} , \{b , c \} , \{ a , b , c \}\}$$

where for example we define the *basic probability assignment*

$$\begin{gathered} m(\varnothing) = 0 , \\ m(\{ a \}) = 0 , \\ m(\{ b \}) = 0 , \\ m(\{ c \}) = 0 , \end{gathered}$$

$$\begin{gathered} m(\{ a , b \}) = 0.25 , \\ m(\{ a , c \}) = 0.25 , \\ m(\{ b , c \}) = 0.25 , \\ m(\{ a , b , c \}) = 0.25 \end{gathered}$$

The non additive property can be easily shown. We have

$$m(\{ a , b \}) > m(\{ a \}) + m(\{ b \})$$

where

$$\{ a , b \} = \{ a \} \cup \{ b \}$$

and

$$\{ a \} \cap \{ b \} = \varnothing$$

For the case

$$\begin{gathered} m(\varnothing) = 0 , \\ m(\{ a \}) = 0.25 , \\ m(\{ b \}) = 0.25 , \\ m(\{ c \}) = 0.25 , \\ m(\{ a , b \}) = 0 , \\ m(\{ a , c \}) = 0 , \\ m(\{ b , c \}) = 0 , \\ m(\{ a , b , c \}) = 0.25 \end{gathered}$$

we also have a property similar to that Destructive Interference in quantum mechanics for which

$$m(\{ a , b \}) < m(\{ a \}) + m(\{ b \})$$

Associated with the basic probability assignment we have a further two measures. The Belief Measure and the Plausibility Measure defined in this way

$$Bel(A) = \sum_{B \subseteq A} m(B) \ , \ Pl(A) = \sum_{B \cap A \neq \varnothing} m(B) \tag{7.3}$$

Note that in the case for which the basic probability assignment is different from zero it is only for the singleton that is with only one element we have

$$Bel(A) = Pl(A) = P(A)$$

Where P(A) is the traditional additive probability. For example we have for

$$\begin{gathered} m(\varnothing) = 0 , \\ m(\{a\}) = 1/3 , \\ m(\{b\}) = 1/3 , \\ m(\{c\}) = 1/3 , \\ m(\{a, b\}) = 0 , \\ m(\{a, c\}) = 0 , \\ m(\{b, c\}) = 0 , \\ m(\{a, b, c\}) = 0 \end{gathered}$$

we have

$$\begin{gathered} Bel(\{a, b\}) = m(\{a\}) + m(\{b\}) = \\ P(\{a\}) + P(\{b\}) = \\ P(\{a, b\}) \end{gathered}$$

7.3 Geometric Interpretation of the Evidence Theory

Given the data in Table 7.1

Table 7.1 Context H(n,p) of Feature H_k , Objects values $\psi_{i,j}$ in the table

Context H(n,p)	H_1	H_2	....	H_p
Object 1	ψ_{11}	ψ_{12}	....	ψ_{1n}
Object 2	ψ_{21}	ψ_{22}	....	ψ_{2n}
....	....	....	....	......
Object n	ψ_{n1}	ψ_{n2}	....	ψ_{np}

For three objects and two states we have the table Example

Table 7.2 Context H (3,3) of Three Features and three objects

Context H(3,3)	H_1	H_2	X_3
Object 1	ψ_{11}	ψ_{12}	ψ_{13}
Object 2	ψ_{21}	ψ_{22}	ψ_{23}
Object 3	ψ_{31}	ψ_{32}	ψ_{33}

The geometric image of table 7.2

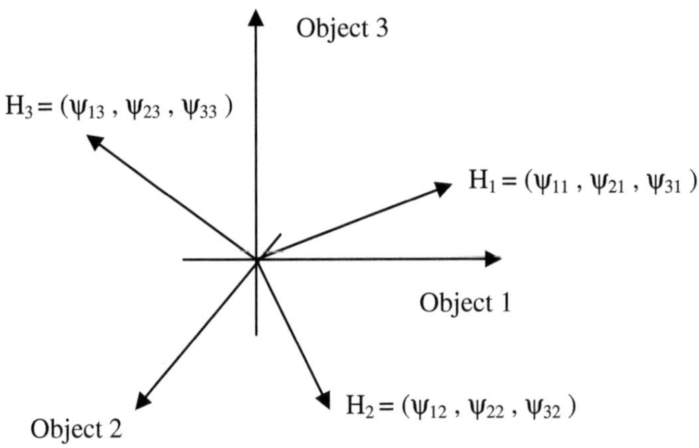

Fig. 7.1 Geometric image of the table 7.2

The table 7.1 can be represented with the matrix H in this way

$$H = \left\{ \begin{bmatrix} \psi_{1,1} \\ \psi_{2,1} \\ \dots \\ \psi_{n,1} \end{bmatrix}, \begin{bmatrix} \psi_{1,2} \\ \psi_{2,2} \\ \dots \\ \psi_{n,2} \end{bmatrix}, \dots\dots, \begin{bmatrix} \psi_{1,p} \\ \psi_{2,p} \\ \dots \\ \psi_{n,p} \end{bmatrix} \right\} = \begin{bmatrix} \psi_{1,1} & \psi_{1,2} & \dots & \psi_{1,p} \\ \psi_{2,1} & \psi_{2,2} & \dots & \psi_{2,p} \\ \dots & \dots & \dots & \dots \\ \psi_{n,1} & \psi_{n,2} & \dots & \psi_{n,p} \end{bmatrix} \tag{7.4}$$

Because the general coordinates

$$H_1 = \begin{bmatrix} \psi_{1,1} \\ \psi_{2,1} \\ \dots \\ \psi_{n,1} \end{bmatrix},$$

$$H_2 = \begin{bmatrix} \psi_{1,2} \\ \psi_{2,2} \\ \dots \\ \psi_{n,2} \end{bmatrix},......,$$

$$H_p = \begin{bmatrix} \psi_{1,p} \\ \psi_{2,p} \\ \dots \\ \psi_{n,p} \end{bmatrix}$$

are not orthogonal the reference system H in which coordinates are the general coordinates H_1 , X_2 ,.....,X_n is a non Eucliden space. Here the distance D among the points is the general form

$$D = \sqrt{g_{11}w_1^2 + g_{22}w_2^2 + ... + g_{pp}w_p^2 + 2g_{12}w_1w_2 + 2g_{13}w_1w_3 + ... + 2g_{n-1,n}w_{n-1}w_n} = \sqrt{\sum_{\alpha=1}^{p}\sum_{\beta=1}^{p} g_{\alpha,\beta}w_\alpha w_\beta} \tag{7.5}$$

Now for

$$\begin{aligned} & g = H^T H \\ & where\ g^T = g \\ & and \\ & W = g^{-1}H^T X \end{aligned} \tag{7.6}$$

We have

$$\begin{aligned} & D^2 = W^T g W = \\ & (g^{-1}H^T X)^T g (g^{-1}H^T X) = \\ & X^T H g^{-1} g g^{-1} H^T X = \\ & (H^T X)^T g^{-1} (H^T X) \end{aligned}$$

When

$$g^{-1} = (H^T H)^{-1} = \\ H^{-1}(H^T)^{-1}$$

The matrix H is a square matrix, we have

$$\begin{aligned} &D^2 = W^T g W = \\ &X^T H g^{-1} H^T X = \\ &X^T H H^{-1} (H^T)^{.1} H^T X = \\ &X^T X \end{aligned}$$

The distance D is invariant for the change of reference or context H. D is always equal to the module of the vector X in the space of the objects. In conclusion **D is an invariant** for any transformation of the feature space H.

When

$$g^{-1} = (H^T H)^{-1} \neq \\ H^{-1}(H^T)^{-1}$$

We compute the projection of X into the space of the features H in this way

$$\begin{aligned} &QX = H(H^T H)^{-1} H^T X \\ &and \\ &QX = HW \end{aligned} \tag{7.7}$$

Or

$$QX = w_1 H_1 + w_2 H_2 + \ldots\ldots + w_p H_p = \\ w_1 \begin{bmatrix} \psi_{1,1} \\ \psi_{2,1} \\ \ldots \\ \psi_{n,1} \end{bmatrix} + w_2 \begin{bmatrix} \psi_{1,2} \\ \psi_{2,2} \\ \ldots \\ \psi_{n,2} \end{bmatrix} + \ldots\ldots w_p \begin{bmatrix} \psi_{1,p} \\ \psi_{2,p} \\ \ldots \\ \psi_{n,p} \end{bmatrix}$$

We remark that

$$Q^2 X = H(H^T H)^{-1} H^T H (H^T H)^{-1} H^T X = H(H^T H)^{-1} H^T X$$

Geometric image of projection

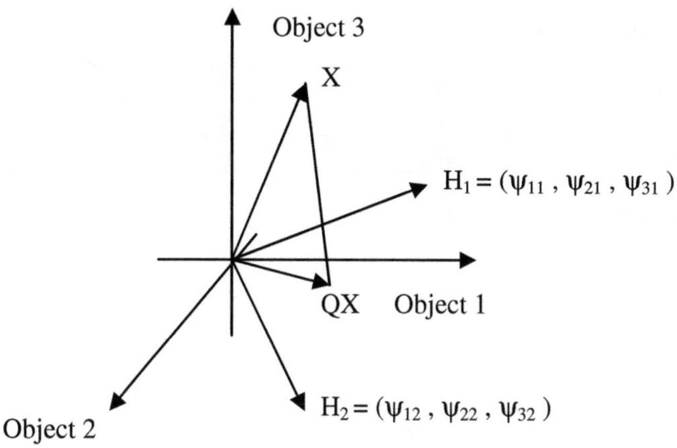

Fig. 7.2 Projection QX of the vector X in the space of the objects into the space of the features (H_1 , H_2)

Now for the projection vector QX we have the distance

$$
\begin{aligned}
& D^2 = W^T g W = \\
& (g^{-1}H^T QX)^T g (g^{-1}H^T QX) = \\
& (QX)^T H g^{-1} g g^{-1} H^T QX = \\
& (QX)^T H g^{-1} H^T QX = \\
& (QX)^T H (H^T H)^{-1} H^T H (H^T H)^{-1} H^T X = \\
& (QX)^T H (H^T H)^{-1} (H^T H)(H^T H)^{-1} H^T X = \\
& (QX)^T H (H^T H)^{-1} H^T X = \\
& (QX)^T QX
\end{aligned}
$$

We note that the distance D is function of the module $\|QX\|$ any transformation for which the module $\|QX\|$ is constant does not change the distance D.

Because $\|QX\| \leq \|X\|$ we have

$$
\begin{aligned}
& D^2 = W^T g W = \\
& (QX)^T QX \leq X^T X
\end{aligned}
$$

When p > n we have table 7.3

Table 7.3 Three Features and two objects

Context H	X_1	X_2	X_3
Object 1	ψ_{11}	ψ_{12}	ψ_{13}
Object 2	ψ_{21}	ψ_{22}	ψ_{23}

And the geometric representation

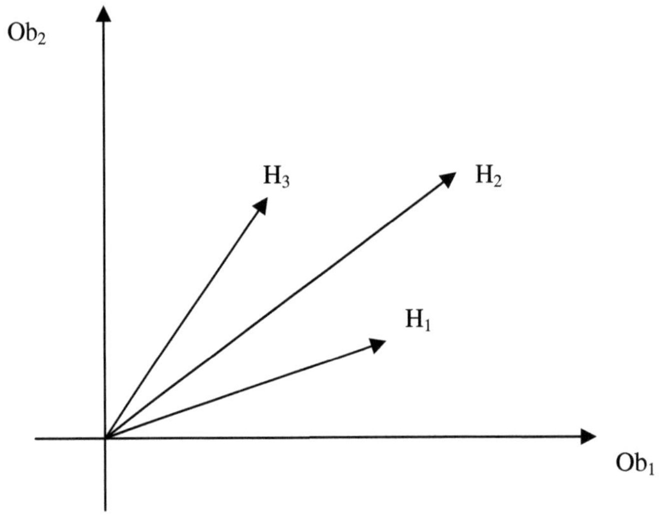

Fig. 7.3 Two objects and three features

Because in figure 7.3 we have

$$H = \begin{bmatrix} \psi_{11} & \psi_{12} & \psi_{13} \\ \psi_{21} & \psi_{22} & \psi_{23} \end{bmatrix}$$

and

$$g = H^T H = \begin{bmatrix} \psi_{11} & \psi_{12} & \psi_{13} \\ \psi_{21} & \psi_{22} & \psi_{23} \end{bmatrix}^T \begin{bmatrix} \psi_{11} & \psi_{12} & \psi_{13} \\ \psi_{21} & \psi_{22} & \psi_{23} \end{bmatrix} =$$

$$\begin{bmatrix} \psi_{11}^2 + \psi_{21}^2 & \psi_{11}\psi_{12} + \psi_{21}\psi_{22} & \psi_{11}\psi_{13} + \psi_{21}\psi_{23} \\ \psi_{11}\psi_{12} + \psi_{21}\psi_{22} & \psi_{12}^2 + \psi_{22}^2 & \psi_{12}\psi_{13} + \psi_{22}\psi_{23} \\ \psi_{11}\psi_{13} + \psi_{21}\psi_{23} & \psi_{12}\psi_{13} + \psi_{22}\psi_{23} & \psi_{13}^2 + \psi_{23}^2 \end{bmatrix} \quad (7.8)$$

Because the determinant of g is equal to zero, we cannot compute W in (7.6).

Here the space of the features cannot be in three dimensional space because X_1, X_2 and X_3 are all in the same two dimensional space of the objects. The number of objects is not sufficient to define the space of the features. The number of features is more then the necessary number of features. The number of the features needed to specify the space of the features is two only.

In conclusion we have three possible situations

1) p < n Here the information is less than that necessary

In this case we cannot built a space of the feature so we have

$$X \neq Q\,X$$

We need to use the projection operator and to obtain the vector

$$Q\,X = H\,W$$

where Q X is the projection of the vector X on the space of the features as we show in figure 7.2.

2) p = n The information supplied is exactly what is necessary and

$$X = Q\,X$$

3) p > n The information is more than the necessary. See figure 7.3 and (7.7), (7.8)

It is noted that The square of the distance D^2 between two points P_1 and P_2 in the space of the features X_j is the basic probability assignment

$$m(H_1, H_2, \ldots\ldots, H_p) = \sum_{\alpha=1}^{p} \sum_{\beta=1}^{p} g_{\alpha,\beta} w_\alpha w_\beta$$

where (7.9)

$$W = \begin{bmatrix} w_1 \\ w_2 \\ \ldots \\ w_p \end{bmatrix} = (H^T H)^{-1} H^T X$$

where

$$g = H^T H = \begin{bmatrix} \psi_{1,1} & \psi_{1,2} & \cdots & \psi_{1,p} \\ \psi_{2,1} & \psi_{2,2} & \cdots & \psi_{2,p} \\ \ldots & \ldots & \ldots & \ldots \\ \psi_{n,1} & \psi_{n,2} & \cdots & \psi_{n,p} \end{bmatrix}^T \begin{bmatrix} \psi_{1,1} & \psi_{1,2} & \cdots & \psi_{1,p} \\ \psi_{2,1} & \psi_{2,2} & \cdots & \psi_{2,p} \\ \ldots & \ldots & \ldots & \ldots \\ \psi_{n,1} & \psi_{n,2} & \cdots & \psi_{n,p} \end{bmatrix} \quad (7.10)$$

For example given the three dimensional space of the features X_1 , X_2 , X_3 in figure 7.2 we have the Basic Assignment Probability

$$m(H_1,H_2,H_3)=g_{11}w_1+g_{22}w_2+g_{33}w_3+2g_{12}w_1w_2+2g_{13}w_1w_3+2g_{23}w_2w_3 \quad (7.11)$$

for

$$H=\begin{bmatrix}1 & 0.5 & 0.3\\ 0.1 & 1 & 0.2\\ 0.3 & 0.4 & 1\end{bmatrix}$$

and

$$P_1=\begin{bmatrix}0.5 & 1 & 0.3\end{bmatrix},$$

$$P_2=\begin{bmatrix}0 & 0 & 0\end{bmatrix}$$

and

$$X=\begin{bmatrix}0.5\\ 1\\ 0.3\end{bmatrix}$$

We have

$$g=H^TH=\begin{bmatrix}1 & 0.5 & 0.3\\ 0.1 & 1 & 0.2\\ 0.3 & 0.4 & 1\end{bmatrix}^T\begin{bmatrix}1 & 0.5 & 0.3\\ 0.1 & 1 & 0.2\\ 0.3 & 0.4 & 1\end{bmatrix}=\begin{bmatrix}1.1 & 0.72 & 0.62\\ 0.72 & 1.41 & 0.75\\ 0.62 & 0.75 & 1.13\end{bmatrix}$$

and

$$W=\begin{bmatrix}w_1\\ w_2\\ w_3\end{bmatrix}=$$

$$(H^TH)^{-1}H^TX=\begin{bmatrix}0.024\\ 1.021\\ -0.116\end{bmatrix}$$

In conclusion we have

$$m(H_1,H_2,H_3)=W^TgW=$$

$$g_{11}w_1^2+g_{22}w_2^2+g_{33}w_3^2+2g_{12}w_1w_2+2g_{13}w_1w_3+2g_{23}w_2w_3=1.34$$

We can prove that the basic probability assignment defined by the geometric structure is not additive. For example when $H = \{H_1, H_2, H_3\}$ we have

$$
\begin{aligned}
&m(H_1) = g_{11} w_1^2 \\
&m(H_2) = g_{22} w_2^2 \\
&m(H_3) = g_{33} w_3^2 \\
&m(H_1, H_2) = g_{11} w_1^2 + g_{22} w_2^2 + 2 g_{12} w_1 w_2 \\
&m(H_1, H_3) = g_{11} w_1^2 + g_{33} w_3^2 + 2 g_{13} w_1 w_3 \\
&m(H_2, H_3) = g_{11} w_2^2 + g_{33} w_3^2 + 2 g_{23} w_2 w_3 \\
&m(H_1, H_2, H_3) = g_{11} w_1^2 + g_{22} w_2^2 + g_{33} w_3^2 + 2 g_{12} w_1 w_2 + 2 g_{13} w_1 w_3 + 2 g_{23} w_2 w_3
\end{aligned}
\tag{7.12}
$$

and we have the non additive property Shafer [4]

$$
\begin{aligned}
&m(H_1, H_2) = g_{1,1} w_1^2 + g_{2,2} w_2^2 + 2 g_{1,2} w_1 w_2 \\
&m(H_3) = g_{3,3} w_3^2 \\
&m(H_1, H_2, H_3) = \\
&g_{1,1} w_1^2 + g_{2,2} w_2^2 + g_{3,3} w_3^2 + 2 g_{1,2} w_1 w_2 + 2 g_{1,3} w_1 w_3 + 2 g_{2,3} w_2 w_3 = \\
&m(H_1, H_2) + m(H_3) + 2\, [g_{1,3} w_1 w_3 + g_{2,3} w_2 w_3]
\end{aligned}
\tag{7.13}
$$

If

$$g_{1,3} w_1 w_3 + g_{2,3} w_2 w_3 \geq 0$$

Then

$$m(H_1, H_2, H_3) \geq m(H_1, H_2) + m(H_3)$$

If

$$g_{1,3} w_1 w_3 + g_{2,3} w_2 w_3 < 0$$

Then

$$m(H_1, H_2, H_3) < m(H_1, H_2) + m(H_3)$$

When

$$g = H^T H = \begin{bmatrix} 1 & 0 & 0 \\ 0 & 1 & 0 \\ 0 & 0 & 1 \end{bmatrix}$$

The space of the features H_1, H_2, H_3 is Euclidean Space. Here case we have

$$\begin{aligned}
m(H_1) &= w_1^2 \\
m(H_2) &= w_2^2 \\
m(H_3) &= w_3^2 \\
m(H_1, H_2) &= w_1^2 + w_2^2 \\
m(H_1, H_3) &= w_1^2 + w_3^2 \\
m(H_2, H_3) &= w_2^2 + w_3^2 \\
m(H_1, H_2, H_3) &= w_1^2 + w_2^2 + w_3^2
\end{aligned} \tag{7.14}$$

and

$$\begin{aligned}
m(H_1, H_2) &= w_1^2 + w_2^2 \\
m(H_3) &= w_3^2 \\
m(H_1, H_2, H_3) &= w_1^2 + w_2^2 + w_3^2 = \\
m(H_1, H_2) &+ m(H_3)
\end{aligned} \tag{7.15}$$

Here the basic probability assignment is an additive measure or probability

7.4 From Evidence Theory to Geometry

Given the basic probability assignment

$m(X_i, X_j)$, $m(X_k)$ and the vector of the coordinates $W = \begin{bmatrix} w_1 \\ w_2 \\ \ldots \\ w_p \end{bmatrix}$

we can compute the metric tensor g in this way

$$g=\begin{bmatrix} g_{1,1} & g_{1,2} & \cdots & g_{1,n} \\ g_{2,1} & g_{2,2} & \cdots & g_{2,n} \\ \cdots & \cdots & \cdots & \cdots \\ g_{n,1} & g_{n,2} & \cdots & g_{n,n} \end{bmatrix}= \tag{7.16}$$

$$\begin{bmatrix} \frac{m(H_1)}{w_1^2} & \frac{m(H_1,H_2)-(m(H_1)+m(h_2))}{2w_1w_2} & \cdots & \frac{m(H_1,H_n)-(m(H_1)+m(H_n))}{2w_1w_n} \\ \frac{m(H_1,H_2)-(m(H_1)+m(H_2))}{w_1w_2} & \frac{m(H_2)}{w_2^2} & \cdots & \frac{m(H_2,H_n)-(m(H_2)+m(H_n))}{2w_2w_n} \\ \cdots & \cdots & \cdots & \cdots \\ \frac{m(H_1,H_n)-(m(H_1)+m(H_n))}{2w_1w_n} & \frac{m(H_2,H_n)-(m(H_2)+m(H_n))}{2w_2w_n} & \cdots & \frac{m(H_n)}{w_n^2} \end{bmatrix}$$

For example we have

$$\begin{aligned} &m(X_1)=0.25 \\ &m(X_2)=0 \\ &m(X_3)=0.25 \\ &m(X_1,X_2)=0 \\ &m(X_1,X_3)=0.25 \\ &m(X_2,X_3)=0.25 \\ &m(X_1,X_2,X_3)=m(X_1,X_2)+m(X_1,X_3)+m(X_2,X_3)-m(X_1)-m(X_2)-m(X_3)=0 \end{aligned} \tag{7.17}$$

Where

$$m(X_1)+m(X_2)+m(X_3)+m(X_1,X_2)+m(X_1,X_3)+m(X_2,X_3)+m(X_1,X_2,X_3)=1 \tag{7.18}$$

when $$W=\begin{bmatrix}1\\1\\1\end{bmatrix} \tag{7.19}$$

we have the metric tensor

$$g=\begin{bmatrix} \frac{1}{4} & \frac{-1}{4} & \frac{-1}{4} \\ \frac{-1}{4} & 0 & 0 \\ \frac{-1}{4} & 0 & \frac{1}{4} \end{bmatrix}$$

The metric tensor g defines the geometry of the evidence for the non additive measure m .

7.5 Quantum Mechanics Interference and the Geometry of the Particles

Given the famous experiments of the two holes where we have interference we know that given the two probability densities h_1 and h_2 the probability density, that is a complex number, of the interference is

$$\begin{aligned} &|h\rangle = |h_1\rangle + |h_2\rangle = \\ &A_1(r,t)e^{iS_1(r,t)} + A_2(r,t)e^{iS_2(r,t)} = \\ &A_1(r,t)\cos(S_1(r,t)) + A_2(r,t)\cos(S_2(r,t)) + i\,[A_1(r,t)\sin(S_1(r,t)) + A_2(r,t)\sin(S_2(r,t))] \end{aligned} \tag{7.20}$$

Where A is the amplitude and S is the phase at specific location r in specific time t. We know that the probability of the interference is

$$\begin{aligned} &P = \langle h|h\rangle = \\ &\langle h_1 + h_2|h_1 + h_2\rangle = \\ &\langle h_1|h_1\rangle + \langle h_2|h_2\rangle + 2\langle h_1|h_2\rangle \end{aligned} \tag{7.21}$$

because the probability densities $|h_1\rangle, |h_2\rangle$ are complex numbers we have

$$\begin{aligned} &P\langle h_1|h_2\rangle = (A_1(r,t)\cos(S_1(r,t)) + A_2(r,t)\cos(S_2(r,t)) + i\,[A_1(r,t)\sin(S_1(r,t)) + A_2(r,t)\sin(S_2(r,t))]) \\ &(A_1(r,t)\cos(S_1(r,t)) + A_2(r,t)\cos(S_2(r,t)) - i\,[A_1(r,t)\sin(S_1(r,t)) + A_2(r,t)\sin(S_2(r,t))]) = \\ &(A_1(r,t)\cos(S_1(r,t)) + A_2(r,t)\cos(S_2(r,t)))^2 + (A_1(r,t)\sin(S_1(r,t)) + A_2(r,t)\sin(S_2(r,t))])^2 = \\ &A_1(r,t)^2 + A_2(r,t)^2 + 2A_1(r,t)A_2(r,t)[\cos(S_1(r,t))\cos(S_2(r,t)) + \sin(S_1(r,t))\sin(S_2(r,t))] = \\ &A_1(r,t)^2 + A_2(r,t)^2 + 2A_1(r,t)A_2(r,t)\cos(S_1(r,t) - S_2(r,t)) = \\ &I_1(r,t) + I_2(r,t) + 2\sqrt{I_1(r,t)I_2(r,t)}\cos(S_1(r,t) - S_2(r,t)) \end{aligned} \tag{7.22}$$

(9)

Where $I_1(x)$ and $I_2(x)$ are the intensity of the electrons beam

Because we have

$$\begin{aligned} &P = P_1 + P_2 + 2\langle h_1|h_2\rangle = I_1(x) + I_2(x) + 2\sqrt{I_1(x)I_2(x)}\cos(\alpha - \beta) \\ &\alpha = S_1(r,t) \\ &\beta = S_2(r,t) \end{aligned} \tag{7.23}$$

Quantum probability P is a non additive probability see Feynman [3]. In this case we assume that P is not a classical probability measure but is the basic probability assignment m. So we have

$$\begin{aligned} P &= m = \\ & m_1 + m_2 + 2\langle h_1 | h_2 \rangle = \\ & I_1(x) + I_2(x) + 2\sqrt{I_1(x) I_2(x)} \cos(\alpha - \beta) \end{aligned} \tag{7.24}$$

For

$$\begin{aligned} w_1(r,t) &= \sqrt{I_1(r,t)} \\ w_2(r,t) &= \sqrt{I_2(r,t)} \end{aligned} \tag{7.25}$$

The (7.24) becomes

$$\begin{aligned} P &= m = \\ & w_1^2 + w_2^2 + 2 w_1 w_2 \cos(\alpha - \beta) \end{aligned} \tag{7.26}$$

In the Gauss plane of the real and imaginary numbers in figure 7.4 we represent the two unitary vectors

$$\begin{aligned} H_1(r,t) &= \frac{h_1(r,t)}{A_1(r,t)} = \\ & \begin{bmatrix} \cos(S_1(r,t)) \\ \sin(S_1(r,t)) \end{bmatrix} \\ H_2(r,t) &= \frac{h_2(r,t)}{A_2(r,t)} = \\ & \begin{bmatrix} \cos(S_2(r,t)) \\ \sin(S_2(r,t)) \end{bmatrix} \end{aligned} \tag{7.27}$$

And the space of the features is

$$H(r,t) = \begin{bmatrix} \cos(S_1(r,t)) & \cos(S_2(r,t)) \\ \sin(S_1(r,t)) & \sin(S_2(r,t)) \end{bmatrix} \tag{7.28}$$

So the metric tensor g is

$$\begin{aligned} g &= H(r,t)^T H(r,t) = \\ & \begin{bmatrix} \cos(S_1(r,t)) & \cos(S_2(r,t)) \\ \sin(S_1(r,t)) & \sin(S_2(r,t)) \end{bmatrix}^T \begin{bmatrix} \cos(S_1(r,t)) & \cos(S_2(r,t)) \\ \sin(S_1(r,t)) & \sin(S_2(r,t)) \end{bmatrix} = \\ & \begin{bmatrix} 1 & \cos(S_1(r,t) - S_2(r,t)) \\ \cos(S_1(r,t) - S_2(r,t)) & 1 \end{bmatrix} \end{aligned} \tag{7.29}$$

The distance is defined in this way

$$D^2 = w_1^2 + w_2^2 + 2w_1 w_2 \cos(S_1(r,t) - S_2(r,t)) \tag{7.30}$$

That is the same of the expression (7.26). With (7.28) we can also compute the contro-variant components of the vector X = X_1 + i X_2 where the coordinates are H_1 and H_2.

$$\begin{bmatrix} w_1 \\ w_2 \end{bmatrix} = \begin{bmatrix} \dfrac{X_2 \cos(S_2) - X_1 \sin(S_2)}{\sin(S_1 - S_2)} \\ \dfrac{X_1 \sin(S_1) - X_2 \cos(S_1)}{\sin(S_1 - S_2)} \end{bmatrix} \tag{7.31}$$

$$\begin{bmatrix} \cos(S_1) & \cos(S_2) \\ \sin(S_1) & \sin(S_2) \end{bmatrix} \begin{bmatrix} w_1 \\ w_2 \end{bmatrix} = \begin{bmatrix} X_1 \\ X_2 \end{bmatrix} \tag{7.32}$$

Geometric image of the probability density, vector X and complex plane.

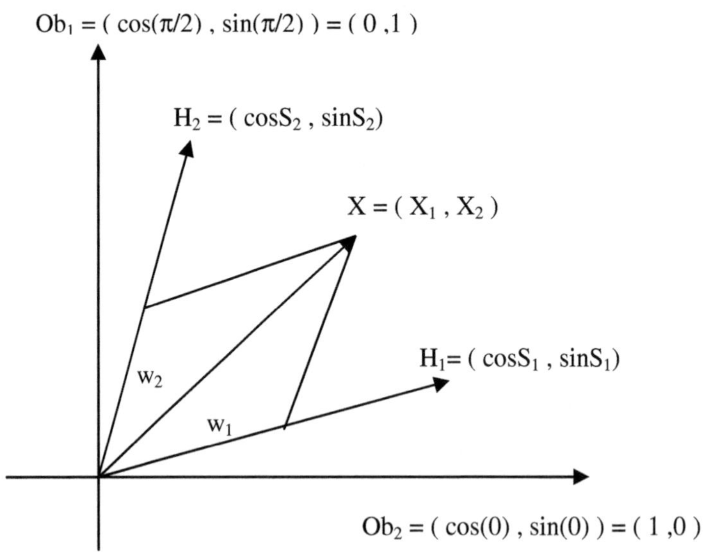

Fig. 7.4 Vector X and vectors H_1, H_2 in the plane of independent set of particles. The phase of the coordinates are zero and π/2.

We remark that with the cotro-variant components in (7.31), compensate the interference formula (7.24). In fact for (7.31) we obtain

$$
\begin{aligned}
&P = P_1 + P_2 + 2\langle h_1 | h_2 \rangle = \\
&I_1(x) + I_2(x) + 2\sqrt{I_1(x) I_2(x)} \cos(\alpha - \beta) = \\
&w_1^2 + w_2^2 + 2 w_1 w_2 \cos(S_1 - S_2) = \\
&\left(\frac{X_2 \cos(S_2) - X_1 \sin(S_2)}{\sin(S_1 - S_2)}\right)^2 + \left(\frac{X_2 \cos(S_1) - X_1 \sin(S_1)}{\sin(S_1 - S_2)}\right)^2 + \\
&2 \frac{X_2 \cos(S_2) - X_1 \sin(S_2)}{\sin(S_1 - S_2)} \frac{X_2 \cos(S_1) - X_1 \sin(S_1)}{\sin(S_1 - S_2)} \cos(S_1 - S_2) = X_1^2 + X_2^2
\end{aligned}
\tag{7.33}
$$

Given the intensities I_1, I_2 of the two beams with quantum interference, we compute the

$$
\begin{bmatrix} w_1(r,t) \\ w_2(r,t) \end{bmatrix} = \begin{bmatrix} \sqrt{I_1(r,t)} \\ \sqrt{I_2(r,t)} \end{bmatrix}
$$

With the geometric transformation (7.32) that can write in this way

$$
\begin{bmatrix} w_1 \cos(S_1) + w_2 \cos(S_2) \\ w_1 \sin(S_1) + w_2 \sin(S_2) \end{bmatrix} = \begin{bmatrix} \sqrt{I_1} \cos(S_1) + \sqrt{I_2} \cos(S_2) \\ \sqrt{I_1} \sin(S_1) + \sqrt{I_2} \sin(S_2) \end{bmatrix} = \begin{bmatrix} X_1 \\ X_2 \end{bmatrix}
\tag{7.34}
$$

We compensate the quantum interference as a pure geometric phenomena.

In figure 7.5 we show another interference experiment

The first crystal in figure 7.5, the black line, split a beam in two beams of particle one with phase zero, red color, the other beam with phase $\frac{\pi}{2}$. The final crystal change the green beam again of $\frac{\pi}{2}$, so the final green beam up is a phase of π and the read beam up has phase zero. The final green beam at the right does not change the phase, the read beam right change the phase by zero to $\frac{\pi}{2}$. The detector 1 in up receive two beams, read and green, with opposite phase so the interference give zero intensity for the beam, the interference is destructive. The detector 2 at the right receive two beams with the same phase so we have the maximum intensity of the interference, the interference is constructive.

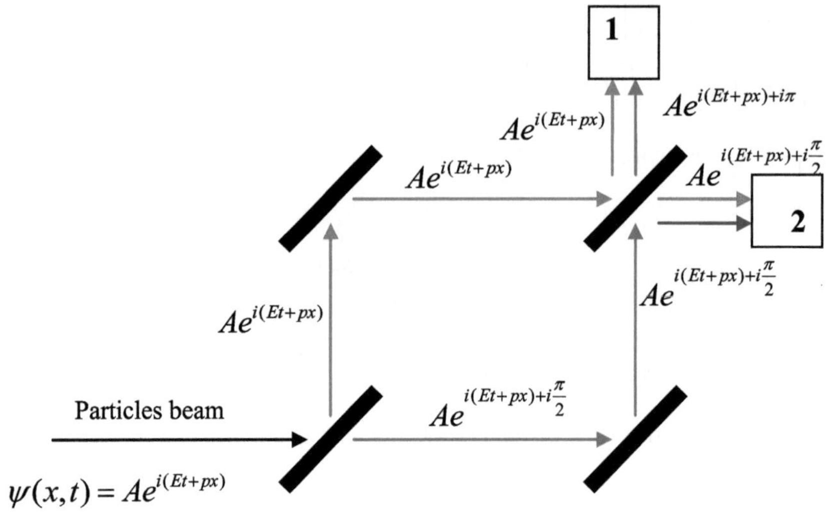

Fig. 7.5 In the first detector we have a destructive interference in the second detector we have the constructive interference

In figure 7.5 we have constructive and destructive interference. From the previous consideration we have that the geometry of the system is different for the first detector than for the second detector.

In the first detector we have the metric P(x) defined in this way:

$$\begin{aligned}
&P(x) = m(x) = w_1^2 + w_2^2 + 2\cos(0-\pi)w_1w_2 = \\
&w_1^2 + w_2^2 - 2w_1w_2 = \\
&(w_1 - w_2)^2 = 0
\end{aligned} \tag{7.35}$$

Because

$$\begin{aligned}
&w_1 = \sqrt{I} \\
&\sqrt{I} = A \\
&w_2 = \sqrt{I} \\
&\sqrt{I} = A \\
&w_1 = w_2
\end{aligned} \tag{7.36}$$

In this case the metric tensor g is

$$g = \begin{bmatrix} 1 & -1 \\ -1 & 1 \end{bmatrix}$$

The matrix g is singular and the interference is destructive. Also if in the input of the system in figure 7.5 the beam is composed by a huge number of the particles, at the detector 1 no particle appear and, as we suggest in this book, this is due to the particular form of the local geometry where the coordinates are the amplitudes of the two beams and the probability is the non Euclidean distance among the beams. In the second detector we have the metric

$$\begin{aligned} m(x) &= w_1^2 + w_2^2 + 2\cos(\frac{\pi}{2} - \frac{\pi}{2})w_1 w_2 = \\ & w_1^2 + w_2^2 + 2w_1 w_2 = \\ & (w_1 + w_2)^2 = 4A^2 \end{aligned} \tag{7.37}$$

The intensity of the interference is the maximum value that is more that the intensity A^2 of the original beam. This again for the influence of the local geometry in the second detector.

The metric tensor in the detector 2 is

$$g = \begin{bmatrix} 1 & 1 \\ 1 & 1 \end{bmatrix}$$

The local geometry in the first detector, the second detector and in the laboratory are different. When the two beams are independent one from the other, we have the traditional expression for the distance

$$m(x) = \psi_1^2 + \psi_2^2 = m_1(x) + m_2(x) = I_1 + I_2 = w_1^2 + w_2^2 \tag{7.38}$$

The Geometry is the Euclidean geometry and no interference is possible.

When the two paths superpose the two beams are considered as one beam. In this case we can have any type of geometry for different values of the phases of the electrons. The phase of the electrons is an intrinsic property that define the local geometry.

In conclusion the square of the distance D or

$$D^2(x) = P(x)$$
$$P(x) = m(x)$$

can have different forms of geometry when we change the phases α and β of the electrons. For the properties of g, we can only detect the difference between the phases this is impossible to know the absolute value of the phases.

We argue that the interference in quantum mechanics does not exist but is only the physical image of a deformation of the local geometry and its metric. The particles are always independent but when are near one with the others the local geometry change and this give us the impression that among the particles there is a quantum interference. We remark that any observation of the particles destroy the superposition state and the local geometry came back to the Euclidean Geometry. In the superposition state the particles are independent but embedded in a deformation of the Euclidean Geometry or a Non Euclidean geometry.

For the relation (7.26), the probability P(x) is given by the contribute of three terms

$$\begin{aligned} P &= P_1 + P_2 + P_3 \\ P_1 &= I_1 \\ P_2 &= I_2 \\ P_3 &= \sqrt{I_1 I_2}\cos(\alpha - \beta) \end{aligned} \tag{7.39}$$

The third probability is the entanglement among the particles due to non Euclidean Geometry. So when two particles are near one with the others the geometry change. When the particles go away and other particles does not generate deformation of the geometry, the local geometry can persist and this can solve the problem of the entanglement. In this model entanglement become an image of the non Euclidean geometry generated when the particles are one very near to the others.

In a more general case we have

$$m(\psi_1, \psi_2, \ldots, \psi_n) = m(\langle \psi | \psi \rangle \tag{7.40}$$

where

$$\begin{aligned} &\psi = h_1 + h_2 + \ldots + h_n \\ &h_1 = \alpha_1 + i\beta_1 \ , \ h_2 = \alpha_2 + i\beta_2, \ldots, h_n = \alpha_n + i\beta_n \end{aligned} \tag{7.41}$$

and

$$\begin{aligned} &\langle \psi | \psi \rangle = \langle h_1 + h_2 + \ldots + h_n | h_1 + h_2 + \ldots + h_n \rangle = \\ &\langle h_1 | h_1 \rangle + \langle h_2 | h_2 \rangle + \ldots + \langle h_n | h_n \rangle + 2\langle h_1 | h_2 \rangle + \ldots + 2\langle h_{n-1} | h_n \rangle = \\ &I_1 + I_2 + \ldots + I_n + 2\sqrt{I_1 I_2}\cos(\alpha_1 - \alpha_2) + 2\sqrt{I_1 I_3}\cos(\alpha_1 - \alpha_3) + \ldots + 2\sqrt{I_n I_{n-1}}\cos(\alpha_{n-1} - \alpha_n) \end{aligned} \tag{7.42}$$

In this case the metric tensor is

$$g = \begin{bmatrix} 1 & \cos(\alpha_1 - \alpha_2) & \dots & \cos(\alpha_1 - \alpha_2) \\ \cos(\alpha_2 - \alpha_1) & 1 & \dots & \cos(\alpha_2 - \alpha_1) \\ \dots & \dots & \dots & \dots \\ \cos(\alpha_n - \alpha_1) & \cos(\alpha_n - \alpha_2) & \dots & 1 \end{bmatrix} \tag{7.43}$$

and

$$w_1 = \sqrt{I_1}, w_2 = \sqrt{I_2}, \dots\dots, w_n = \sqrt{I_n} \tag{7.44}$$

In conclusion we have the distance

$$\begin{aligned} &D^2 = \langle \psi | \psi \rangle = \\ &P = g_{11}w_1^2 + g_{22}w_2^2 + \dots + g_{nn}w_n^2 + 2g_{12}w_1w_2 + 2g_{13}w_1w_3 + \dots + 2g_{n,n-1}w_nw_{n-1} \end{aligned} \tag{7.45}$$

where

$$\begin{aligned} g = &\begin{bmatrix} 1 & \cos(\alpha_1 - \alpha_2) & \dots & \cos(\alpha_1 - \alpha_n) \\ \cos(\alpha_2 - \alpha_1) & 1 & \dots & \cos(\alpha_2 - \alpha_n) \\ \dots & \dots & \dots & \dots \\ \cos(\alpha_n - \alpha_1) & \cos(\alpha_n - \alpha_2) & \dots & 1 \end{bmatrix} = \\ &\begin{bmatrix} g_{11} & g_{12} & \dots & g_{1n} \\ g_{n1} & g_{22} & \dots & g_{2n} \\ \dots & \dots & \dots & \dots \\ g_{n1} & g_{n2} & \dots & g_{nn} \end{bmatrix} \end{aligned} \tag{7.46}$$

We can say that a complex interference process can be represented by a Non Euclidean n dimensional space of the beams. In the situation of the n dimensional Euclidean particles the beams are independent one from the other. But one we have a superposition of the beams, the multiple interference is a deformation of the n dimensional Euclidean Space into a non Euclidean n dimensional space with multiple entanglements among the beams. We argue that this can be a complete geometric image of the superposition and entanglement. We also assume that there is a deeper connection among the basic assignment probability and quantum probability with a common geometric image.

7.6 Conclusion

In this chapter we reformulate quantum mechanics in line with the recent theory of evidence. In the Evidence Theory we assume that the probability is non additive and takes the name of basic probability assignment. With geometric image of evidence theory and of the basic probability assignment we assume that general coordinates of intensity of beams can substitute the density probability which values are complex numbers. We use only real numbers as density of the beam particles in general coordinates to model interference and entanglement in quantum mechanics. Without entanglements every density of a beam is one independent from the other and the classical Euclidean Geometry is sufficient to represent the quantum mechanics. But with superposition and entanglement the Euclidean Geometry is reshaped in a Non Euclidean Geometry.

References

[1] Brody, D.C., Hughston, L.P.: Statistical Geometry in Quantum Mechanics, arXiv: gr – qc/9701051 v2 (October 22, 1997)
[2] Resconi, G., Klir, G., Pessa, E.: Conceptual foundation of quantum mechanics the role of evidence theory, Quantum sets, and Modal Logic. International Journal of Modern Physics C 10(1), 29–62 (1999)
[3] Feynman, R.P.: Lectures On Physics, vol. III. Addison Wesley Publishing Company (1965)
[4] Shafer, G.: A mathematical theory of Evidence. Princeton University Press (1976)
[5] Resconi, G., Klir, G.J., Clair, U.S.: Hierarchical Uncertainty Metatheory Based Upon Modal Logic. Int. J. of General Systems 21, 23–50 (1992)
[6] Harmanec, Klir, G.J., Resconi, G.: On Modal Logic Interpretation of Dempster-Shefer Theory of Evidence. Int. J. Intelligent Systems 9, 941–951 (1994)
[7] Resconi, G., Jain, L.C.: Intelligent Agents. Springer, Heidelberg (2004)
[8] Resconi, G., Türkşen, I.B.: Canonical forms of fuzzy truthfulnesss by meta-theory based upon modal logic. Information Sciences 131, 157–194 (2001)
[9] Resconi, G., Pessa, E., Mignani, R.: Non-Conservative Gravitational Equations. General Relativity and Gravitation 29(8) (1997)

Chapter 8
The Logic of Uncertainty and Geometry of the Worlds

Abstract. The classical logic is built in an axiomatic way without containing any uncertainty. Over the years, many attempts have been made to extend the classical logic framework to incorporate uncertainties. This work has produced different and apparently conflicting definitions of uncertainties. Previously we have argued that an extended modal logic framework can provide a *unifying formal language for many formalizations of uncertainty* such as probability, approximate probability, evidence theory, fuzzy sets, and rough sets.

In this chapter, we have shown how such an extended modal of logic framework provides a unifying formal language for formalizations of probability and fuzzy sets theories. We also show that this originally pure theoretical approach can be used to describe a mathematical model for linguistic uncertainty in a unified framework. The approach combines concepts of modal logic together the linguistic context space previously proposed. This combined approach is illustrated with an application to the question of economical preference between customers and goods.

A common language able to represent different types of uncertainties is useful because it allows us to define the uncertainties using simple entities and to compare different types of uncertainties as being different aspects of the same fundamental structure. The justification of different theories of uncertainty using a unified language is beyond the scope of this paper.

Our goal is to rebuild a foundation of the uncertainty concept by using a new interpretation of the modal logic structure. Using this new foundation we discover that disparate types of uncertainties and the idea of uncertainty itself can be understood. This new foundation opens opportunities to discover hidden connection between different types of uncertainties.

8.1 Introduction

In this chapter, we describe a new unified approach to the logic of uncertainty. The logic of uncertainty is an Umbrella Term [Gaines, 1976; Hawthorne, 2004, Kovalerchuk, 1990] that covers Fuzzy Logic, Probability Logic, Lukasiewicz Infinitely Valued Logic and other forms. The classical true/false logic and set theory do not use the Umbrella Terms. However, both are connected with the

G. Resconi: Geometry of Knowledge for Intelligent Systems, SCI 407, pp., pp. 229–279.
springerlink.com

probability theory that deals with uncertainty. Similarly Modal Logic and Non-Monotonic Logic also do not use the term uncertainty, but are connected with fuzzy logic and fuzzy measure respectively. Gaines [1976] argued that there are close links, both formal and semantic, between fuzzy logic and probability logics, but it is still difficult to identify their distinct areas. Gaines developed a multivalued logic in terms of a Truth Function over a lattice of propositions that encompasses a wide range of the logic of uncertainty. He also introduced quantification and modal operators to the basic logic. This approach expanded the set of logic and showed links between them. It left untouched the *issue of actually getting and of assigning values of truth functions* over a lattice of propositions. In essence, all modern logics of uncertainty lack such internal components. In this paper we show that adding this component serves two goals:

(1) Making uncertainty theories more complete with a clear semantic meaning and
(2) Creating a base for comparison of uncertainty theories semantically.

Known theories of uncertainties start from the assumption that *uncertainty values have been assigned somehow outside the theory*. The goal of these theories is to provide logic as a *manipulating tool* with already available values of uncertainty. That is combining these uncertainty values. This approach has a fundamental flaw as it is difficult and often impossible to justify logic operations with uncertainty values without knowing how the uncertainty values have been generated. In physics it would be similar to theories without operational measurement procedures for the phenomena. In physics the fundamental concept of operationalism was introduced a long time ago by Percy Bridgman (1882-1961) who was inspired to use this approach of Einstein. Among theories of uncertainties only the probability theory has mathematical statistics that provides a systematical way of getting probability values operationally. Mathematical statistics is not a part of the Kolmogorov's axiomatic probability theory.

The difference between our approach and Gaines' approach is in the introduction of modal logic and context spaces as a way of getting uncertainty values in theories such as fuzzy logic theory, Dempster-Shaffer evidence theory, and probability theory in a unified way. This is done by introducing deeper structures of Modal Logic, Agent Theory, and Context Spaces to the logic of uncertainty. This solves the important problem of making the uncertainty theories consistent with the operational approach used by Bridgman's Operationalism used in physics.

To explain the impact of the Operational Approach on Uncertainty Theories we need to separate *conceptual ideas* of these theories and *actual logic operations* they currently employ. Our modal logic and context space approach preserves conceptual intent of these theories, but not necessarily their current logic operations. Note, that Zadeh never insisted that operations he originally proposed for fuzzy logic such as Min and Multiplication for "AND" operation should be considered as the only possible operations, viewed them rather as an illustration of the concept.

The conceptual appeal of the fuzzy logic theory was two-fold:

(1) Recognition of difficulties of control theory to built models when a physical phenomenon was not known sufficiently well to derive *differential equations*, and
(2) The availability of successful human control expertise that could be expressed informally in a *natural language*. The uncertainty of the situation was clear, but it was not clear how it could be formulated as a probability theory challenge. In essence it included many *more challenges than merely combining uncertainties* that the logics of uncertainties intend to formalize.

The *problem of a task formulation/formalization* is often missed in the discussions about differences and similarities between fuzzy logic and the probability theory. The situation is somewhat similar to the development of the utility theory for economics that incorporated formalized uncertainty values as probabilities, but also combines them with other entities such as the utility of goods. In essence, such tasks provide context for the logic of uncertainty and justification for operations that potentially may not even be probabilistic. This poses an important question:

"Is it correct to compare different uncertainty theories without a knowledge of their intended context, tasks and measurement procedures? "

In general the answer is: "No", but having too many informal potential contexts, tasks and measurement procedures it is often *impractical* to implement the context based approach on the full scale.

A practical approach is to define uncertainty values *procedurally and operationally* explicitly and in a unified way. This is the approach we have used in this paper. It means that uncertainty values and operations previously introduced in Modal Logic, Fuzzy Logic and other theories without explicit procedures must be revised.

The operation that we revise in modal logic is the Complement/Negation. In the classical Modal Logic a set of worlds where the proposition is true is a complement of the set of the worlds where the same proposition is false. This closed world assumption may not always be true. We also introduce the average value of truth. This is a new operation that did not exist previously in modal logic.

We also change the interpretation of the modal logic *concept of the world.* In classical modal logic any modality or any world is a stand alone part of the real or conceptual universe. Any modality is completely *separate and independent* from the others, in say space and time. We may ask the same question for one part of the universe or another. Here agents that decide if the proposition is true or false are not important because such agents can be substituted by a single evaluation function that provides the logic value of a proposition in different situations and worlds in the same way.

In our case agents are fundamental because for the same situation, position, time, and type of object the agents as worlds decide if the proposition is true. The *state of the agent may alter the truth value* provided by the same agent for the same situation. This resembles a Turing machine concept of the machine state that may change the output. In this case an agent may conflict with herself for the same proposition about the same reality if the state is changed. In addition different truth values provided by different agents can conflict too for the same reality. In our case the world is a conceptual world of the agent which decides if one proposition is true or false. This interpretation allows us to talk about agents as worlds.

The *definition of a world* in classical modal logic and in our theory are quite different because the agent gives us the possibility to obtain a set of conflicting worlds where different logic values can be given to the same attribute.

Introduction of agents gives a new flavor to standard Min and Max operations used in fuzzy logic by defining these operations for sets of agents. Let A and B be sets of agents where A is a subset of the agents B. In this case A= A $\cap$ B and the number of propositions |g| that are true for agents A $\cap$ B is defined by the Min formula, |g|=min(|a|,|b|). Similarly |g|=max(|a|,|b|) if A $\cup$ B =B.

We begin with the description of the Kolmogorov's classical approach to probability which is based on the set theory. Due to isomorphism between the set theory and the classical logic it is possible to associate the ordinary probability calculus with classical logic reasoning. We also clarify the meaning of the True and False Logic value for this goal. We discover that to define the true and false logic value of a proposition, we must introduce an agent or a group of agents that provide a true or false value for the proposition. Such agents are models of possible worlds in the modal logic. When we move from an ordinary set to a set of worlds we can create a new type of logic. The new type of logic is responsible for a new type of uncertainties. The transition from the set of elements to a set of worlds can change the properties of the set theory itself. An example is such as tautology and contradiction. We show below that we can have an Irrational Possible World where the contradiction is true or where the tautology is false.

In modal logic settings some worlds are not isolated one and other but one world can see and accept the logic value of a proposition in another world. This type of vision and communication between worlds is formalized by the concept of Accessibility Relation. The Accessibility Relation holds between worlds w and w′ if and only if w′ is **possible** given the **facts** of w [Garson, 2003]. This concept is used here for building the Logic of Uncertainty.

The logic operations are not always the same but depend on the distribution of the logic values inside the set of worlds that form associations of incoherent objects. In the paper we show that such incoherent objects form a foundation of fuzzy logic propositions.

The overall goal of this chapter is to show that a natural extension of the set theory to a set of possible Rational and Irrational worlds provides a new type of logic that represents and gives **a uniform foundation** for uncertainty models such as fuzzy sets, fuzzy measures, imprecise probability, and evidence theory. In this way, we base the emerging uncertainty logics on a more simple and stable foundation. We also show that the concept of Possible Worlds naturally extends the traditional probability calculus.

This chapter is organized as follows. Section 2 is devoted to concepts of modal logic and possible worlds. It describes Kripke Framework, Definitions, and Meaning of the possible worlds, and the Tarski Truth definition. Section 3 discusses links between concepts of the possible world and probability. After that section 4 shows how the concept of possible worlds opens a way to break the symmetry in probability Calculus and Evidence Theory. In essence, this opens a way to introduce new types of uncertainties to these theories. Section 5 is devoted to Fuzzy Sets and Logic Theory, its operations and the concept of irrationality is introduced to provide an interpretation of this theory. Further sections provide deeper understanding the meaning of rational and irrational worlds, classical logic, and linguistic uncertainty for fuzzy logic. Section 6 discusses concepts of irrational and rational worlds in a more general framework. Section 7 defines transformation of set of worlds. Section 8 discusses invariant expressions in fuzzy logic. Section 9 discusses the linguistic context space of the worlds. The last section (section 10) discusses economic model of worlds with irrational and rational customers in connection with the Fuzzy Sets Theory and communication among customers in connection with the rough sets. In conclusion we summarize results.

8.2 Modal Logic and Meaning of Worlds

There is a wide-spread acceptance that Kripke models [Chellas, 1980] offer the fundamental semantics for modal logic. Further there are several generalizations of Kripke models [Chellas, 1980] that have the most universal characteristics of such models. In this section, we give a brief description of Kripke model.

8.3 Kripke Modal Framework

Let L_{ML} be a language for modal logic formed from a set P of **elementary (atomic) sentences** q_i Q={ q_i } using propositional connectives such as $\neg$ (negation), $\wedge$ (conjunction), $\vee$ (disjunction), $\Rightarrow$ (material implication), $\Leftrightarrow$ (equivalence), and modal operators such as (necessity) and $\Diamond$ (possibility).

A *Kripke model* is defined as the following triple:

$$M = < W, R, V >, \tag{8.1}$$

where W={w} is a non-empty set of w entities called **Possible worlds**, $R \subseteq W \times W$ is a relation on W called an **accessibility relation** on W, and V is a **Valuation Operator** for atomic sentences at each possible world, $w \in W$:

$$V: \; Q \times W \rightarrow \{T, F\}. \tag{8.2}$$

The accessibility relation has the following interpretation: "In ordinary speech, the claim that A is necessary does not require the truth of A in *all* possible worlds, but rather only in a certain class of worlds which I have in mind, for example, worlds where I avoid penalties for failure to pay. In order to provide a generic treatment of **necessity**, we must say that A is true in w iff A is true in **all worlds that are related to w in the right way**. So for an operator interpreted as necessity, we introduce a corresponding relation R on the set of possible worlds W, **traditionally called the accessibility relation**. The accessibility relation R holds between worlds w and w′ iff w′ **is possible** given the **facts of w**" [Garson, 2003]. Thus the term accessibility means that if world w′ has "access" to the information about internals of world w (that is statement A is true in w), then it is **possible** that A in w′ can be true also. In essence, R formalizes possible truth output from the communication between worlds w and w′ in W.

The valuation operator is extended in the usual way and, in particular, for modal operators, - necessity and ◊ - possibility:

$$V(\quad q, w) = T \Leftrightarrow \forall\, w'(wRw' \Rightarrow V(q, w')) = T \Leftrightarrow \Xi_w \subseteq ||p||^M, \tag{8.3}$$

$$V(\Diamond q, w) = T \Leftrightarrow \exists w'\ (wRw' \text{ and } V(q, w')) = T \Leftrightarrow \Xi_w \cap ||q||^M \neq \varnothing \tag{8.4}$$

where $\Xi_w = \{w' \mid wRw'\}$ and $||q||^M = \{w \mid V(q, w) = T\}$. Note that $||q||^M$ is called the **truth set** associated with the proposition of an atomic sentence p in a Kripke model M in the literature of philosophical logic. It is well known that different conditions on accessibility relations correspond to different axiom schemas.

Using common logic default notation we can simplify (8.3) and (8.4) by writing

$$V(\quad q, w) \Leftrightarrow \forall\, w'\ (wRw' \Rightarrow V(q\ w')) \quad \Leftrightarrow \Xi_w \subseteq ||q|^M,$$

$$V(\Diamond q\ w) \Leftrightarrow \exists w'\ (wRw' \text{ and } V(q\ w')) \Leftrightarrow \Xi_w \cap ||q|^M \neq \varnothing$$

where, for instance, $V(\quad q\ w)$ stands for $V(\quad q\ w) = T$. Similarly if $V(\quad q\ w)$ would be false,

$V(\quad q\ w)=F$ then we could write $\neg V(\quad q\ w)$.

Property (8.3) states that q is necessarily true in the world w then it is true in every world that w can "access". Property (8.4) states that if p is possibly true in world w then it is true at least in one world that w can "access". Thus, (8.3) and (8.4) set up a stronger communication structure between worlds than only access to the information about another world.

8.4 Definitions of the Possible Worlds

In [Resconi, 2004] we defined the *possible world* as an *entity by which we can procedurally assign a logic value to a sentence q*.

For example given the sentence p = "The color of the house is red" we want to have a *procedure* to know if p is true or false. To know this the procedure could be to ask an agent to assign the logic value to p with an explicit or implicit construction process by which they compares internal knowledge with the external knowledge. In this interpretation, the world is an agent or a group of agents that has sufficient information to build the logic value in specific situations.

Summarizing we think that the **logic value** is not a *simple function* as in the classical modal logic , but is a *constructive explicit or implicit complex process* inside of the agent by which we can define the logic valuation function V. In this paper we show how a possible world can be defined in different situations. One of such situations is described by the topological logic given in a topological space

[Rescher, 1968]. We identify the **topological logic** in a compact way by defining a parameterized operator $P_{\alpha}(q)$ in the propositional logic using Resher notation:

$$P_{\alpha}(q), \text{<substitute p to q>} \tag{8.5}$$

along with a standard notation for a parameterized predicate, $P_{\alpha}(q)$, which means that " The proposition p is realized at the position α", where α is an element in a **topological space**. The **position** α can be a special position in the Cartesian coordinates or can be defined by any positional scheme as an index of the object, e.g., a seat number in the lecture room. The position can be also the time coordinate. Let (8.5) be constructed to mean "the proposition p is true in a possible world located in α". Thus, the Rescher's topological logic can improve the meaning of the possible world concept. In this way, a possible world is defined inside the conceptual space in the Wittgenstein – Carnap logic theory [Wittgenstein, 1953, Carnap, 1971]. An extensive discussion on the *conceptual space* is provided in [Gardenfors, 2000]. The conceptual space is a framework for representation of the concepts in the space of the features.

We frequently compare the experience we are currently having with memory of an early episode. This cognitive capacity shows that we can judge, various relations among our experiences. In particular, we can tell how *similar* a new phenomenon is to an old one. To establish the similarity relation we use the *space of the features as the conceptual space*. Depending on the nature of the features, we may be able to introduce similarity measures and physically meaningful distances. This is typically possible if features have the same physical modality, for example lengths, but if feature 1 is temperature and feature 2 is a pressure the meaning of the distance between points and meaning of geometrical constructs lines, planes is not obvious in such conceptual space. If these difficulties are overcome then introducing the distances we can know the similarity between concepts.

We continue our discussion on the possible worlds. Rescher [1975] defined The Possible World using the concept of **possible individuals** or agents as primitive entities. In Rescher's interpretation the theory of possible worlds and the theory of essentialism discussed by [Rescher, 1975]) are intimately intertwined because possible worlds and *essential properties* that stand in mutually reciprocal coordination. More exactly a property is essential if it takes place in every possible world. Thus, this is the property that can be used to define a given set of possible worlds.

A *possible world* can be defined as a single possible individual or agent, or as a collection of duly combined agents. In fact, every agent can be considered as an entity that can give a logic value to every sentence in that world.

8.5 Meaning of the Possible World

In the classical modal logic [Chellas, 1980] a set of **possible (atomic) sentences** $p_0, p_1, p_2, \ldots$ is given. For each natural number n, the set S_n collects just those worlds at which an **atomic sentence** p_n is true. A **model** M consists of two elements, W and S,

$$M=< W, S >,$$

When W is a set of possible worlds and S s an abbreviation of a sequence S_0 , S_1, S_2,…. of subsets of W.

In other words a *model* M is a pair <W, Ω>, where Ω is a set of Predicates $\{Q_i\}$, such that

$$\forall\, i\ Q_i(w) \Leftrightarrow w\in S_i.$$

According to the model theory [Robinson, 1963] M is a **model** for the set of sentences when all the sentences p_k are true in M through the correspondence C. In the model M each sentence p in language L is matched to a statement that affirms or denies sentence p, this statement could be one of the propositions "p is true" or "p is false". In Figure 8.1 we show correspondence C between the language L and the model M.

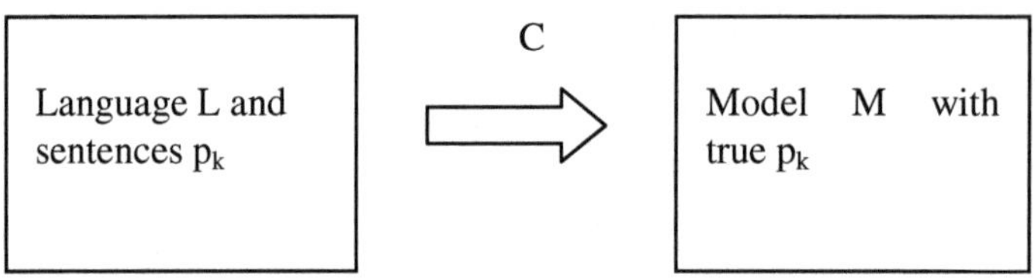

Fig. 8.1 Abstract language and model M where p_k are true

In accordance with the model theory correspondence C provides semantics for language L and expressions p_k because it identifies if p_k is true. The model M can be considered as a set of worlds where all the propositions p_k are true. Thus, a set of worlds provides a **context** where p_k are all true. In the model theory terms, this context is called a **model** for $\{p_k\}$ by C.

On the other hand, model theory is a theory of truth. It aims to characterize all true sentences of a certain *object–language*. Object language is the language in which we write the sentences. The object-language is normally a logic language based on the meaning of the proposition and the meaning of their meta-language translation. The meaning of the object-language and sentences are the logic values true and false. The meaning of the *meta-language* is the map of the object language expressions with other domains such as mathematics, geometry and so on.

According to Tarski [1956] the notion that "x is a true sentence" is contained in the following schema

$$\text{x is a true sentence if and only if q} \tag{8.6}$$

Thereby x is a meta-language name of a sentence in the object-language L and q is a meta-language translation by C in the model M of the object-language sentence. Tarski provided an example of semantic definition for x that is "it is snowing":

"it is snowing" (x) in the object-language L is a true sentence (q) if and only if it is snowing (q) in the model M of the world.

8.6 Discussion of Tarski's Truth Definition

In mathematical tasks expressions such as "not", "and", "for some", is included in" have clear meaning as *mathematical concepts*. The model theory naturally interprets the model structure as a *representation of reality*. From mathematical viewpoint Tarski's definition of "True" is well defined, however, in many cases expression (8.6) is vacuous if there are *no explicit criteria* as to when the meta-language sentence holds or does not hold in the model M. In fact, Tarski's logic theory assumes that

1) The meaning is defined as a *direct* relation C between the expressions and their referents and
2) The *meaning relation* C is an *external* relation between language symbol L and its referent M that is "out there in the world" [Tarski, 1956].

The relation C is defined *a priori* without any reference to the process by which we can build the relation C. Such process can be identified by using concepts of agents and ontologies. The *standard ontology* is the description of the concepts and relation among concepts. The standard model theory ontology offered by Tarski and others raises the problem:

"Where do agents fit within standard model theory?"

One of the possible answers is that the agent provides a *theoretical interpretation* of the language in the model. That is correspondence C identified without giving a procedural (operational) mechanism for assigning true/false values in the Model. If this answer is not accepted then it is necessary to resolve the following problems associated with use of standard model theory for a natural language:

1) How can we *interpret* language terms such as the personal pronouns "I" , "you", and "we" without the introduction of agents? In the Montague interpretation, "I," "you", and "we" are interpreted as positions of the agent in a particular space-time. The coordinate approach in the topological logic [Reference?] does nothing more than *assign* referents to context dependent expressions (world).

In traditional logic this assignment is by definition containing any map between symbols and world entities is viewed as an interpretation. In this paper we want to introduce another element that is not present in the ordinary logic. This element is the *agent* which generates the map that. In the ordinary logic, the *map* is defined without any explicit reference to a particular agent which takes the *responsibility* to define this map.

The goal of this paper is to show that with the introduction of the concepts of a world and an agent we are able to introduce the concept of uncertainty to the classical logic structures. In contrast, Tarski's approach uses an external referent or a model without the agent concept. The same takes place in Montague's approach, where every word such as "I" and "you" are only positions. In Montague's approach we cannot assume that an agent exists that decides to assign a logic value to the sentence p. Thus, we cannot have any type of uncertainty associated with possible errors in the agent judgment.

2) Another problem in the standard ontology is in the analysis of *non-literal uses* of words and sentences, such as irony, metaphor, etc. Since there is only one notion of meaning, defined as a direct relation between expressions and the model theory reality, the only way to treat such non-literal uses is by postulating syntactic-semantic *ambiguity*. More exactly this means that without expanding an ontology by introducing an agent that can give two different meanings to a sentence we may not be able to analyze irony and metaphors. With an agent we can introduce an external meaning and an internal meaning that somebody wants to express by irony and metaphors.

3) The next problem is to relate the standard model theory approach to the *subjective belief worlds*. So the model structure must describe not only the objective that is the real and possible worlds, but also the subjective world.

4) The fourth problem is the *treatment of vagueness* in the standard model theory. One may want to treat the vagueness of natural language concepts in terms of *different degrees of truth*.

5) Finally we consider the possible treatments of non-logic meaning in the standard model theory. The words "red", "green" and "blue" behave alike syntactically and semantically in that they denote functions with the same domain range structure on the same semantic type. What can we say about the *difference* of their meanings? The words such as "red" and "green" are color attributes of objects. These words are of the same type; because they play the same adjective role in sentences. Internal images of "red" or "green" can be a completely different one from one an another for individual agents. Thus, introduction of the agent generates a new degree of uncertainty related to the *internal perception* of the attributes such as a color of the object.

Kripke [1974] defines concepts of the world and a society of individuals using *individual entities connected* with one and other. Individual entities are the basic elements to introduce pragmatics. Thus, to solve the classical ontology problems 1 to 5 listed above we apply a pragmatic approach to the natural language [Hausser, 1987] and intelligent agents [Resconi, 2004].

The purpose of the natural-language processing is to model the communication between the speaker and the hearers. Communication may be viewed as sharing and extending speaker-hearers internal models of beliefs. These may or may not reflect the external "real" world). Referential semantics [Hausser, 1987] is based on the idea that knowing the meaning of a sentence may be equated with knowing under what circumstance (possible world) the sentence would be true. In the referential semantic an approach which aims at modeling the information processing of the agent speaker and the agent hearer is called a Paradigm II. In this case, the model is treated as part of the speaker and the hearer.

Below we discuss concepts of **context** and **adaptive agents**. It is noted that the *agent of the first order* changes the environment by using the input sensors. For example, the agent can move the object and thus change the environment.

An *agent of the second order* or an *adaptive agent* can act to understand rules of the external environment, and to simulate the perceived environment inside agent's mind before acting. This agent adapts the internal rules to the external rules and adapts the internal images to the external images.

With a second order agent we can know if a sentence is true or false. Using this approach, **worlds** and **agents** are **synonymous** with each other. The definition of the world concept should not be separated from the concept of adaptive agents of the second order.

Below we fuse two approaches presented in [Hausser, 1989] and [Resconi, 2004]. In [Resconi, 2004] a *second order intelligent agent* is modeled using the concept of communication. This means we can say if a sentence is true or false and interpret and understand the sentence. The communication involves two points:

(1) "The external sentence and its context" and
(2) "The internal sentence and its context".

The first one of these includes

(a) The *sentence* of the sender S and the
(b) *Physical facts* for example boxes with their colors.

Here the sentence is a physical entity, for example either a sentence written on paper or a sequence of recorded sounds.

In the second one sentence and physical facts become concepts of *sensory perception*:

1. The comprehension of the sentence by the agent which knows the language and
2. Comprehension of the physical facts by the agent for example the concept of the color is the agent interpretation of the physical color in the context of the facts.

The architecture of such an adaptive agent is shown in Figure 8.2.

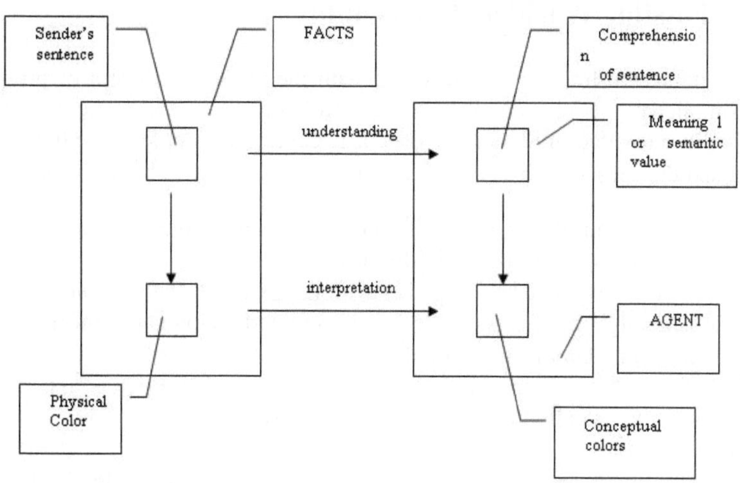

Fig. 8.2 Architecture of the adaptive agent that interacts with the context of the facts

Given an object, we can say that the object is red only when the internal concept of the agent matches the physical interpretation of the color. True and False values represent a result of the matching process inside the agent. The agent has two different channels in which to say that a color is true. Within the internal channel, the agent relates the internal meaning of Symbols, Icons, Indexes, and Pictures with the concepts in the context of the facts in order to provide understanding. The external channel provides an interpretation of the external color using the sensory organs. When we see the color of an object and we match the internal color with the external color, we define the value of the color attribute for the object. A matching process provides the True color. In this way, we can solve many difficulties inside the Tarski's concept of semantic values and define the world by using an individual active entity or agent. In Figure 8.3 we show the connection between the real world and the agent.

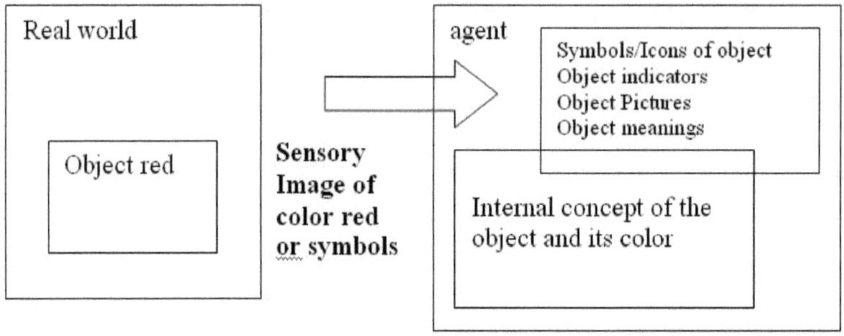

Fig. 8.3 Adaptive agent that knows when the sentence "The color of an object is red" is true

8.7 Possible World and Probability

At this point, we link a possible world with the classical concept of probability. For simplicity, we rewrite the classical Kolmogorov's axiomatic definition of probability provided in Shafer, Vovk, 2001. Given the sample space the Universal Set Ω and the set F of subsets of Ω, which we call *events* we have five axioms:

1. F is a field of sets, that is if F contains E and G then F also contains their Intersection, Union and Difference,
2. F contains the set Ω Together with the axiom 1. This axiom states that F is an algebra of sets. When F is closed under Countable Infinite Intersections and Unions it is called A σ Algebra,
3. To each set E in F a non-negative real number P(E) is assigned. This number is called the *probability of the event E.*
4. $P(\Omega) = 1$
5. If E and G are disjoint ($E \cap G = \varnothing$) then $P(E \cup G) = P(E)+P(G)$

The set theory is a model for the classical logic because there is an isomorphism between them. In fact given the propositions p and q in the proposition logic, we can generate other propositions by the three elementary operations of $\neg$ (negation), $\wedge$ (conjunction), and $\vee$ (disjunction).

In this chapter we use notation $s \rightarrow S$ to indicate that a proposition s is true for every element in the set S. Similarly we may have $\neg s \rightarrow S^C$. Here S^C is the complement set of S.

Similarly $s \wedge q \rightarrow S \cap Q$

$$s \vee q \rightarrow S \cup Q$$

Using the Kolmogorov's definition of the probability, when $S, Q \in \Omega$ and $S \cap Q = \varnothing$.

we have $S \cup S^C = \Omega$ and

$$\neg s \rightarrow S^C, \quad P(S^C) = 1 - P(S)$$

$$s \wedge q \rightarrow S \cap Q, \quad P(S \cap Q) \rightarrow P(S \mid Q)\, P(Q)$$

$$s \vee q \rightarrow S \cup Q, \quad P(S \cup Q) = P(S) + P(Q) \quad \text{with } S \cap Q = \varnothing.$$

Briefly, we can associate one and only one expression in probability calculus for every expression in the logic. In a similar way when the set S is a set of different objects, we can say that S has a set of adaptive agents or worlds. One for any object in S, by which we can know whether it is true or false for a particular property of the object.

The sample space Ω is a collection of different independent objects (worlds). Inside Ω no logic conflict is possible because the objects (worlds) in S for which s is true and the objects in S^C where s is false or $\neg s$ is true are different. Therefore, we have

$$s \wedge \neg s = F \quad \text{and } S \cap S^C = \varnothing$$

When s is true for one object it cannot for the same object be false. The set of worlds S associated with s has no object in common with the set of the worlds S^C.

For the equal probability hypothesis, the probability of the event S is equal to the fraction of the objects in Ω for which the proposition s is true. In Ω there are objects in for which s is true and other objects for which $\neg s$ is true. The probability measure does not carry the information on the **individual object as**, it is lost during the process of its computation. That is a generalization. Only the logic value (false, true) of s for the property of the object will be remembered. For example, given six objects

$$\Omega_{ob} = \{ Ob_1, Ob_2, \ldots\ldots, Ob_6 \}$$

associated with a possible world for each object

$$\Omega_w = \{ w_1, w_2, \ldots\ldots, w_6 \}$$

and an adaptive agent C_i that decides if the sentence p is true or false at every object

$$\Omega_C = \{ C_1 , C_2 ,......,C_6 \}$$

There may a set of logic values in every world.

$$\Omega_T = \{ \text{True, False, True, True, False, True} \}.$$

In conclusion, the probability to have the property or the **event** p is equal to

$$P(q) = P(E) = 4/6$$

where $E = \{ Ob_1 , Ob_3 , Ob_4 , Ob_6 \}$

When we permute the objects in this way

$$\Omega^* = \{ Ob_2 , Ob_1 , Ob_3 , Ob_5 , Ob_6 , Ob_4 \}$$

we have the set of worlds

$$\Omega^* = \{ w_2 , w_1 , w_3 , w_5 , w_6 , w_4 \}$$

and

$$\Omega^*_T = \{ \text{False , True , True , False , True , True} \}$$

and

$$P (q) = P (E^*) = 4/6$$

where $E^* = \{ Ob_2 , Ob_3 , Ob_5 , Ob_6 \}$

Remark

We remark that in the Kolmogorov's definition of probability with the world interpretation we have *symmetry.* For all permutations of the objects in the set Ω the number of worlds, where p is true is always the same. Thus, P(q) is **invariant** to the permutation operator.

Summing up, the uncertainty in the probability calculus is related to the loss of individuality of the objects. In fact, we are interested only in the number of worlds or objects where p is true without any remark regarding the individual object. The information is connected with the set of events where p is true, but a huge number of individual objects can generate the same set of events where p is true. No information is known on the structure of object inside the set Ω .

8.8 Break of Symmetry in Probability Calculus and Evidence Theory by Using Possible Worlds

As we have seen above in the classical definition of the probability, P(p) is **invariant** to the permutation operator. It is assumed that the worlds are **independent** and **are not connected** with others. However, introduction of *relations between worlds* via an accessible relation R can change the Kolmogorov's axioms as follows:

1 F is a *field* of sets. This means that if F contains E and G then F contains their intersection, union, and difference.

2 F contains the set Ω (that is the Ω or the universal set). Together with axiom 1, this says that F is an *algebra of sets*. When F is closed under countable infinite intersection and union it is called a *σ algebra.*

3 Another set $W=\{w_i\}$ is also given. Each element of W is called a *world*. Every element Ob_i of Ω is associated with one and only one world w_i in W and with an *accessible relation* R in Ω. If $R(Ob_i,Ob_j)=T$ then element Ob_i can "access" element Ob_j. That is it can get information about element Ob_j. Having each Ob_i associated with some world w_i, it is possible to say that world w_i can "access" world w_j, that is it can get information about world w_j. In the simplest case Ob_i can be the same as w_i. Elements Ob_i also sometimes called objects or elementary events. Similarly set E of elementary events Ob_i is called a set of *events*. The predicate $Q_j(Ob_i)$ is defined for each elementary event Ob_i. If $Q_j(Ob_i)$=True then we say that event Ob_i is true (takes place). In essence, predicate Q_j defines the concepts elementary events. The predicate Q_j creates a structure of events. More generally, a set of predicates Q_j and axioms that set up relations between these predicates provide a structure for the set of the worlds. An *Ontology Agent (OA)* can set up this structure and *evaluation agents* (EAs) can evaluate if $Q_j(Ob_i)$ is true or not for a world w_i that is associated with Ob_i.

4 A non-negative real number P(E) is assigned to each set E in F. This number P(E) is called the *probability of the event (set) E.* Note: traditionally P(E) reflects the number of elementary events in E relative to total number of events in Ω. <Probabilities in models, Karnap, make reference, Karnap, 1962 >

5 $P(\Omega) = 1$. This axiom reflects the intuitive idea that number of elementary events in Ω relative to itself is 1.

6 If E and G are disjoint ($E \cap G = \varnothing$) then $P(E \cup G) = P(E)+P(G)$.

7 Two non-negative real numbers $P_1(E)$ and $P_2(E)$ are assigned to each set E in F such that $P_1(E) \le P(E) \le P_2(E)$, where $P_1(E)$ is the *relative number of worlds* in E that are at least *necessary* true *for one element in E* , $P_2(E)$ is the relative number of worlds that are at least *possibly true for one element in E.* Thus, $P_1(E)$ and $P_2(E)$ are lower and upper limits of probability p(E), also $[P_1(E), P_2(E)]$ is an interval probability.

<we need to define here concepts of possibly true and necessary true using accessibility relation R or make a reference to the part of the paper where it was done>.

8 The measure m(E) is the relative number of worlds that are possible true *for all the elements in E and cannot possibly be true for other elements of F.*

Statements.
If axioms 1-8 take place then we have the classical relation

$$\sum_{2^\Omega} m(E) = 1 \,,\; P_1(E) = \sum_{H \subseteq E} m(H) \,,\; P_2(E) = \sum_{H \cap E \neq \varnothing} m(H)$$

where the computation of m, P_1 and P_2 is based on use of accessibility relation. We can also show that

$$P_1(E) \leq P(E) \leq P_2(E)$$

and if E and G are disjoint ($E \cap G = \emptyset$) then $P(E \cup G) = P(E)+P(G)$,

$$P_1(E \cup G) \geq P_1(E)+P_1(G)\ ,\ P_2(E) = 1 - P_1(E^C)$$

The difference between m and probability P can be represented by a measure of breaking symmetry in m. Function m depends on relation R that breaks the symmetry of worlds or elements involved. In probability calculus the elements have the same value for all the possible permutations, that is these elements are symmetrical. For P_1 and P_2 the permutation changes the meaning of necessary and possible true values.

It is remarked that ;

It can be proved that ,

$$P_1(E_1) + P_1(E_2) \leq P_1(E_1 \cup E_2)$$

In $P_1(E_1)$ we have all the worlds in which the relations are internal to E_1 and also for $P_1(E_2)$. In $P_1(E_1 \cup E_2)$ we have the worlds which relations are internal to E_1 and E_2 plus the worlds $P(E_1, E_2)$, which one relations are between E_1 and E_2 so we have

$$P_1(E_1 \cup E_2) = P_1(E_1) + P_1(E_2) + P(E_1, E_2)$$

Can be proved that

$$P_2(E) = 1 - P_1(E^C),$$

where E^C is the complementary event of E.

For example:

For $E = \{ C_1, C_2, C_3 \}$ and the relation R in Figure 8.4 only the world w_1 is possibly true for every element C_1 , C_2 , C_3 of E.

For example:

Given three types of objects with three colors we have

The set of objects (worlds) with red color is $\{ w_2 \}$,
The set of objects (worlds) with green color is $\{ w_1, w_5, w_6 \}$,
The set of objects (worlds) with white color is $\{ w_4, w_3 \}$,

Probability for the red color P (red) = 1/6
Probability for the green color P (green) = 3/6
Probability for the white color P (white) = 2/6

Accessibility relation R is presented in Table 8.1 and illustrated in Figure 8.4.

Table 8.1 Logic Communication which accessibility is the relation among Semantic agents or worlds

	w_1	w_2	w_3	w_4	w_5	w_6
w_1	1	1	1			
w_2		1		1		
w_3			1	1	1	
w_4				1		1
w_5					1	1
w_6						1

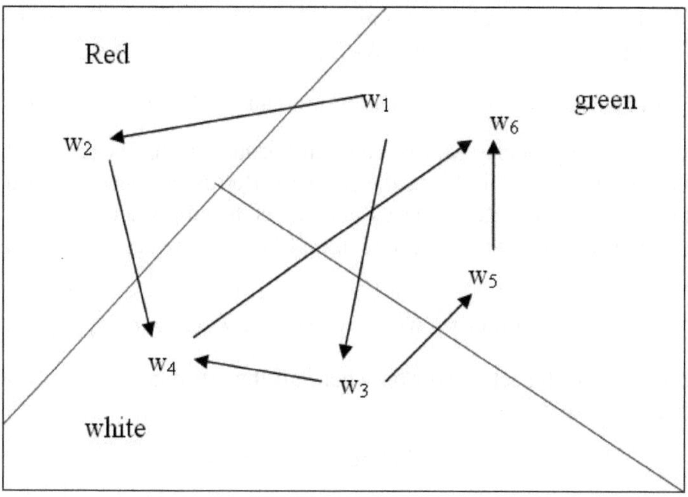

Fig. 8.4 Graph of the relation R

Given Π = {Red , Green, White } the power set is

$$2^{\Pi} = \{ \{Red , Green, White \} , \{Red , Green \} , \{Red ,White \} , \{ Green, White \},$$

$$\{ Red \} , \{ Green \} , \{ White \} , \varnothing \}$$

We remark that P ({ Red }) = 1/6 , P ({ White }) = 2/6 and P ({ Green }) = 2/6 with the additive property

$$P (\{Red ,White \}) = 3 / 6 .$$

To compute P_1 we must enumerate the set of worlds where this property is necessarily true in all the accessible worlds.

The set of worlds where "Red" is necessary true is ∅
The set of worlds where "White" is necessary true is ∅
The set of worlds where "Green" is necessary true is { w_5 , w_6 }

Let us comment the difference between the classical probability with 4 without the accessibility relation A_r. We have P({Red})=1/6 because only w_2 has this property out of six worlds. This property {Red} is not a necessary property because w_2 it is not expanding from w_2 if we follow the accessibility relations A_r ,shown by the arrows in Figure 8.4. We can access w_4 with A_r , but {Red} is not a property of w_4, thus, property {Red} cannot be necessary. In fact there is no a single world beyond w_2 that is connected with w_2 by the accessibility relation A_r that has the {Red} property. Thus P_1({Red}=0.

Because in the worlds w_5 and w_6 the property "Green" is necessary true. In fact Green is true in w_5 and in w_6 but is also true for all the worlds accessible to w_5 and w_6 . We have that the relative number of worlds for which Green is possible true is 2/6 and P_1 ({ Green }) = 2/6. We have

$$P_1 (\{ Red \}) = 0 , P_1 (\{ White \}) = 0 \text{ and } P_1 (\{ Green \}) = 2/6$$

The world w_2 the property Red is true, but because w_2 has access to the world w_4 where the property Red is false, the property Red is not necessary true in w_2. So the set of worlds where Red is necessary true is the empty set.

Because the set of worlds where Green and White are necessary true is { w_4 , w_3 , w_5 , w_6 } we have

$$P_1 \{ Green , White \} = 4/6 > P_1 (\{ White \}) + P_1 (\{ Green \})$$

We also remark that the set of worlds where all the attributes is necessary true is the Universal set. We have

$$P_1 (Red , Green , White) = 1$$

The measure P_1 is a non additive measure. For the measure P_2 we have

$$P_2 (\{ Red \}) = 2/6 , P_2 (\{ White \}) = 3/6 \text{ and } P_2 (\{ Green \}) = 4/6$$

We have the attribute "Red" is possible true for the set { w_2 ,w_1 }. In w_2 "Red" is true and is false in w_1 but it is true in the accessible world w_2

The attribute "White" is possibly true for the set { w_4 , w_3 , w_2 }.

In w_4 and w_3 "White" is true , in w_2 "White" is false but in the accessible worlds w_4 "White" is true, so we have that "White" is possible true for the set { w_4 , w_3 , w_2 }. The attribute "Green" is possibly true for the set { w_6 , w_5 , w_4 , w_3 } because "Green" is true in w_5 and also in w_6 but is false in w_4 and w_3 are true in the accessible worlds w_6 and w_5

In w_1 the attribute "Red" is false but because has access to the world w_2 where "Red" is true. In the world w_1 the property "Red" is false but it is possible true. In conclusion we have that

Red or Green are possible true in { w_2 , w_1 , w_6, w_5 , w_4 , w_3 }

$$P_2 (Red , Green) = 1$$

Red or White are possible true in $\{ w_2 , w_4 , \quad w_3 , w_1 \}$

$$P_2 (\text{Red} , \text{White}) = 4/6$$

Green or White are possible true in $\{ w_2 , w_1 , \quad w_6 , w_5 , w_4 , w_3 \}$

$$P_2 (\text{Green} , \text{White}) = 1$$

$$P_2 (\text{Red} , \text{Green} , \text{White}) = 1$$

the measure m(E) is the relative number of worlds that are possible true *for all the elements in E.*

We have only one world w_1 for which all the elements Red, Green and White are all possible true. In w_1 Green is true , in w_1 "Red" is false but is possible true. In w_1 "White" is false but is possible true. So we have

$$m (\text{Red} , \text{Green} , \text{White}) \quad = 1/6$$

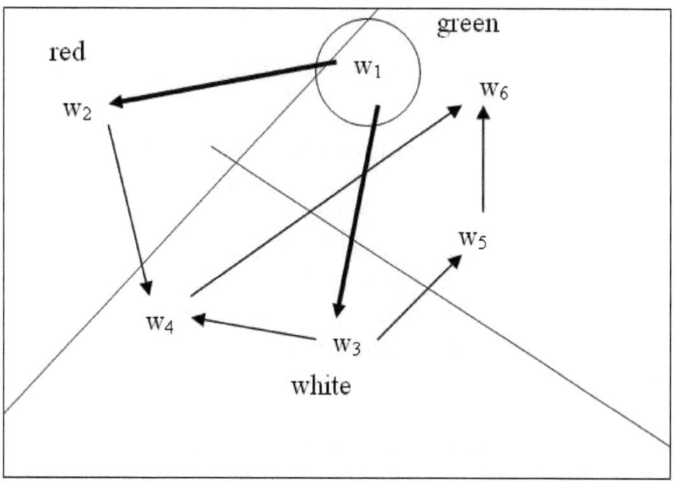

Fig. 8.5 World w_1 that has access to the tree elements Red,White and Green. In w_1 Green is true, Red and White are possible true.

We have that

The set of worlds where only "Red" is possible true is the empty set $\varnothing$
The set of worlds where only "Green" is possible true is $\{ w_5 , w_6 \}$
The set of worlds where only "White" is possible true is $\varnothing$

So we have

$$m (\text{Red}) = 0$$

$$m (\text{Green}) \quad = 2/6$$

$$m\,(\text{White}) = 0$$

Because no words are true for all the elements Red and Green we have

$$m\,\{\text{Red}, \text{Green}\} = 0$$

In the word w_2 Red is true and White is possible true and no other elements are possible true, we have

$$m\,(\{\text{Red}, \text{White}\}) = 1/6$$

In the words w_4 , w_3 White is true and Green is true and no other elements are true, we have

$$m\,(\{\text{Green}, \text{White}\}) = 2/6$$

In the word w_1 Green is true and Red , White are possible true and no other element are possible true, so we have

$$m(\{\text{Green}, \text{White}, \text{Red}\}) = 1/6$$

In conclusion for the measure m that is basic probability assignment in the evidence theory is

$$\begin{aligned}
&m\,\{\text{Red}, \text{Green}, \text{White}\} = 1/6\\
&m\,\{\text{Red}, \text{Green}\} = 0\\
&m\,\{\text{Red}, \text{White}\} = 1/6\\
&m\,\{\text{Green}, \text{White}\} = 2/6\\
&m\,\{\text{Red}\} = 0\\
&m\,\{\text{Green}\} = 2/6\\
&m\,\{\text{White}\} = 0
\end{aligned}$$

So we have

$$m\,(\{\text{Red}, \text{Green}, \text{White}\}) + m\,(\{\text{Red}, \text{Green}\}) + m\,(\{\text{Red}, \text{White}\}) + m\,(\{\text{Green}, \text{White}\}) + m\,(\{\text{Red}\}) + m\,(\{\text{Green}\}) + m\,(\{\text{White}\}) = 1$$

We note that the property

$$P_1(E) = \sum_{H \subseteq E} m(H)$$

is true because we have the sets inside { Green , White } are { Green , White }, { Green }, { White } For the point 9 at the page 15 we have P_1 (Green , White) = m (Green , White) + m (Green) + m (White) = 4/6

We see in the definition 5.

The measure m(E) is not additive and the conditions

$$m\,(A \cup B) > m\,(A) + m\,(B),\ m\,(A \cup B) = m\,(A) + m\,(B) \text{ and } m\,(A \cup B) < m\,(A) + m\,(B).$$

are context dependent.

In one context we have

$$m(\{Red, Green\}) = 0 \text{ and } \quad m(\{Red\}) + m(\{Green\}) = 2/6$$

Then

$$m(\{Red, Green\}) < m(\{Red\}) + m(\{Green\})$$

In other contexts we have

$$m(\{Green, White\}) = m(\{White\}) + m(\{Green\})$$
$$m(\{Red, White\}) > m(\{White\}) + m(\{Red\})$$

Remark

The modal logic representation of the probability P and of the interval probability (P_1 , P_2) by accessible relation R , can be used to model the imprecise probability, evidence theory (Dempster- Shafer theory), where P_1 = Belief measure and P_2 = Plausible measure and rough sets and granulation theory [Pawlak, 1991].

Given the approximate space (U , θ) , where θx is the equivalent class of x ∈ U , the rough set is a pair $(\underline{X}, \overline{X})$ where $\underline{X}$ is the lower approximation of X and $\overline{X}$ is the upper approximation of X . When in Kolmogorov's definition of probability we compute the probabilities for $\underline{X}$, X and $\overline{X}$, we have

$$P(\underline{X}) \leq P(X) \leq P(\overline{X}).$$

When at every equivalent class θx we associate a world and an accessible relation between the equivalent class θx and all the equivalent classes in the environment of θx , we have that $P(\underline{X}) = P_1$ and $P(\overline{X}) = P_2$.

8.9 Fuzzy Set Theory

8.9.1 Modified Probability Axioms Approach

One of the major motivations to introduce fuzzy sets was the need to deal in a formal way with imprecision in *natural languages*. Often imprecision means that a word representing an "entity" such as age, temperature, velocity … cannot have a crisp logic evaluation. The meaning of a word in a proposition may usually be evaluated in different ways for different assessments of an entity by different agents, that is worlds. In an attempt to respond to these challenges the fuzzy sets theory made significant progress, but met fundamental conceptual difficulties. Fuzzy sets operations are heuristic and often ad hoc.

Our approach to build a stronger foundation for the fuzzy sets theory is to expand it by *incorporating the concepts of a population of observers and a population of possible worlds* into the algebra of fuzzy subsets, the universe of discourse and to make any explicit linkage between these concepts.

Different and in some cases *conflicting evaluations* for the same proposition may occur for the same entity when the property is evaluated in *different worlds in different way*. We have no agreement among all the worlds or agents. For example, suppose the entity is a particular temperature of a room, and we ask for the property *cold*, when we have no instrument to measure that property. The meaning of the entity, "temperature", is composed of assessments that are the *opinions* of a population of observers that evaluate the predicate "cold". Without the population of observers, and their assessments, we cannot have the entity "temperature" and the predicate "cold". When we move from crisp sets to fuzzy sets, we move from a simple way of producing aggregate assessments such as (A and B) =true. If A and B are true a much more complex way of producing such an assessment. In an abstract way, the population of the assessments of the entity becomes the population of the worlds that is associated with crisp logic evaluations of the set of propositions.

If we know only the name of a person (entity), we cannot know if s/he is "old". Additional observation-based information (context) could be that s/he is a manager, married, and plays with toys. The last information on toys conflicts with information that he/she is "old" that is providing a contradictory context.

For the fuzzy sets the Kolmogorov's axioms are changed in this way:

1. F is a field of sets which means that whenever F contains E and G, it contains their intersection, union, and difference).
2. F contains the set Ω. Together with axiom 1, this says that F is an algebra of sets on Ω.
   When F is closed under countable infinite intersection and union, it is called a σ algebra.
3. Every element (object Ob_i) of Ω is associated with one and only one world.
4. A non negative real number μ(E) is assigned that is equal to the relative number of worlds that are true in E.
5. If E and G are *disjoint* ($E \cap G = \varnothing$) then $\mu (E \cup G) = \mu (E) + \mu (G)$. Note that the traditional fuzzy sets theory *does not distinguish* between disjoint and overlapping sets. It defines $\mu (E \cup G) = \max(\mu (E), \mu (G))$ for both disjoint and overlapping situations, in essence being *context free*.
6. Let set E be in the field of sets F and *property* C is assigned true inside E and is false outside of E in F for the object Ob, C(Ob)=F. <BK think about axioms for the individual object only>
7. Let set E be in the field of sets F and *property* D is assigned true inside E and is false outside of E in F for the object Ob, D(Ob)=T.

8. Properties C and D can be assigned independently in axioms 6 and 7 to object Ob, but can be contradictory for this object Ob: C(Ob) $\Rightarrow \neg$ D(Ob), that is C(Ob) causes that D(Ob) should be false. The classical Kolmogorov's axioms do not allow such contradiction.

Thus, this object Ob has two or more different *contradictory properties* in F at the same time. Such object can be called *incoherent object with fuzzy properties.* For example, a pencil with two colors is an incoherent object, because we cannot answer the question: What is the color of the pencil?". We can only say that for a fraction it is of one color and for the other fraction it is of another color. Often incoherence is a result of a restricted set of terms (red, green) without a term "partially red and partially green" rather than a property of the object Ob itself.

Comment: The axiom 5 that disjoint sets exist is a cornerstone of the Kolmogorov's axiomatic probability theory. This is a very natural assumption when we talk about card games ("7" cannot be "6" at the same time). The property C of the object Ob to be "7" and property D of the object to be "6" are mutually exclusive (contradictory for the same card Ob): C(Ob) $\Rightarrow \neg$ D(Ob). But for a more open natural language processing situations this assumption often is not true. For instance, sets of "young people" (property C) and "medium age people" (property D) are mutually exclusive for age 15 but overlap at the ages of 30-32. Thus, C(32) $\Rightarrow$ D(32) but C(15) $\Rightarrow \neg$ D(15).

There is a possibility to make disjoint sets by introducing concepts such as "young and not medium age", "medium age and not young", but these concepts are rather *artificial* and communication with such long terms hardly can be called *natural* language communication. Thus, we have a fundamental question" How to expand Kolmogorov's theory for such not-disjoint situations?" The axioms 6-8 introduce properties C and D to deal with both contradictory and non-disjoint situations.

Example

Given the set of colors Ω = {Red, Green, White } and one object which color is the superposition of Red, Green and White. The color of the object is unknown but is the superposition of three colors. In fact, the object has the three colors in different parts. A group of agents or worlds S_1 see only the red part of the object, another group S_2 sees only the green part, and third group S_3 sees only the white. However, the object itself has all the three color together so the color of the object is fuzzy. If we have this particular situation

$$\text{Red} \rightarrow S_1 = \{ w_2 \}$$
$$\text{Green} \rightarrow S_2 = \{ w_1 , w_6 , w_5 \}$$
$$\text{White} \rightarrow S_3 = \{ w_4 , w_3 \}$$

Finally, the object has three colors in the same time, but with different *weights*.

Table 8.2 Possible world and cases relation

	Red	Green	White	Expression
w_1	0	1	0	Green
w_2	1	0	0	Red
w_3	0	0	1	White
w_4	0	0	1	White
w_5	0	1	0	Green
w_6	0	1	0	Green

The three cases are one independent from the others and describe the conflicting color of the object.

The color of the object can be represented as a superposition of different colors with different weights proportional to the number of agents that recognize the property as true.

The *relative number of the agents* is

$$\mu(q) = \frac{\|W(q)\|}{\|W\|}$$

where $\mu(P)$ is the membership function for a fuzzy set of the color q. Using this formula, we have

$$\mu(\text{Red}) = 1/6, \quad \mu(\text{Green}) = 3/6, \quad \text{and} \quad \mu(\text{White}) = 2/6$$

for the previous example. This example is based on the assumption that worlds (agents) are *independent.* It fits well to the classic probabilistic axiomatic formalism with $\mu(\text{Red}) + \mu(\text{Green}) + \mu(\text{White}) = 1$. Fuzziness here is in the fact that agents (observers) see only different part of the object but judge the color of the *whole object.* In essence they observe different objects but assign the color to the whole object. To resolve uncertainty in this situation we have two options: (a) to introduce another set of objects (parts of the original object), or (b) expand the set of colors allowing color combinations such as Red and Green, that will indicate that the object has both these colors. In many cases both options may not be practical: alternative (a) may produce too many new different objects and alternative (b) may produce too many color combinations. In essence here fuzziness, uncertainty is caused by our willingness to limit the set of entities considered as objects and a set of colors and their combinations. Such willingness can be motivated by our inability to operate with a large number of objects and colors.

8.9.2 *Fuzzy Logic Situations*

Given the set W of worlds with a set of properties true for subsets of W, and which superposition gives the property of the object, we can have different situations:

1. The set W of the worlds or agents is divided in equivalent classes of worlds one for every property
2. The set W is the union of subsets where intersection is empty

In the first case, the object property is the simple superposition of the properties in the set W as we show in Table 8.2. These properties may or may not contradict each other in agent perception. If there is a *contradiction* then we have a *probabilistic case*. If there is no contradiction between agents perception of properties we have a deterministic *classic logic case*.

In Table 8.3 and in Figure 8.4 we show the *second situation* where the first agent w_1 sees *both red and green parts* of the object, but other agents see *only one color* parts. All agents are forced to express their perception in a predefined *restrictive language* of colors and object concepts. The agents can use only the three fixed colors and can assign the color to the whole object without right to assign it to the part of the object Ob. This creates the situation shown in Table 8.3, where the agent assigned two colors to the object, instead of instructing that part A is Red and part B is Green of the object Ob.

Example

Table 8.3 Cases where multicolor parts of the object observed by some agents

Agent/world	Red	Green	White	Expression	Parts of the object
w_1	1	1	0	Red ∧ Green	A, B
w_2	1	0	1	Red ∧ White	A, C
w_3	0	1	0	Green	B
w_4	0	0	1	White	C
w_5	0	1	0	Green	B
w_6	0	1	0	Green	B

In a graphical way we have

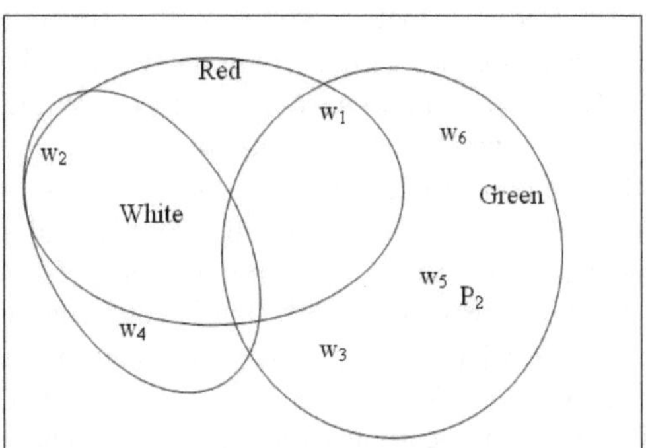

Fig. 8.6 Complex superposition of logic expressions

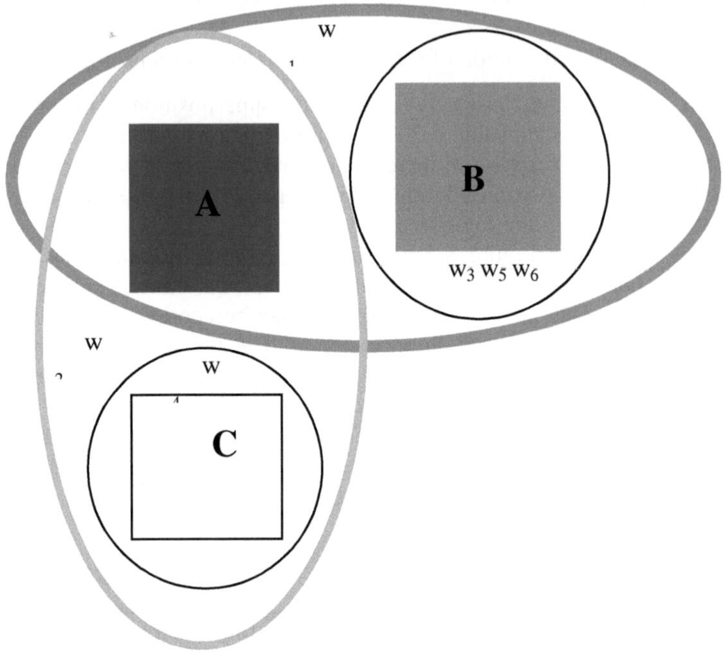

Fig. 8.7 Complex superposition of object's parts

We can compute $\mu(q) = \dfrac{\|W(q)\|}{\|W\|}$ for q = red using Table 8.3. It is 2/6. We also can compute

μ (Green)=4/6 and μ(White)=2/6. The total sum 2/6+4/6+2/6=8/6>1, that is to violate probability axioms and reproduces the situation known in *fuzzy logic*.

If we compute μ(A=Red) =||W(A=Red)||/||W(A)||=2/2=1 and μ(A≠Red)=0, thus μ(A=Red)+ μ(A≠Red)=1, that reproduces the situation known in the *probability theory and classic logic*. Similarly we have μ(B=Green) =||W(A=Green)||/||W(B)||=4/4=1 and μ(A=White) =||W(A=White)||/||W(C)||=2/2=1 and reproduce probabilistic and classic logic situation.

Let us analyze why the fuzzy logic situation produced the sum=8/6. We have Ω = {Red, Green, White } as a base (atomic elements) of the algebra of properties then in the axiomatic probability theory we can have only sets: {R}, {G},{W}, {R or G}, {R or W}, {G or W},{R or G or W}, that is equivalent to {R}, {G},{W}, {R ∪ G}, {R ∪ W}, {G ∪ W},{R ∪ G ∪ W}. We cannot produce sets {R & G}, {R & W}, {G & W}, {R & G & W}, that is {R ∩ G}, {R ∩ W}, {G ∩ W}, {R ∩ G ∩ W} within this algebra. We can only join atomic elements, we cannot divide them.

To be able to divide them and convert fuzzy logic situation to probabilistic one we need to have a new algebra, Ω\` with elements such as R ∩ G ∩ W}, {R ∩ ¬G∩ W}}, {R ∩ ¬G∩ ¬W}{R ∩ ¬W} for example. In this algebra the base red is {R ∩ ¬G∩ ¬W}. It is not old red, R from Ω.

This means that the **context is changed**. In Ω we had context {Red, Green, White } as elementary (atomic) units. There are no smaller elements, thus we have no R&G, R&W, R&G&W and cannot use them in Ω. In Ω\` all of them are permitted and what was R in Ω is now R & ¬G & ¬W in Ω\`.

This change has an important implication, {R}, {G}, and {W} are not disjoint anymore and formula $\mu(q) = \frac{\|W(q)\|}{\|W\|}$ does not count occurrences of disjoint events.

If Property q is red it is now R & ¬G & ¬W. There is no single such case in table 8.3 and we cannot compute μ (R & ¬G & ¬W) from Table 8.4 without analyzing the last column that present parts A,B,C and D of object Ob.

Let us discuss the implications of considering non-disjoint events with the sum of measures μ of all alternatives greater than 1 as it is done fuzzy logic. We can assign a measure in any possible way, including any t-norms and t-conorms if we do not combine those measures later, but only compare them directly with each other to find out which is greater. In fact in most interesting and important tasks we want to combine measures. Combining of measures of non-disjoint events is a challenge. Non-disjoint events can happen in a variety of situations. It is possible that a world (agent) is partially blind, cannot recognize red or recognizes it unreliably, can quickly change color judgment or an object actually consists of two objects, one is red and another one is white. If we know for disjoint events μ(Red) and μ(Green) we cannot compute μ(Red or Green) from them.

In probability theory μ(R or G) =μ(R)+ μ(G) for disjoint events, but in our situation it is not the case. We still can use formula

$$\mu(\text{R or G}) = \|W(\text{R or G})\|/\|W\|,$$

but this formula will provide results that differs from both probability and context independent t-co-norms provide. The last formula is truly *context dependent*, which is very desirable, but it does not give any computational shortcuts as probability and fuzzy theories do. In (8.2) we need to compute μ(R or G) directly. We can not use formulas like μ(R)+ μ(G), but we can use a general probability formula μ(R&G)= μ(R)+ μ(G)- μ(R&G). Thus, from pure technical viewpoint the last formula means that we actually created a new probability space with elementary events such as R&G.

What is a relation of a modified probability axioms presented in previous section with fuzzy logic? New axioms open the way of practical use of the concept of an *irrational world* that is beyond the classical modal logic formalism. Note that the philosophic description of the modal logic has a concept of an impossible possible world. The classical model logic is in essence a propositional modal logic without abilities to deal with uncertainties and individual objects. In the classical modal logic the world w is limited by the propositions q and t such as q = "color red" and t = "color green" without any reference to a particular object for which these propositions are true. In fact in the same world w we can have two objects one red and another one green. Thus, in the same world two propositions q = "the object is red", t = "the object is green" can be true without any contradiction, irrationality or uncertainty because objects on q and t can be different. The first order modal logic [Fitting, Mendelsohn, 1999] can resolve this issue.

Such irrationality is not only the result of a limited propositional language, but it is also as a property of the "measurement procedure" that an agent uses to get the truth value of the proposition. For instance, an agent may have only a straight wood stick to measure the length. This stick does not bend and should be applied only once to measure the length of the curve. With this limitation we cannot measure the arc length accurately. We may select different start and end points on the curve close to its ends and have hundreds of different "lengths" of the curve, none of which is correct.

+++++++++++++

When we limit the language that the agent can use judging statements and identified objects of the statements' truth we are limiting accuracy of our judgment ("measurement"). We force an agent to tell only "red" or "green" similarly to the measurement instruction "do not bend a stick". In this case it is not an irrational world itself. It is a world with an *irrational and limited observer.* We can call this world irrational too, but in essence the observer brought irrationality to this world. Thus the irrationality can have quite different sources. If we cannot separate the observer from the instrument we can get irrationality and difficulty of measurement known in quantum mechanics.

Now we want to clarify the relation of our modal logic approach with probability theory and fuzzy logic relative to the concept of the *individual object.* Our modal logic approach allows to introduce uncertainty measures $\mu(Ob)$ for each individual object by constructing a *set of possible worlds for that individual object.* In contrast probability $p(Q(Ob))$ typically measures the number of objects that have property Q relative to the total number of objects. Thus, probability rather captures the distribution of property Q in a set of objects, than really measuring uncertainty that the specific individual object Ob has property Q. To be able to judge the individual object in such probabilistic approach we need to assume that all objects are alike and chances all of them to have property Q are the same. Thus, the probabilistic approach fits less to judging individual objects than the modal logic approach. The fuzzy logic approach has the same goal as we have to introduce a measure of uncertainty of the property of the individual object. The modal logic with possible worlds makes this approach rigorous. It can avoid using heuristic operations often used in fuzzy logic.

Note that there is also another possibly to introduce a measure of uncertainty of the individual object that is termed as context space approach [Kovalerchuk, 1996] that we discuss in the next section. This approach has a natural link to the modal logic approach. Let $W_i=\{w(Ob_i)\}$ be a set of possible worlds for the individual object Ob_i with the same set of predicate terms $\Omega_i= \{P_{i1},P_{i2}, \ldots , P_{in}\}$ for every world from W_i. The set of predicate terms Ω_i can be build in a such way that we can introduce a probability space over these predicate terms. We call this probability space an exact complete context for the object Ob_i.

8.10 Context Space Approach

An alternative way to accommodate contradictory and non-disjoint situations is building an interconnected set of probability spaces. This approach called linguistic context space approach [Kovalerchuk, 1996] is outlined below and is presented in more detail in section 9. We introduce a probability space for an individual age, say 35, and assume that a population of agents assign one of the terms {young, middle-age, old} for this age. Exactly this was done in terms of modal logic above. The world of each agent A for object x is {young, middle-age, old} with agent's estimates of P(x, young), P(x, middle-age), and P(x, old).

In this example the base elements of the probability spaces are numeric ages [0, 150] that overlap for some linguistic age categories. Having age categories "young", "middle-age", and "old" we cannot assign age 35 the only label "young". Age 35 also can be viewed as "middle age" too. Thus we cannot build a single mapping f:

$$f: [0, 150] \rightarrow \{young, middle\text{-}age, old\},$$

<Renumber formulas>

where f(35) will be young only, f(35)="young". It is also meaningful to have f(35)="middle-age".

In contrast, in card games it was possible unambiguously call a card "6" or "7" and then explore probabilities $P_g(6)$ and $P_g(7)$ that a player has them in the game g. However, we cannot build and explore probabilities, P(young) and P(middle-age) that a player has a young or middle age partner in the same way, because we have difficulties to identify these age categories for the partner. Nevertheless, we can build two classes of probabilities P_f and μ, where the first one will be a probabilistic version of function f in (8.1) that we describe below and another one will be similar to $P_g(6)$ and $P_g(7)$.

The probabilistic version of function f can be played by three conditional probability distributions (and spaces):

$$P_{young}(x)= P(x \mid young),\ P_{middle\text{-}age}(x)= P(x \mid middle\text{-}age),\ P_{old}(x)= P(x \mid old)$$

with x from [0,150]. For instance, we may have $P_{young}(35)=0.7$, $P_{middle\text{-}age}(x)=0.2$, and $P_{old}(x)=0.05$. These three probabilities are from different probability spaces and thus their sum may not be equal to 1. The first space may have only two elements {young, not young}, the second one may have two other elements {middle-age, not middle age}, and the third one may consists of {old and not old}. Moreover, the sum operation does not have meaning for probabilities from different spaces. Next we need to build new probabilities p_g(young) and P_g(middle-age) to evaluate game partner's age in a particular game room using $P_{young}(x)$, $P_{middle\text{-}age}(x)$, $P_{old}(x)$ Let's assume that we know age distribution of the people in the game room, $\mu(x)$, where x is a numeric age value. We need to evaluate P_g(young) and P_g(middle-age) using this information. We can use Bayesian approach

$$P_g(\text{young})=\sum_{x=1\text{-}150} p(x| \text{ young})p(\text{young}).$$

Using our previous notation it means

$$P_g(\text{young})=\sum_{x=1\text{-}150} P_{young}(x)\mu(x).$$

Age distribution of the people in the game room, $\mu(x)$ can be computed using standard frequencies. A conditional probability $P_{young}(x)$= P(x | young) can be evaluated by counting for every fixed x, say 35 the number of worlds (cases) where this age is called young, $|W_{young}(35)|$, relative to the total number of "young" worlds $|W_{young})|$ for all ages, that is

$$P_{young}(x)= P(x \mid \text{young})= |W_{young}(35)|/|W_{young}|$$

Here we assume that the age young is one of the three age category alternatives {young, middle-age, old}, that is we have a dependency

$$P_{young}(35) + P_{middle\text{-}age}(35) +P_{old}(35) =1,$$

$$\text{e.g., } P_{young}(35)=0.7, P_{middle\text{-}age}(35)=0.2, P_{old}(35)=0.1.$$

If age young is a part of the space {very young, young, middle-age, old, very old} then | the number worlds where people of age 35 called young, $|W_{35}(\text{young})|$, can be different with property

$$P_{very\ young}(35) + P_{young}(35)+ P_{middle\text{-}age}(35) +P_{old}(35) + P_{very\ old}(35) =1.$$

We also discuss this issue in section 8.9.

In an alternative way we can write

$$P_g(\text{young})=\sum_{x=1\text{-}150} p(\text{young} \mid x)P.(x).$$

A conditional probability p(young |x) can be evaluated by counting how often word "young" is used for a fixed age x, say 35, that is

$$P(\text{young} \mid x)= |W_{35}(\text{young})|/|W_{35}|,$$

where $|W_{35}|$ is the number of worlds (people of age x=35) and $|W_{35}(\text{young})|$ is the number worlds where people of age 35 are called young.

Similarly we also can compute

$$P_g(\text{middle-age})=\sum_{x=1\text{-}150} P(x| \text{ middle-age })p(\text{middle-age}).$$

What is the conceptual difference between p(young |x) and P(x| young)? In p(young |x) the base set is a set of *numeric ages* [0, 150] and in the p(x| young) the set of *linguistic categories* {young, middle age, old}. In the second case we switched the base to words, where the base of the probability space consists of a set of linguistic terms {young, middle-age, old}.

8.11 Comparison of Two Approaches

Both approaches presented in sections 5.1 and 5.2 are intuitively natural and representation of real tasks based on them is also intuitive. The major differences are coming from the procedures to identify uncertainties. Both approaches assume that probabilities of linguistic alternatives can be computed, that is for computing $p_{young}(x)= p(x \mid young)$ and $p_{red}(x) = p(x \mid red)$ we may need to know the meaning of young and red for a particular (or typical) context. Without such context young can be almost any age and red can range from very dark red that is almost black to very light red that as almost white. There are many successful experiments that were able to identify probabilities of linguistic ages for typical contexts [e.g., Hall, Kandel, 1986].

Both approaches can naturally accommodate agents in two ways:

- an agent provides probabilities $P_{young}(x)= P(x \mid young)$, $P_{middle\text{-}age}(x)= P(x \mid middle\text{-}age)$, $P_{red}(x) = P(x \mid red)$ and $P_{blue}(x) = P(x \mid blue)$ for each object x and
- an agent provides probabilities $P_x(young)= P(young \mid x)$, $P_x(middle\text{-}age)= P(middle\text{-}age \mid x)$, $P_x(red) = P(red \mid x)$ and $P_x(blue) = P(blue \mid x)$.

For instance, we can compute P(young |35) as $|W_{35}(young)|/|W_{35}|$, where $W_{35}(young)$ is a set of agents or worlds for which young is true and W_{35} is a set of worlds (people) of age 35. At first glance $W_{35}(young)$ differs from traditional crisp sets, that is it may not be easy to identify that an element belongs to it from the agent's viewpoint, but for the common language use context people do this without much difficulties.

8.12 Irrational World or Agent

Given the sentences P = White and ¬P = Black we may have worlds with ignorance which of these sentences is true for object Ob. A logically rational agent (world) may just acknowledge ignorance and will not assign any value to P and ¬P. An irrational agent (world) can assign logical values illogically, stating that both P and ¬P are true, that implied that the contradiction $P \wedge \neg P$ is true. Alternatively the irrational agent (world) can claim that both P and ¬ P are false, that implies that the tautology $P \vee \neg P$ is false too. See such worlds w_1, w_4 and w_5 in Table 8.4.

An example: A human can be asked: "Mark colors that you see in the object Ob". If the experiment setting allows answering this question two times, but with only two choices: "White" and "Black" a human that see both colors will mark both answers. See this in Table 8.4 for world w_1. There is no contradiction in the context of this experiment setting. World (agent) w_1 is quite logical, but if we do not know the context of this experiment world w_1 looks illogical only looking at the truth table.

We may also consider another context (another experimental setting) that allows answering only once and again with the same "White" or "Black" answers only. In this case a rational agent (world) will not provide any answers, but irrational can randomly answer "white" or "Black". If we would not know this experimental context we will consider the provided random illogical answer quite logical just looking at the truth table.

This brings us to the interesting conclusion: irrationality of the world cannot be judged from the truth tables themselves. In addition we need to know the **context.**

Table 8.4 Irrational world

	P	¬P	Expression
w_1	1	1	$P \wedge \neg P = T$
w_2	0	1	$\neg P$
w_3	1	0	P
w_4	0	0	$P \vee \neg P = F$
w_5	1	1	$P \wedge \neg P = T$
w_6	1	0	P

Allowing contradictory values for P and ¬ P in some worlds and keeping classical logic interpretation of & and $\wedge$ operators leads us to expressions shown in the last column in Table 8.4, where classical contradiction $P \wedge \neg P$ is true in the irrational world w_1 and classical tautology $P \vee \neg P$ is false in the irrational world w_4.

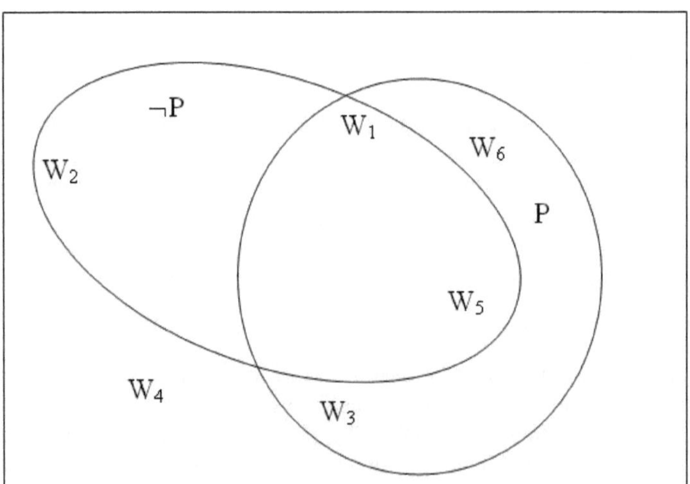

Fig. 8.8 Scheme of worlds or irrational agent in the worlds w_1 and w_5, where $P \wedge \neg P$ is true and w_4 where $P \vee \neg P$ is false

Gaines's logic of uncertainty is based on lattices [Gaines, 1976]. The irrational worlds we discuss here can be described using Gaines's lattices with truth values defined as follows: for any A and B: A&B=TT, A&B=TF; A&B=FT; A&B=FF. If A =P and B= $\neg$ P we have a situation shown in Table 8.4.

8.13 Fuzzy Set, Zadeh Min Max Composition Rules and Irrationality

In an **irrational world** w we may have two contradictory statements A and B that claim, respectively that P is true and not P is true, A = (P = T) and B ≡ ($\neg$ P = T). Such irrationality may have many different reasons. One of them we already discussed above. The object has part R that has P and part Q that has no property P , but the agent was forced to use a restrictive language that does not have concepts of object parts, thus agent's judgments were assigned to the whole object. Alternatively the agent may not know that he/she observed different parts of the same objects, he/she may assume that each time the same whole object was observed. Another reason could be that the agent was forced to use a restrictive language about properties. Property P, say color Red may not be the best term for the situation. It may be difficult to judge a color of a particular multicolor object using just two categories, {Red, not Red}. In such irrational situation the agent has several options: (a) assigning color randomly that will result in irreproducible truth values, (b) always assigning True in the case of doubt, or do opposite – always assign False in the case of doubt. The third alternative is to assign true to $P \wedge \neg P$ which can be interpreted that there are difficulties to assign value rationally, consistently.

Assigning $P \wedge \neg P$ = True can be a temporary assignment before resolving ambiguity by expanding and correcting the set of predicate and object terms of the language used or/and improving observation and measurement process. Before knowing the logic value of a proposition we have confusion and semantic uncertainty that can be expressed by assigning True value to $P \wedge \neg P$.

An option with expanding the language by changing the language of truth values themselves was proposed by Lukasiewicz who introduced value "do not know" (½) as a third value. This avoids direct contradiction in logic. The Lukasiewicz's value ½ can be studied by the set of two statements in a single world or in two worlds that are evaluated in two different ways the same property of one object as p is True and not p is True. Thus, we assign $\mu = ½$. In essence $P \wedge \neg P$ = True expresses the same idea as $\mu = ½$. Note that $\mu = ½$ is less descriptive than $P \wedge \neg P$ = True when we have more than two contradictory worlds. $P \wedge \neg P$ = True means that there are at least two worlds that contradict.

Below we consider the case of irrational worlds presented in Table 8.5 and illustrated in Figure 8.9. Table 8.5 differs from Table 8.4, but similarly to Table 8.4 we identify expressions shown in the last column applying the classical implication operation ($A \wedge B \Rightarrow A$) for irrational worlds.

Using this classical operation in the **irrational world** w_1 we have A = (P = T), that is P is True, and

B ≡ (¬ P = T), that is not P is T. Using classical implication we produce P=T from P∧¬ P=T. Similarly, ¬ P=T can be inferred in w_1. Al alternative approach would be not to apply classical implication to irrational worlds, but this alternative is beyond the scope of this paper.

Expressions in the last column in Table 8.5 are

Table 8.5 Fuzzy Rules

	P	¬P	Expression
W_1	1	1	P∧¬ P=T
W_2	0	0	P∨¬ P = F
W_3	1	0	P∧¬ P= F
W_4	0	0	P∨¬ P = F
W_5	1	1	P∧¬ P
W_6	1	1	P∧¬ P

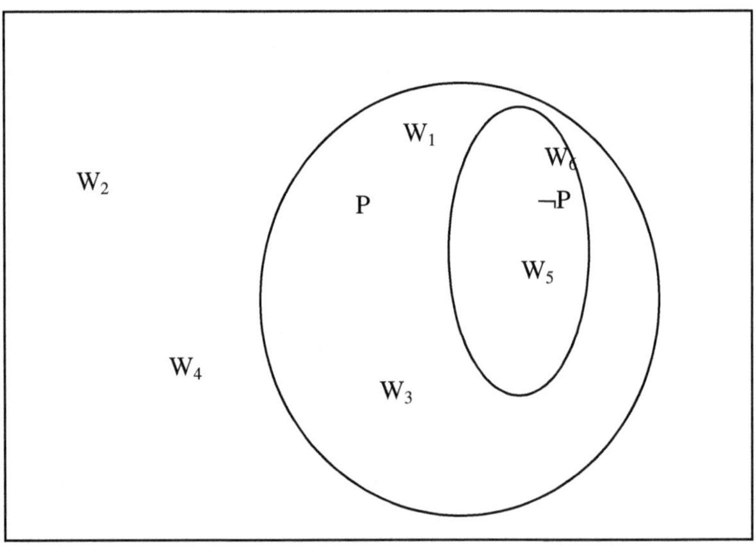

Fig. 8.9 Set P and ¬P one included in the other

In this chapter we use notation P → $\{w_1,w_3,w_5,w_6\}$ to indicate that a property holds in the set $\{w_1,w_3, w_5,w_6\}$. Similarly we have ¬P → $\{w_5,w_6\}$.

Note that having six worlds $\{w_1, w_2, w_3, w_4, w_5, w_6\}$ the compliment to worlds $\{w_5, w_6\}$ is $\{w_1, w_2, w_3, w_4\}$: $\{w_5, w_6\}^C=\{w_1, w_2, w_3, w_4\}$ which differs from $\{w_1,w_3,w_5,w_6\}$ where P is true.

Consider worlds $\{w_5,w_6\}$ P ∧ ¬P is true and worlds $\{w_1,w_3,w_5,w_6\}$ where P ∨ ¬P is true:

$P \wedge \neg P \rightarrow \{w_5, w_6\}$ and $P \vee \neg P \rightarrow \{w_1, w_3, w_5, w_6\}$. Table 8.5 indicates that P is true in four worlds and $\neg P$ is true in two worlds, that provides fuzzy logic membership function values $\mu(P) = 4/6$, $\mu(\neg P) = 2/6$ and $\mu(P \vee \neg P) = 4/6$.

Here $\mu(P) + \mu(\neg P) = 1$, that is consistent with classic logic and fuzzy logic rule $\mu(\neg P) = 1 - \mu(P)$, but value $\mu(P \vee \neg P) = 4/6$ differs from $\mu(P \vee \neg P) = 1$.

Let us clarify the reason for such difference. The number 4/6 for $\mu(P \vee \neg P)$ has an empirical meaning, it counts the number of irrational worlds where true value was assigned to expression $P \vee \neg P$. Fuzzy logic complement rule is an axiomatic rules that was selected because it is consistent with classical logic when μ takes only 0 or 1 values, but for μ that differs 0 or 1 the complement rule is heuristic and may not match empirical data as it was shown for fuzzy logic min/max composition rules [Zimmerman, Zysno 1980; Kovalerchuk, Talianski, 1991].

In our approach with irrational worlds it is allowed to assign logic values T/F independently for expressions $P \wedge \neg P$, P and $\neg P$. Thus we have eight possible world types for each object.

$P \wedge \neg P$ =T, P=T, $\neg P$=T
$P \wedge \neg P$ =T, P=T, $\neg P$=F
$P \wedge \neg P$ =T, P=F, $\neg P$=T
$P \wedge \neg P$ =T, P=F, $\neg P$=F

$P \wedge \neg P$ =F, P=T, $\neg P$=T
$P \wedge \neg P$ =F, P=T, $\neg P$=F
$P \wedge \neg P$ =F, P=F, $\neg P$=T
$P \wedge \neg P$ =F, P=F, $\neg P$=F

Now we will try to clarify applicability of fuzzy logic composition rules for P and $\neg P$

$$\mu(A \wedge B) = \min(\mu(A), \mu(A))$$

$$\mu(A \vee B) = \max(\mu(A), \mu(B))$$

These rules provide $\mu(P \wedge \neg P) = 2/6$ and $\mu(P \vee \neg P) = 4/6$ for Table 8.5. Fuzzy logic theory does not apply these rules for A=P and B= $\neg P$. In such case fuzzy logic uses the complement rule $\mu(\neg P) = 1 - \mu(P)$.

In the case of Table 8.4 and Figure 8.8 (when a set of worlds include irrational worlds) property P is true in the set $\{w_1, w_6, w_3, w_5\}$ and $\neg P$ is true in the set $\{w_5, w_2, w_1\}$ and $P \wedge \neg P$ is true in the set $\{w_5, w_1\}$. Thus

$$\mu(P) = 4/6, \quad \mu(\neg P) = 3/6, \quad \text{and} \quad \mu(P \wedge \neg P) = 2/6.$$

We can also compute $\mu(P \wedge \neg P) = 2/6$ and $\mu(P \vee \neg P) = 5/6$ using the formulas that involve fuzzy logic min, max operators [Resconi , Klir, 1992, Resconi, Turksen, 2001]:

$$\mu(P \wedge \neg P) = 2/6 = \min(\mu(P), \mu(\neg P)) - |\{w_1, w_6, w_3, w_5\}^C \cap \{w_5, w_2, w_1\}|/6 = 3/6 - 1/6$$

$$\mu (P\vee \neg P) = 5/6 = \max (\mu (P) , \mu (\neg P)) + |\{w_1,w_6 ,w_3,w_5\}^C \cap \{w_5, w_2 , w_1\}| /6 \quad = 4 / 6 + 1 /6.$$

This follows from $P\wedge \neg P \rightarrow \{w_5,w_1\}$ and $P\vee \neg P \rightarrow \{w_5,w_1 , w_2 , w_3 , w_6\}$ because $\{w_1,w_6 ,w_3,w_5\}^C \cap \{w_5, w_2 , w_1\} = \{w_2\}$, where w_2 is rational world where $\neg P$ is true (see Table 8.4 and Figure 8.8),

In the formulas that involve min composition rule the nested part is computed using min and non-nested part is computed using $\{w_1,w_6 ,w_3,w_5\}^C \cap \{w_5, w_2 , w_1\}| /6$.

8.14 Irrational and Rational Worlds

When $\neg P$ is true in the set $\{w_5,w_6\}$ and P is true in the set $\{w_1 , w_3 , w_5,w_6\}$ no world is irrational as we can see in Table 8.6.

Table 8.6 No irrational worlds

	P	¬P	Expression
w_1	1	0	P
w_2	0	1	¬ P
w_3	1	0	P
w_4	0	1	¬ P
w_5	1	0	P
w_6	1	0	P

Figure 8.10 shows a graph of Table 8.6.

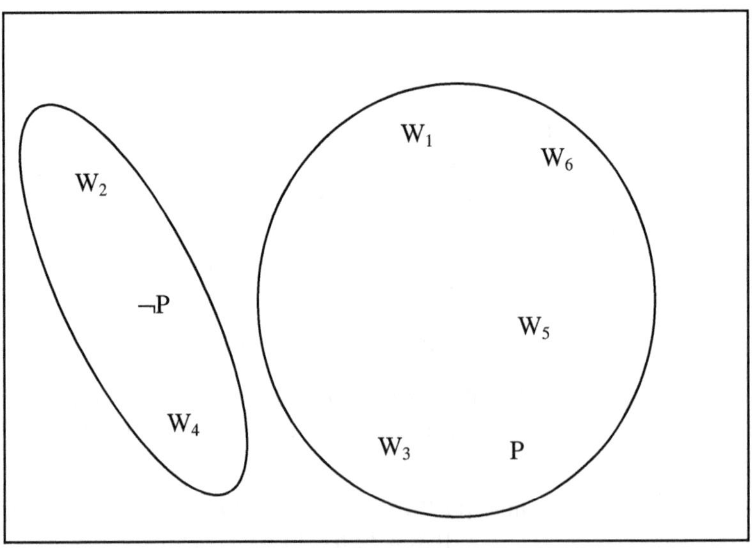

Fig. 8.10 Set of worlds without any irrational situation

For the table 8.6 we have

$$\mu(P) = 4/6 \quad \text{and} \quad \mu(\neg P) = 2/6$$

and

$$\mu(P \wedge \neg P) = 0 = \min(\mu(P), \mu(\neg P)) - |\{w_1, w_6, w_3, w_5\}^C \cap \{w_5, w_2, w_1\}| /6 = 2/6 - 2/6$$

$$\mu(P \vee \neg P) = 1 = \max(\mu(P), \mu(\neg P)) + |\{w_1, w_6, w_3, w_5\}^C \cap \{w_5, w_2, w_1\}| /6 = 4/6 + 2/6$$

In this case we came back to the classical situation where the set of worlds where $P \wedge \neg P$ is true is empty and the set of the world where $P \vee \neg P$ is true is the universal set W. The contradiction $P \wedge \neg P$ is always false and the tautology $P \vee \neg P$ is always true.

Remark
The situations with irrational worlds are beyond the situations considered in traditional sets of rational worlds. This, the concept of irrational worlds extends the traditional Kolmogorov's axioms to irrational situations.

In Zadeh's fuzzy logic rule $\mu(A \wedge B) = \min(\mu(A), \mu(B))$ is applied for all A and B including $B = \neg A$.. Thus for conjunction operation fuzzy logic follows the *principle of irrationality*, which means truth of contradiction $\mu(A \wedge \neg A)$ can be above 0.

In contrast rule $\mu(A \vee B) = \max(\mu(A), \mu(B))$ is applied only if B is not a negation of A ($B \neq \neg A$). It means that the fuzzy logic is partially context dependent. For negation case when A=P and B= $\neg P$ fuzzy logic uses rules $\mu(P \vee \neg P)=1$ and $\mu(\neg P) = 1 - \mu(P)$, That is requiring that truth of tautology should be equal to 1. However, in automatic fuzzy logic reasoning we may not have tools to know that B= $\neg A$. Both A and B can be given by their numeric values only, say $\mu(A)=0.7$and $\mu(B)=0.3$. As a result we may compute $\mu(A \vee \neg A)$ as $\max(\mu(A), \mu(\neg A))$.=0.7 <1 that is irrational situation $\mu(A \vee \neg A) < 1$.

The truth table of a set of worlds that contains $\mu(A \wedge \neg A) > 0$ and $\mu(A \vee \neg A) < 1$ indicates that some worlds satisfy principle of irrationality, and can be irrational. However more exactly it indicates presence of an *irrational formalization* (knowledge representation) of real entities. The irrational formalization may come in the form the superposition of two *different contexts*. For example, in one context we can say that a book is good, but the same book in another context is bad. This is not a question of different and conflicting logic evaluations provided by the same person in the same context. In the irrational situation we have no conflict for the logic evaluation but we have *different contexts* in which the same world or person give irrational judgment. The person simply may not know the correct context for making the judgment.

Remark
The classical logic provides the fundamental axioms [Mendelson, 1964]

$$A \rightarrow (B \rightarrow A)$$

$$(A \rightarrow (B \rightarrow C) \rightarrow ((A \rightarrow B) \rightarrow (A \rightarrow C))$$

$$(\neg B \rightarrow \neg A) \rightarrow (\neg B \rightarrow A) \rightarrow B)$$

that are always true (tautology). In the irrational world these axioms may fail. Thus, the concept of the irrational world (where the contradiction is true or the tautology is false) separates fuzzy logic in our possible world formalization from the classical logic and the classical set theory.

The same concept of irrational world separates the proposed approach from the classical modal logic. The classical modal logic is context dependent (topologic logic) but it does not allow an irrational world and superposition of conflicting logic propositions for the same object.

The concept of irrationality explains deeper meaning of the fuzzy logic. This allows us to rebuilt the fundamental rules in fuzzy logic and extend the fuzzy logic to more general and abstract forms of uncertainties that is impossible to study with the traditional fuzzy logic.

8.15 Mapping Set of Worlds

When P is true in the set of worlds $\{w_1, w_3, w_5, w_6\}$ and $\neg P$ is true in the set $\{w_2, w_4\}$ no world is irrational and the classical axioms of the previous chapters are true. To get irrational worlds we need to allow that set $\{w_2, w_4\}$ where $\neg P$ is true be any set of worlds from $\{w_1, \ldots, w_6\}$ not only the compliment to $\{w_1, w_3, w_5, w_6\}$.

We use vector V ($\neg P$) as a representation of truth value of $\neg P$ in worlds $\{w_1, w_2, w_3, w_4, w_5, w_6\}$

If we have a set of worlds with vector $V_1(\neg P) = [F, T, F, T, F, F]$ and another set of worlds where $V_2(\neg P) = [F, F, F, F, T, T]$ then the mapping Γ between these worlds can be defined by the mapping (transformation) matrix Γ, $\Gamma V_1^T = V_2^T$, where V_1^T and V_2^T are transposed vectors $V_1(\neg P)$ and $V_2(\neg P)$:

$$\Gamma = \begin{bmatrix} 1 & 0 & 0 & 0 & 0 & 0 \\ 0 & 0 & 0 & 0 & 1 & 0 \\ 0 & 0 & 1 & 0 & 0 & 0 \\ 0 & 0 & 0 & 0 & 0 & 1 \\ 0 & 1 & 0 & 0 & 0 & 0 \\ 0 & 0 & 0 & 1 & 0 & 0 \end{bmatrix}$$

and

$$\begin{bmatrix} 1 & 0 & 0 & 0 & 0 & 0 \\ 0 & 0 & 0 & 0 & 1 & 0 \\ 0 & 0 & 1 & 0 & 0 & 0 \\ 0 & 0 & 0 & 0 & 0 & 1 \\ 0 & 1 & 0 & 0 & 0 & 0 \\ 0 & 0 & 0 & 1 & 0 & 0 \end{bmatrix} \begin{bmatrix} F \\ T \\ F \\ T \\ F \\ F \end{bmatrix} = \begin{bmatrix} F \\ F \\ F \\ F \\ T \\ T \end{bmatrix}$$

In a graphic way this is presented in Figure 8.11.

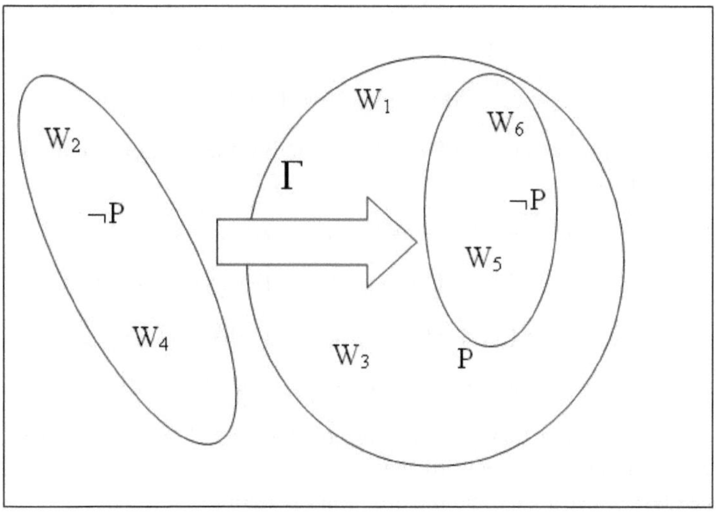

Fig. 8.11 Mapping Γ of the classical complement $\{w_2,w_4\}$ to non-classical complement $\{w_5,w_6\}$

We will also use notation $\Gamma(\neg P)$ for ΓV_1^T when vector V^T represents truth values $\neg P$ in all worlds $\{w_1, w_2, w_3, w_4, w_5, w_6\}$ after mapping Γ is applied to V^T. Similarly we will use notation $\Gamma(Q)$ for any Q.

For vector V_1^T we have

$$\mu(P \wedge \neg P) = 0, \quad \mu(P \vee \neg P) = 1,$$

and after mapping Γ have been applied to V_1^T we have

$$\mu(P \wedge \Gamma(\neg P)) = 2/6, \quad \mu(P \vee \Gamma(\neg P)) = 4/6.$$

Here $P \wedge \Gamma(\neg P)$ means a conjunction of two truth vectors for P and $\Gamma(\neg P)$ respectively.

Applying mapping Γ means that we change **context** of computing truth values. With this in mind membership function $\mu(P \wedge \Gamma(\neg P))$ *evaluates truth values of two different contexts mixed together.*

Having equivalence $A \rightarrow B$ and $\neg A \vee B$ we can rewrite three classical axioms in this way

$$A \rightarrow (B \rightarrow A) \equiv \neg A \vee (\neg B \vee A)$$

$$(A \rightarrow (B \rightarrow C) \rightarrow ((A \rightarrow B) \rightarrow (A \rightarrow C)) \equiv$$

$$\equiv \neg(\neg A \vee (\neg B \vee C)) \vee (\neg(\neg A \vee B) \vee (\neg A \vee C))$$

$$(\neg B \rightarrow \neg A) \rightarrow (\neg B \rightarrow A) \rightarrow B) \equiv$$

$$\neg(B \vee \neg A) \vee \neg(B \vee A) \vee B$$

Now we can see how mapping Γ can be applied to these axioms and provide their generalization.

$$\Gamma A \rightarrow (B \rightarrow A) \equiv \Gamma(\neg A) \vee (\neg B \vee A)$$

$$(A \rightarrow (B \rightarrow C) \rightarrow ((A \rightarrow B) \rightarrow (A \rightarrow C)) \equiv$$

$$\equiv \neg(\Gamma(\neg A) \vee (\neg B \vee C)) \vee (\neg(\Gamma(\neg A) \vee B) \vee (\Gamma(\neg A) \vee C))$$

$$(\neg B \rightarrow \neg A) \rightarrow (\neg B \rightarrow A) \rightarrow B) \equiv \neg(B \vee \Gamma(\neg A)) \vee \neg(B \vee A) \vee B.$$

In conclusion we note that the operators of the type Γ give us the possibility to move from the classical logic to many different fuzzy logics that are generated from the classical structure and the different operators Γ.

8.16 Invariant Expressions in Fuzzy Logic

The following property (fuzzy logic invariant) holds in fuzzy logic

$$\max[\mu(P), \mu(\neg P))] + \min(\mu(\neg P), \mu(P)) = 1.$$

Below we illustrate this invariants under mapping Γ:

$$\max[\mu(P), \mu(\Gamma(\neg P))] + \min(\mu(\neg P), \mu(\neg \Gamma(\neg P)) = 1.$$

Given the fuzzy logic expression

$$A = [P \vee \Gamma(\neg P)],$$

we can define $B = A \vee \neg A$,

where $\neg A$ is true in the worlds where A is false.

Let P is true in $\{w_1, w_3, w_5, w_6\}$, $\neg P$ is true in $\{w_2, w_4\}$ and $\Gamma(\neg P)$ is true in $\{w_5, w_6\}$. If we apply fuzzy logic operators to compute $\mu[P \vee \Gamma(\neg P)]$ we obtain $\mu[P \vee \Gamma(\neg P)] = \max[\mu(P), \mu(\Gamma(\neg P))] = \max[4/6, 2/6] = 4/6$.

Next we have $\neg(P \vee \Gamma(\neg P)) = \neg P \wedge \neg \Gamma(\neg P)$ and $\mu(\neg P) = 2/6$ because $\neg P$ is true in $\{w_2, w_4\}$. In addition $\neg \Gamma(\neg P)$ is true in $\{w_2, w_4, w_1, w_3\}$ that includes $\{w_2, w_4\}$, thus $\mu(\neg \Gamma(\neg P) = 4/6$ and $\mu[\neg P \wedge \neg \Gamma(\neg P)] = \min(\mu(\neg P), \mu(\neg \Gamma(\neg P)) = 2/6$

In conclusion we have the fuzzy logic invariant

$$\max[\mu(P), \mu(\Gamma(\neg P))] + \min(\mu(\neg P), \mu(\neg \Gamma(\neg P)) = 1$$

8.17 Linguistic Context Space of the Worlds

Linguistic uncertainties play an important role in human reasoning. Previously we introduced the concept of a linguistic context space [Kovalerchuk, 1996, 2000] to deal with such uncertainties in both probabilistic and fuzzy logic settings. The context space framework clarified the difference and interrelations between fuzzy logic and probability theories and identified the area where fuzzy logic has a solid foundation beyond its heuristic base. It is shown that well-justified interconnected membership functions can be build as cross-sections of probability distributions (spaces). This was briefly outlined in section 5.3. Below we show how the context space concept related to the concept of uncertainty in possible worlds that we discussed in the previous sections.

The combination of modal logic and context space approaches makes *explicit* the process of membership functions and probability computations. Traditional probability and fuzzy logic theories treat these issues as *external* to both theories. These axiomatic theories start from the assumption that uncertainty values are given (already computed somehow) and should satisfy axioms of probability theory or fuzzy logic theory, respectively.

We have made computing uncertainties values internal to the theory on uncertainty based on possible worlds in the previous sections. However the modal logic accessibility relation used sets up a very *simple structure* of relations between worlds that only permits to say "If A is true in the world w and world w` can access world w than A can be true in w`." The context space approach allows us to deal with more complex situations than the modal logic accessibility relations allow.

For instance the context space will allow to deal with cases like this "If P(a) is true in the world w and world w` can access world w and predicates than Q(b) can be true in w`.". Thus, the context space approach allows works in much richer *first order logic* instead of propositional logic. In this way can operate explicitly with different individuals a, b and predicates P and Q. This deeper approach is more practical and useful to real world situations and applications.

Let us given a set of terms or **term context**

$$C = \{ \lambda_1 , \quad \lambda_2 , \ldots\ldots\ldots, \quad \lambda_n \}$$

and an object defined by the values of parameters x_i from the parameter set

$$X = \{ x_1 , \quad x_2 , \ldots\ldots\ldots, x_m \}.$$

Each term λ_i is a linguistic term (fuzzy degree), λ_i can be one of the terms "very small", "small", "medium", "large", "very large", and so on.

As we discussed in section 6, if worlds rational in the modal logic interpretation of the fuzzy set, then Kolmogorov's probability can be used to model the hidden structure of uncertainties in possible worlds. Now we associate each property x with uncertain linguistic degree λ_k. The frequency $p_x(\lambda_k)$ of use of different terms λ_k from C for the same age x gives us an instrument to model incoherent properties of elements in X . We can also define a set of function $\mu_{\lambda k}(x)$ for each λ_k as follows $\mu_{\lambda k}(x)= p_x(\lambda_k)$. The difference between p_x and $\mu_{\lambda k}$ is that each p_x has fixed x as a parameter and λ_k as variable, but each $\mu_{\lambda k}$ has λ_k as a fixed parameter and x as a variable. If we introduce an **uncertainty function** $F(\lambda_k ,x)$ of two variables λ_k and x then fuzzy membership functions $\mu_{\lambda k}(x)$ and pseudo probability distributions $p_x(\lambda_k)$ are two different cross-sections of F.

The set of terms C is a base of the **context space.** We are especially interested in the situation when properties are interrelated, that is typical, in real world situations. For example, if person x is identified by the parameter x_i from space X = { Age } such that x_i = 32 years, then we can use a term context { λ_1 = Not Old, λ_2 = Almost Old, λ_3 = Old}=C(32).

Below we use abbreviations NO for Not Old, AO for Almost Old and OL for Old. Alternatively the term context can be:

{ λ_1 = Very Young, λ_2 = Old} that is not a right context for age 32.

There is no reason much to use this term context for age 32. Thus, we need to design the term context *individually* for each specific age, not as a single one for every age.

Now we can introduce a concept of worlds to the context spaces to be able to compute the uncertainty function F. We will call such world's **context worlds**.

For the same age = 32, we may have the **context worlds**:

$$\lambda_1 \rightarrow \{ w_2 \} \quad , \quad \lambda_2 \rightarrow \{ w_1 \ , \quad w_5 , w_6 \} , \quad \lambda_3 \rightarrow \{ w_4 , w_3 \}$$

where w_2 is a world that uses term λ_1 for age 32, { w_1 , w_5 , w_6 }are worlds that use term λ_2 for the same age, and { w_4 , w_3 } are worlds that use term λ_3. Graphically we show these worlds in Figure 8.12.

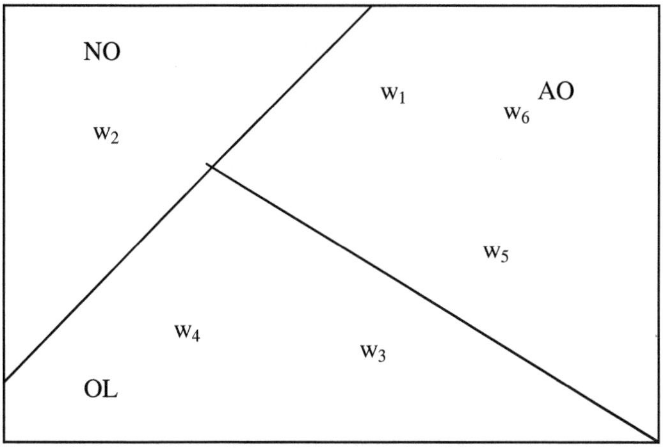

Fig. 8.12 Equivalent classes of worlds with three elements in the linguistic context

Therefore, we have *measures of uncertainty* for age 32

NO(32) = 1/6 , AO(32) = 3/6, OL(32) = 2/6, and

NO(32) + AO(32) + OL(32) = 1

There is a clear relation between the context terms and the context of associated worlds (world context). In our example with three-term context the complement of term NO is the set of terms { AO, OL }, NO^C = {AO, OL}. This is matched to a property of associated worlds w(NO^C) = w(AO $\cup$ OL)={ w_1 , w_5 , w_6 , w_4 , w_3 }.

Let us consider another example, where

NO corresponds to { w_1 , w_3 , w_4, w_5 , w_6 } with NO(32) = 5/6,

AO corresponds to { w_3 , w_5 } with AO(32) = 2/6, and

OL corresponds to { w_2 , w_4 } with OL(32) = 2/6.

These worlds can be represented as shown in Figure 8.13.

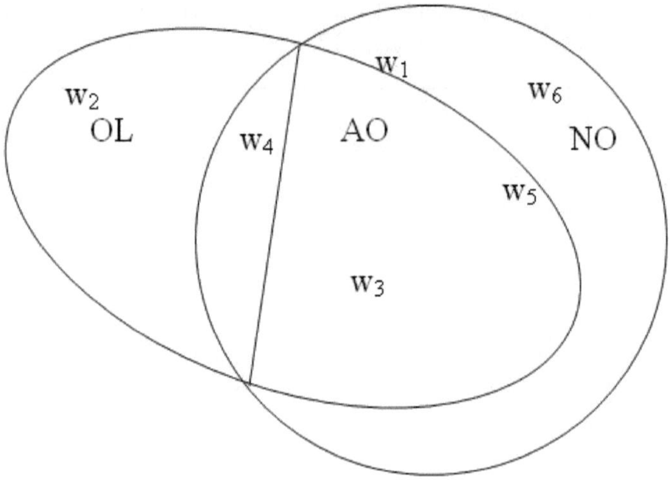

Fig. 8.13 The complement set of NO is NO^C = (AO , OL) in this case $w_4 \in NO$, $w_4 \in NO^C$ the same for w_5 and w_3. So the worlds w_4 , w_3 , w_5 are counted two time because are irrational. When we eliminate the internal irrationality, we have the traditional pseudo probability calculus U = OL(32)+AO(32)+NO(32)- IN(32) =1 where IN = OL$\cap$NO = (w_4 , w_3 , w_5).

In this example both OL and NO are true in w_4. On the other hand the complement to OL is {NO, AO}, because our complete context U is limited by these three terms in this example,

$$U = NO \cup AO \cup OL.$$

Having OL true in w_4 we should have its complement {NO, AO} false in w_4, that is $\neg$ NO and $\neg$ AO should be true in w_4. Thus, in world w_4 true both NO and $\neg$ NO that is w_4 is an irrational world. Worlds w_3 and w_5 are also irrational because having No and AO for the same age 32.

We define the measure of uncertainty of the whole term context U is

$$U(32) = NO(32)+AO(32)+OL(32) – IN(32)$$

where intersection IN(32) is formed by irrational worlds w_3 , w_4 , and w_5 .

Sets of worlds for AO and NO interest in two worlds w_4 and w_5 giving

$$In(32, AO \cap NO)=2/6.$$

Similarly sets of worlds for OL and NO intersect in a single world w_4 with

$$In(32, OL \cap NO)=1/6.$$

Thus, IN(32)= In(32, AO $\cap$ NO)+ In(32, OL $\cap$ NO)=2/6+1/6 and

Finally U(32) = NO(32)+AO(32)+OL(32) – IN(32)=5/6 + 2/6 + 2/6 – 2/6-1/6= 1.

This U(32)=1 indicates that the set of context terms U is *complete* for evaluating age 32. In contrast if U would consist of two terms OL (old) and VO(Very Old) we may get for worlds (agents) { w_1 , w_2 , w_3 , w_4 , w_5 , w_6 }

OL corresponds to { w_2 , w_4 } with OL(32) = 2/6 and

VO = OL^C corresponds to {w_4} with NO(32) = 0/6,

and U(32)=3/6. This indicated that such U is very incomplete context for these agents/worlds.

In this case we have. In this case we have that the tautology is false for the irrational worlds

w_1 , w_3 , w_5 , w_6. The irrational worlds are outside the universal set of worlds. This is beyond any modal logic or set theory.

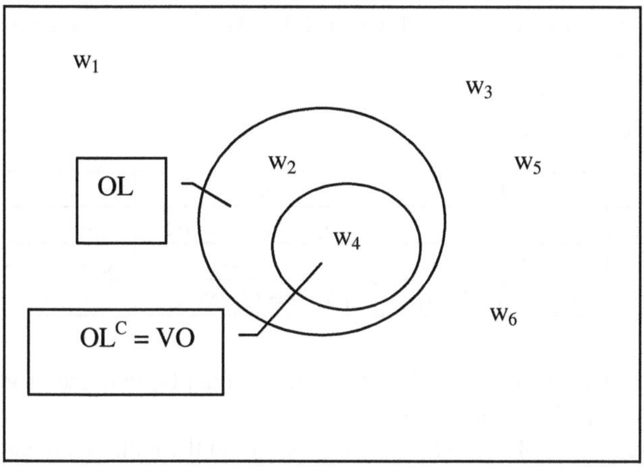

Fig. 8.14 The complement set of OL = (w_2, w_4) is OL^C = (w_4) in this case $w_4 \in OL$, $w_4 \in OL^C$ So the worlds w_4 is counted two time because is irrational. When we eliminate the internal irrationality , we have the traditional pseudo probability calculus

U = OL(32)+VO(32) - IN(32) =2/6 where IN = OL∩NO = (w_4). But because U < 1, we have worlds out of the universal set. In this case the worlds our of U are irrational because are out of the universal set U and this is impossible for the traditional set theory. For the worlds out of U the tautology is false. For the set of the words at the intersection of OL and OL^C the contradiction is true.

We may also get an *over complete term context* with UJ(32)>1 if we do not subtract intersection In(32) that may happen if we do not know that worlds are irrational. Remark that the over complete context is a result of interplay of two components: (1) the set of worlds (agents), and (2) the set of context terms. If a set of context terms is too small then agents may not be able to classify ages without overlap. The same can happen if terms are nested (Old can be interpreted as a part of Very Old).

8.18 Economic Model of Worlds

Given three products G_1, G_2, and G_3 and six customers $C_1,C_2,\ldots,C_6$ we have nine sentences (see Table 8.7).

Table 8.7 Sentences P <Make q>

	G_1	G_2	G_3
G_1	$G_1 = G_1$	$P_1 \equiv (G_1 > G_2)$	$P_2 \equiv (G_1 > G_3)$
G_2	$\neg P_1 \equiv G_1 \le G_2$	$G_2 = G_2$	$P_3 \equiv= (G_2 > G_3)$
G_3	$\neg P_2 \equiv G_1 \le G_3$	$\neg P_3 \equiv= G_2 \le G_3$	$G_3 = G_3$

where $G_k = G_k$ are tautologies for k=1,2,3. Another set of sentences is presented in Table 8.8.

Table 8.8 Sentences Q

	G_1	G_2	G_3
G_1	$G_1 = G_1$	$Q_1 \equiv (G_1 < G_2)$	$Q_2 \equiv (G_1 < G_3)$
G_2	$\neg Q_1 \equiv G_1 \ge G_2$	$G_2 = G_2$	$Q_3 \equiv G_2 < G_3)$
G_3	$\neg Q_2 \equiv G_1 \ge G_3$	$\neg Q_3 \equiv= G_2 \ge G_3$	$G_3 = G_3$

Let P = "The customer C likes more the product G_k than the product G_h" and in a formal notation $P = (G_k > G_h)$.

Because $S \equiv \neg P \wedge \neg Q \equiv (C_k = C_h)$ in the natural language we have

S = "The customer C_k equally likes products G_k and G_h",

The customers (agents) are mutually independent and can give the logic value (true/false) to each sentence. Any customer can be viewed as a world in a classical modal logic. Therefore,

$$C_1 \to w_1 , C_2 \to w_2 ,\ldots\ldots, C_6 \to w_6 ,$$

where we associate each possible world with a customer.

Next, we associate customers (agents) with preference sentences P_1 P_2 and P_3. Thus, we may have

$$P_1 \to \{ w_2 \} = \{ C_2 \} \quad \text{where } P_1 = \ (G_1 > G_2)$$

This means that customer C_2 prefers product G_1 to product G_2.

Similarly we may have $P_2 \rightarrow \{ w_1, w_6, w_5 \} = \{ C_1, C_6, C_5 \}$ where $P_2 = (G_1 > G_3)$

Customers C_1, C_6, and C_5 prefer product G_1 to product G_3,

$$P_3 \rightarrow \{ w_4, w_3 \} = \{ C_4, C_3 \} \quad \text{where} \quad P_3 = (G_2 > G_3)$$

Customers C_4 and C_3 prefer product G_2 to product G_3. In a graphic way we have

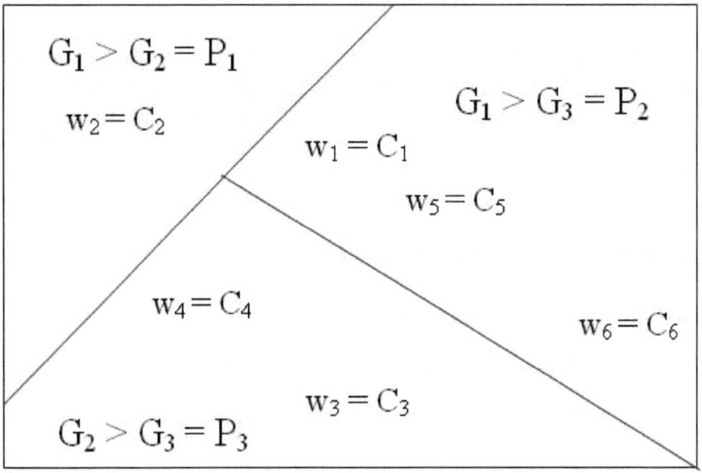

Fig. 8.15 Sets of agents (words) and sentences $P_1 = G_1 > G_2$, $P_2 = G_1 > G_3$, $P_3 = G_2 > G_3$

The chain $G_2 \geq G_1 > G_3$ can be written in this way

$$G_2 \geq G_1 > G_3 = (G_1 \leq G_2) \wedge (G_1 > G_3) = \neg P_1 \wedge P_2 .$$

This is because, we know that P_1 is equal to $G_1 > G_2$ thus $\neg P_1$ is $G_2 \geq G_1$ and P_2 is equal to $G_1 > G_3$. When all the worlds (customers) are rational, we have

$$\neg P_1 = G_1 \leq G_2 = (G_1 > G_3) \wedge (G_2 > G_3) \wedge (G_1 = G_2)$$

because we have $P_2 \rightarrow \{ C_1, C_6, C_5 \}$, $P_3 \rightarrow \{ C_4, C_3 \}$, $S = \varnothing$ we have $\neg P_1 \rightarrow \{ C_1, C_3, C_4, C_5, C_6 \}$ and

$$\neg P_1 \wedge P_2 \rightarrow \{ C_1, C_6, C_5 \} , \quad \mu(\neg P_1 \wedge P_2) = 3/6.$$

Here the *arrow* means that we *associate a sentence* with a set of customers (worlds) that agree with the sentence. For the customers C_1, C_6, C_5 prefer product G_2 to product G_1 and product G_1 to product G_3. Similarly for the logic expression (sentence) $P_1 \vee P_2$ we have

$P_1 \vee P_2 \rightarrow \{ C_1, C_2, C_6, C_5 \}$ and the bifurcation $G_1 > G_2$, $G_1 > G_3$ or

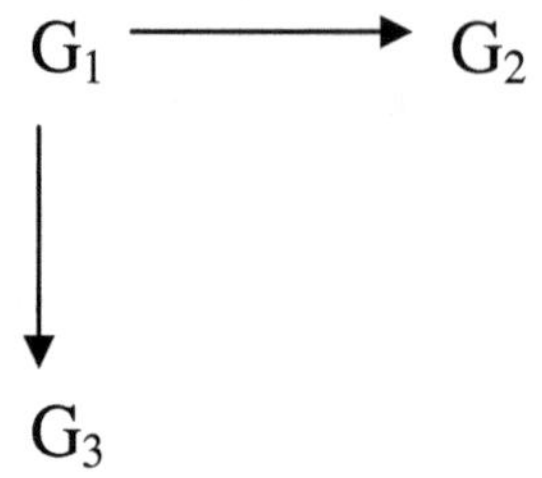

At the logic expression $P_1 \vee P_2$ we associate the graph bifurcation. We remark that a complex graph can be represented with the previous consideration by a logic expression. At the same logic expression we can associate also the measure

$$\mu (P_1 \vee P_2) = 4/6.$$

8.19 Irrational Customers and Fuzzy Set

For irrational customers we may have

$$P_1 \rightarrow \{ C_1 , C_3 , C_5 , C_6 \}, \text{ where } P_1 = G_1 > G_2$$

$$P_2 \rightarrow \{ C_1 , C_5 \}, \text{ where } P_2 = G_1 > G_3$$

$$P_3 \rightarrow \{ C_2 \}, \text{ where } P_3 = G_2 > G_3$$

In a graphic way this case is shown in Figure 8.16.

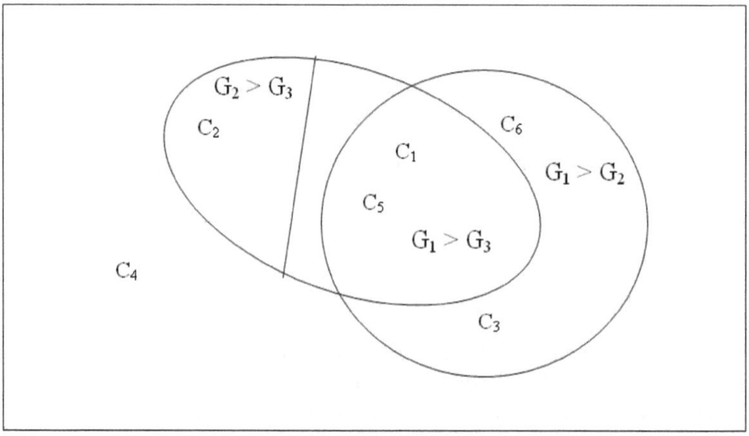

Fig. 8.16 A set of customers where customer C_4 is irrational (it is out of the context)

Because we have $\neg P_1 = G_1 \leq G_2 = (G_1 > G_3) \wedge (G_2 > G_3) \wedge (G_1 = G_2) = \{C_1, C_5, C_2\}$

For the previous considerations we have

Contradiction $= \neg P_1 \wedge P_1 = \{C_1, C_5\}$.

The customers C_1, C_5 are irrational costumers. On the other hand we have

Tautology $= \neg P_1 \vee P_1 = (G_1 > G_3) \wedge (G_2 > G_3) \wedge (G_1 = G_2) \wedge (G_1 > G_2) =$

$$\{C_1, C_2, C_3, C_5, C_6\},$$

Boris: How propositions can be equal to customers?

but the universal set is $U = \{C_1, C_4, C_2, C_3, C_5, C_6\}$. In this case we have that

$$\neg P_1 \vee P_1 \text{ is false for } \{C_4\}.$$

The customer C_4 is an irrational agent for him all the possible sentences and the negations of the possible sentences are all false. In the ordinary logic if one sentence is false the negation is true.

For the previous customer C_4 has all the sentences and the negations are not complete set of sentences. Other possible sentences exist but we have not considered. The set of the sentences is open.

8.20 Communication among Customers and Rough Sets

When customers communicate one with the other, we can have the structures shown in Figure 8.17.

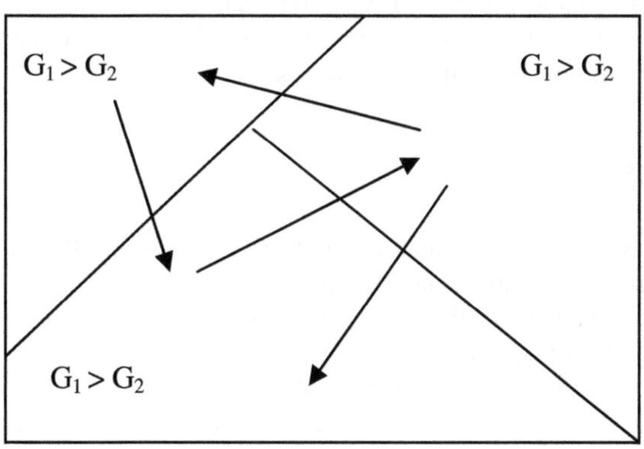

Fig. 8.17 Communicating agents

In figure 17 we show a set of customers that can communicate one with the other. In this case we have three different probabilities $P_1 \leq P \leq P_2$, where P_1 is the relative number of worlds for which, for example, $G_1 > G_3$ is necessary , P is the relative number of worlds where $G_1 > G_3$ is true and P_2 is the relative number of worlds where $G_1 > G_3$ is possible. We study the accessible communication among worlds. We can repeat the same concepts in the case of the customers and products. In this case, we have set of customers for which a preference is necessary, set of customers for which the preference is true and set of customers for which the preference is possible.

8.21 Conclusion

In this paper, we presented the hidden structure of the uncertainty by using the modal logic extended to irrational worlds or a scheme of worlds. Appear that the different forms of uncertainties can be captured by the **scheme of the worlds and communication among worlds.** In a future work we can introduce new type of uncertainty by composition of the principles of uncertainty due to the probability, imprecise probability, and irrationality in fuzzy set theory.

The paper is also dedicated to describing the fuzzy set and fuzzy measure (evidence) by only one model of worlds. This simplifies the complexity of the fuzzy set operations as t-norm and t-conorm. In the same time we put in evidence the difference between the classical probability based on a set theory without conflict and incoherence among worlds or point of view with the fuzzy set theory and fuzzy logic as conflicting and incoherent (irrational) structures.

In conclusion in the new model at the traditional uncertainty in probability calculus based on the set theory, we adjoin a new type of uncertainty based on set of conflicting and incoherent worlds. The set of worlds is not comparable with the classical set of elements. In fact the set of worlds is the instrument by which we evaluate if one element has or has not a property, the classical set takes as obvious the evaluation of the property for a given element. In this way we open a new domain of research in the uncertainty and logic by which we can take care of more complex conceptual types of uncertainties.

References

1. Carnap, R., Jeffrey, R.: Studies in Inductive Logics and Probability, vol. 1, pp. 35–165. University of California Press, Berkeley (1971)
2. Fagin, R., Halpern, J.: Reasoning about Knowledge and Probability. Journal of the ACM 41(2), 340–367 (1994)
3. Edmonds, B.: Review of Reasoning about Rational Agents by Michael Wooldridge. Journal of Artificial Societies and Social Simulation 5(1) (2002), `http://jasss.soc.surrey.ac.uk/5/1/reviews/edmonds.html`
4. Ferber, J.: Multi Agent Systems. Addison Wesley (1999)
5. Gigerenzer, G., Selten, R.: Bounded Rationality. The MIT Press, Cambridge (2002)
6. Halpern, J.: Reasoning about uncertainty. MIT Press (2005)

7. Hisdal, E.: Logical Structures for Representation of Knowledge and Uncertainty. Springer, Heidelberg (1998)
8. Resconi, G., Jain, L.: Intelligent agents. Springer, Heidelberg (2004)
9. Resconi, G., Kovalerchuk, B.: The Logic of Uncertainty with Irrational Agents. In: Proc. of JCIS-2006 Advances in Intelligent Systems Research. Atlantis Press, Taiwan (2006)
10. Kahneman, D.: Maps of Bounded Rationality: Psychology for Behavioral Economics. The American Economic Review 93(5), 1449–1475 (2003)
11. Kovalerchuk, B.: Analysis of Gaines' logic of uncertainty. In: Turksen, I.B. (ed.) Proceeding of NAFIPS 1990, Toronto, Canada, vol. 2, pp. 293–295 (1990)
12. Kovalerchuk, B.: Context spaces as necessary frames for correct approximate reasoning. International Journal of General Systems 25(1), 61–80 (1996)
13. Kovalerchuk, B., Vityaev, E.: Data mining in finance: advances in relational and hybrid methods. Kluwer (2000)
14. Wooldridge, M.: Reasoning about Rational Agents. The MIT Press, Cambridge (2000)
15. Montero, J., Gomez, D., Bustine, H.: On the relevance of some families of fuzzy sets. Fuzzy Sets and Systems (2007) (in print)
16. Atanassov, K.T.: Intuitionistic Fuzzy Sets. Physica Verlag, Springer, Heidelberg (1999)
17. Flament, C.: Applications of graphs theory to group structure. Prentice Hall, London (1963)
18. Ruspini, E.H.: A new approach to clustering. Information and Control 15, 22–32 (1969)
19. Priest, G., Tanaka, K.: Paraconsistent Logic. Stanford Encyclopedia of Philosophy (2004), http://plato.stanford.edu/entries/logic-paraconsistent/

MIX
Papier aus verantwortungsvollen Quellen
Paper from responsible sources
FSC® C105338

If you have any concerns about our products,
you can contact us on
ProductSafety@springernature.com

In case Publisher is established outside the EU,
the EU authorized representative is:
Springer Nature Customer Service Center GmbH
Europaplatz 3, 69115 Heidelberg, Germany

Printed by Libri Plureos GmbH
in Hamburg, Germany